VOLUME FIVE HUNDRED AND NINETY FOUR

METHODS IN ENZYMOLOGY

A Structure-Function Toolbox for Membrane Transporter and Channels

METHODS IN ENZYMOLOGY

Editors-in-Chief

ANNA MARIE PYLE
Departments of Molecular, Cellular and Developmental Biology and Department of Chemistry
Investigator, Howard Hughes Medical Institute
Yale University

DAVID W. CHRISTIANSON
Roy and Diana Vagelos Laboratories
Department of Chemistry
University of Pennsylvania
Philadelphia, PA

Founding Editors

SIDNEY P. COLOWICK and NATHAN O. KAPLAN

VOLUME FIVE HUNDRED AND NINETY FOUR

METHODS IN ENZYMOLOGY

A Structure-Function Toolbox for Membrane Transporter and Channels

Edited by

CHRISTINE ZIEGLER

Department of Structural Biology,
Max Planck Institute of Biophysics,
Frankfurt am Main, Germany

ELSEVIER

ACADEMIC PRESS

An imprint of Elsevier

Academic Press is an imprint of Elsevier
50 Hampshire Street, 5th Floor, Cambridge, MA 02139, United States
525 B Street, Suite 1800, San Diego, CA 92101–4495, United States
The Boulevard, Langford Lane, Kidlington, Oxford OX5 1GB, United Kingdom
125 London Wall, London, EC2Y 5AS, United Kingdom

First edition 2017

ISBN: 978-0-12-812353-9
ISSN: 0076-6879

For information on all Academic Press publications
visit our website at https://www.elsevier.com/books-and-journals

Publisher: Zoe Kruze
Acquisition Editor: Zoe Kruze
Editorial Project Manager: Helene Kabes
Production Project Manager: Magesh Kumar Mahalingam
Cover Designer: Greg Harris

Typeset by SPi Global, India

CONTENTS

CONTRIBUTORS

Abdul Aziz-Qureshi
Stockholm University, SE-106 91 Stockholm, Sweden

Maria Barthmes
Nanion Technologies GmbH, Munich, Germany

Andre Bazzone
Max Planck Institute of Biophysics, Frankfurt/Main; Nanion Technologies GmbH, Munich, Germany

Andreas Bøggild
DANDRITE, Nordic-EMBL Partnership for Molecular Medicine, Aarhus University, Aarhus, Denmark

Byron Carpenter
MRC Laboratory of Molecular Biology, Cambridge, United Kingdom

David Drew
Stockholm University, SE-106 91 Stockholm, Sweden

Patricia C. Edwards
MRC Laboratory of Molecular Biology, Cambridge, United Kingdom

Rouslan G. Efremov
Vrije Universiteit Brussel, Brussels, Belgium

Klaus Fendler
Max Planck Institute of Biophysics, Frankfurt/Main, Germany

Jens Frauenfeld
Salipro Biotech AB, Stockholm, Sweden

Christos Gatsogiannis
Max Planck Institute of Molecular Physiology, Dortmund, Germany

Eric R. Geertsma
Goethe University Frankfurt, Frankfurt/Main, Germany

Joris M.H. Goudsmits
University of Groningen, Groningen, The Netherlands

Katharina Holzhüter
Goethe University Frankfurt, Frankfurt/Main, Germany

Benedikt Kuhn
Goethe University Frankfurt, Frankfurt/Main, Germany

Chiara Lee
Stockholm University, SE-106 91 Stockholm, Sweden

Joseph A. Lyons
DANDRITE, Nordic-EMBL Partnership for Molecular Medicine, Aarhus University, Aarhus, Denmark

Jacopo Marino
University of Zurich, Zurich, Switzerland

Pascal F. Meier
Stockholm University, SE-106 91 Stockholm, Sweden

Joseph A. Mindell
Membrane Transport Biophysics Section, Porter Neuroscience Research Center, National Institute of Neurological Disorders and Stroke, National Institutes of Health, Bethesda, MD, United States

Christopher Mulligan
School of Biosciences, University of Kent, Canterbury, United Kingdom

Poul Nissen
DANDRITE, Nordic-EMBL Partnership for Molecular Medicine, Aarhus University, Aarhus, Denmark

Christos Pliotas
University of St Andrews, St Andrews, Fife, United Kingdom

Stefan Raunser
Max Planck Institute of Molecular Physiology, Dortmund, Germany

Dirk J. Slotboom
University of Groningen, Groningen, The Netherlands

Annette Strege
MRC Laboratory of Molecular Biology, Cambridge, United Kingdom

Christopher G. Tate
MRC Laboratory of Molecular Biology, Cambridge, United Kingdom

Antoine M. van Oijen
University of Groningen, Groningen, The Netherlands

PREFACE

Membrane transport is fundamental in living cells. However, limited molecular details on function and structure left us with more open questions than answers on this enigmatic biological process. Since solving the first structure of a membrane protein in 1985, X-ray crystallography and single particle cryo-electron microscopy have yielded constantly increasing number of atomic and near-atomic structures, however their amount lingered behind expectations. Similar, dynamical methods like NMR, EPR and single-molecule FRET faced multiple experimental challenges, e.g., the low amount of expressed membrane proteins and the lack of adequate membrane-mimicking environment. In addition, functional descriptions of active and passive membrane transport processes essential to complement and interpret structure and dynamics are likewise difficult to obtain. Undoubtedly, working with membrane transport proteins remains challenging. The overwhelming struggles that might be encountered at each step have figuratively stigmatized this fascinating research field.

New approaches in structure-function methodology and membrane protein handling have recently surfaced, raising hope that we will enter a new era of unprecedented molecular detail in membrane transport. This includes new aspects in membrane protein expression, in thermo-stabilization, in providing more native membrane environment as well as new technical advances in obtaining molecular and functional information.

In this regard, we are proud to present in this volume of Methods in Enzymology a timely overview on cutting-edge techniques that have advanced the structure-function toolbox of membrane transporters and channels. The individual chapters describe innovations from expression over structure determination to dynamical characterization providing creative experimental solutions to typical problems when working with membrane proteins. Hence, this volume is a compilation of knowledge of long-standing membrane protein experts, who have contributed for years with their work to the field of membrane transport.

Stefan Raunser and co-workers begin the volume with a chapter describing a method to study membrane proteins by single particle analysis in a nearly native lipid bilayer using lipid nanodiscs. In case studies, they recount cryo-EM-specific benefits and challenges of this method focusing on membrane proteins.

The second chapter, by Klaus Fendler and colleagues, introduces an electrophysiology method, the Solid Supported Membrane method, for a cell-free functional characterization of electrogenic transporters and channels. The chapter provides in "hands-on" format detailed information on sample and buffer preparation as well as the interpretation of the results.

The groups of Poul Nissen and Jens Frauenfeld report on a complementary method to prepare bilayer nanoparticles using saposin A, a protein that is involved in lipid transport, as flexible and adaptive scaffold protein. The chapter by Dirk Slotboom and co-workers focuses on single-molecule imaging of liposome-reconstituted membrane transporters. Here membrane-reconstituted transporters with fluorescent labels are used to visualize the transport-associated dynamics by total internal reflection fluorescence microscopy at the single-molecule level.

The fifth chapter by members of the group of David Drew covers the important issue of screening for overexpression condition of membrane proteins. In a high-throughput setup in *Escherichia coli* the authors document the advantages of their "one-shot expression approach" approach on several well-known transport proteins.

Eric Geertsma and colleagues take a different approach on the same problem, the overexpression of membrane proteins in in *Escherichia coli*. Their procedure involves transcriptional fusions of small additional RNA combined with a tuneable promoter adjusting the membrane protein expression rate to the downstream folding capacity.

The following chapter by Joseph Mindell and Christopher Mulligan is an outstanding review of the very successful techniques that exploit the chemistry of cysteine to test the structure and mechanism not only of membrane proteins. The authors describe numerous applications reaching from cross-linking and labelling for monitoring conformational changes, to cross-linking strategies coupled to structural studies.

Cysteine modifications are also an important tool in Christos Pliotas's chapter; this chapter provides a detailed description of pulsed-EPR methodology using mechanosensitive ion channels as model to investigate oligomerization, conformation, and the effect of lipids on the protein.

The last chapter offers insights into strategies from Christopher Tate's lab for the thermo-stabilization of a GPCR–mini-G protein complex. This chapter summarizes theoretical and practical aspects of the methodology used for stabilizing this complex and provides examples of thermo-stability shifts.

We are convinced that your research will be stimulated by these new approaches and hope that you will enjoy this collection.

Christine Ziegler

CHAPTER ONE

Lipid Nanodiscs as a Tool for High-Resolution Structure Determination of Membrane Proteins by Single-Particle Cryo-EM

Rouslan G. Efremov*[,1], **Christos Gatsogiannis**[†,1], **Stefan Raunser**[†,1]
*Vrije Universiteit Brussel, Brussels, Belgium
[†]Max Planck Institute of Molecular Physiology, Dortmund, Germany
[1]Corresponding authors: e-mail address: rouslan.efremov@vib-vub.be; christos.gatsogiannis@mpi-dortmund.mpg.de; stefan.raunser@mpi-dortmund.mpg.de

Contents

Abstract

The "resolution revolution" in electron cryomicroscopy (cryo-EM) profoundly changed structural biology of membrane proteins. Near-atomic structures of medium size to large membrane protein complexes can now be determined without crystallization. This significantly accelerates structure determination and also the visualization of small bound ligands. There is an additional advantage: the structure of membrane proteins can now be studied in their native or nearly native lipid bilayer environment. A popular lipid bilayer mimetic are lipid nanodiscs, which have been thoroughly characterized and successfully utilized in multiple applications. Here, we provide a guide for using lipid nanodiscs as a tool for single-particle cryo-EM of membrane proteins. We discuss general methodological aspects and specific challenges of protein reconstitution into lipid nanodiscs and high-resolution structure determination of the nanodisc-embedded complexes. Furthermore, we describe in detail case studies of two successful applications of nanodiscs in cryo-EM, namely, the structure determination of the rabbit

Methods in Enzymology, Volume 594
ISSN 0076-6879
http://dx.doi.org/10.1016/bs.mie.2017.05.007

ryanodine receptor, RyR1, and the pore-forming TcdA1 toxin subunit from *Photorhabdus luminescens*. We discuss cryo-EM-specific hurdles concerning sample homogeneity, distribution of reconstituted particles in vitreous ice, and solutions to overcome them.

1. INTRODUCTION

It is predicted that ~23% of human genes code for membrane proteins (Uhlén et al., 2015). Because of their surface accessibility and central importance in many cellular mechanisms, they are attractive drug targets. Thus, in 2006, membrane proteins constituted more than 60% of all known drug targets (Overington, Al-Lazikani, & Hopkins, 2006).

Structural information is key for understanding the interaction between membrane proteins, in particular receptors, and small drug molecules in atomic detail. However, structural investigations on membrane proteins are also important to understand essential cellular mechanisms, including signaling, transport, and energy conversion at the molecular level.

While some structures have been obtained by electron crystallography (Raunser & Walz, 2009) and solution NMR (Hiller & Wagner, 2009), the majority of membrane protein structures has been determined by X-ray crystallography (Vinothkumar & Henderson, 2010). Although the structure of the first membrane protein was solved in 1985 (Deisenhofer, Epp, Miki, Huber, & Michel, 1985), membrane proteins have remained a challenge for structural biology over the past decades. Besides difficulties in obtaining ordered crystals in the case of crystallography, major bottlenecks have been the expression, solubilization, and purification of sufficient amounts of recombinant and biologically active membrane proteins. In recent years, significant progress has been made to overcome these chalenges, and many structures of important membrane proteins have been solved. These include transporters, respiratory complexes, G-protein-coupled receptors (GPCRs), ion channels, and adventitious membrane proteins (Vinothkumar & Henderson, 2010).

However, in many cases, in particular for mammalian membrane proteins, crystals of sufficient quality have not been obtained. Recently, the introduction of a new generation of direct electron detectors (McMullan, Faruqi, Clare, & Henderson, 2014) and the parallel development of image-processing software (Brilot et al., 2012; Li et al., 2013) increased the resolution limit of single-particle electron cryomicroscopy (cryo-EM) to atomic resolution. Cryo-EM of the post-"resolution revolution" era, as Werner Kühlbrandt coined it (Kühlbrandt, 2014), allows now for the direct structure determination of proteins to near-atomic resolution (between 3 and 4Å) without crystallization. Consequently, this new development has had a major impact on structural biology, and many structures of mammalian protein complexes that resisted crystallization so far have been determined in the last 3 years. Good examples are γ-secretase (Bai et al., 2015), respiratory complex I (Fiedorczuk et al., 2016; Zhu, Vinothkumar, & Hirst, 2016) and supercomplexes (Letts, Fiedorczuk, & Sazanov, 2016; Wu, Gu, Guo, Huang, & Yang, 2016), ATP synthase (Allegretti et al., 2015), different ions channels (Liao, Cao, Julius, & Cheng, 2013; Wu, Yan, et al., 2016), and transporters (Kim et al., 2015).

As for X-ray crystallography, pure protein samples are needed for single-particle cryo-EM studies. Typically, purification protocols for membrane proteins start with extracting the transmembrane proteins from membranes by solubilizing them in detergents. Protein purification by means of different techniques, mainly affinity and size-exclusion chromatography (SEC), are also performed in the presence of detergent to prevent the precipitation of the protein. Detergent-solubilized membrane proteins can be directly used for single-particle cryo-EM. Since cryo-EM allows observing individual unrestrained protein molecules embedded in a thin layer of amorphous ice, there are no restrictions on choosing a suitable detergent, which is usually not the case in crystallography. For the same reason, detergent-free systems can also be used to study the structures of membrane proteins. Among these systems are polymers such as amphipols (Popot et al., 2011) and styrene–maleic acid (SMA) copolymers (Scheidelaar et al., 2015), which can substitute detergents, or lipid-based systems, such as lipid nanodiscs (Schuler, Denisov, & Sligar, 2013), a saposin-lipoprotein nanoparticle system (e.g., Salipro®, EP 2745834 B1) (Frauenfeld et al., 2016), and even liposomes (Wang & Sigworth, 2010).

Since detergents must be used at concentration levels above the critical micelle concentration (CMC) to prevent precipitation of the protein, their concentration in the solution is high enough to decrease the contrast of cryo-EM images (Schmidt-Krey & Rubinstein, 2011). This is not the case when reconstituting membrane proteins in amphipols. Therefore, amphipols have been successfully used for the determination of the first high-resolution structures of the TRPV1 channel (Liao et al., 2013) and γ-secretase (Bai et al., 2015). In general, amphipols are popular polymers for reconstituting membrane proteins in single-particle cryo-EM (Popot et al., 2011). Although this process is more straightforward than reconstituting proteins into lipid environments, amphipols as well as detergents have the disadvantage that the system does not contain lipids. There are a number of reasons why studying the structure of membrane proteins in a lipid environment has advantages over examining them in detergent.

First of all, the lipid bilayer is the native environment for membrane proteins, and it has significantly different physical properties than its mimetic detergent micelle (Zhou & Cross, 2013). In contrast to planar lipid bilayers, detergent micelles typically have a spherical shape and differ considerably in their dielectric constant and electrostatic potential. In some cases these differences can have severe consequences, resulting in structures of nonnatively folded membrane proteins when crystallized from detergent solution (Cross, Murray, & Watts, 2013; Zhou & Cross, 2013). However, the effects are normally milder and while the proper fold is preserved when the membrane protein is solubilized in detergent or reconstituted in polymers, the protein dynamics, which are more sensitive to the environment than the structure, are very often influenced or altered. For example, the absorption spectrum of bacteriorhodopsin shifts upon solubilization in detergent and its photocycle kinetics are changed (Lam & Packer, 1983), even if its structure remains virtually unaltered. A more recent vivid example is related to the functional dynamics of the rabbit ryanodine receptor 1 (RyR1). Solubilized in Tween-20 (Yan et al., 2015) or CHAPS (Zalk et al., 2015), the structure of the channel is similar to the one determined in a lipid environment (Efremov, Leitner, Aebersold, & Raunser, 2015). However, in Tween-20 it is locked in a closed state and does not open under conditions that induce the full opening of the channel in lipid membranes (Bai, Yan, Wu, Li, & Yan, 2016).

The function of membrane proteins often depends on the direct interaction with lipids (Phillips, Ursell, Wiggins, & Sens, 2009). A layer of annular lipids surrounds the proteins. These lipids are in direct contact with the hydrophobic surface of the protein and tightly seal the membrane at the protein–membrane interface. Often they take distorted arrangements to adapt to the rough surface of the protein (Lee, 2004) and play an important role in stabilizing membrane protein oligomers (Gupta et al., 2017). However, most of the annular lipids are removed during the purification process, and in most cases they are not reintroduced afterward for crystallography or single-particle cryo-EM with detergents or polymers. In many crystal structures (Yeagle, 2014) and also recent cryo-EM structures (Liao et al., 2013) though, some very tightly bound lipid molecules remain bound during purification and can be resolved in the structures.

The optimal way to purify a membrane protein together with its natural annular lipids is to use detergent-free purification procedures that make use of SMA copolymers (Dörr et al., 2014, 2016) or nanodiscs (Civjan, Bayburt, Schuler, & Sligar, 2003). The development of these techniques is still in its infancy, and it has not yet been broadly applied. The most common approach is thus to purify the protein using detergents and reintroduce lipids afterward. In most cases, lipid composition can be varied and adjusted to mimic the natural membrane environment or to identify key lipids and lipid compositions. This has been done, for instance, for bacteriorhodopsin (Lee et al., 2015), the ABC transporter MbsA (Kawai, Caaveiro, Abe, Katagiri, & Tsumoto, 2011), or cytochrome P450 reductase (Das & Sligar, 2009).

Among the different available options for reconstituting membrane proteins in a lipid bilayer, the lipid nanodisc system currently presents the best-understood and -characterized method. Lipid nanodiscs were developed in the laboratory of Stephen Sligar (Bayburt & Sligar, 2003). They are composed of a patch of lipid bilayer, which is surrounded by membrane scaffold protein (MSP). MSP is a derivative form of apolipoprotein A-1 (Apo-A1) (Ritchie et al., 2009) and composed of short amphipathic helices. In an assembled nanodisc, two MSPs arrange in a parallel or antiparallel manner, forming a belt around the hydrophobic region of the lipid bilayer that includes a few hundred lipid molecules (Fig. 1). Constructs of MSP with

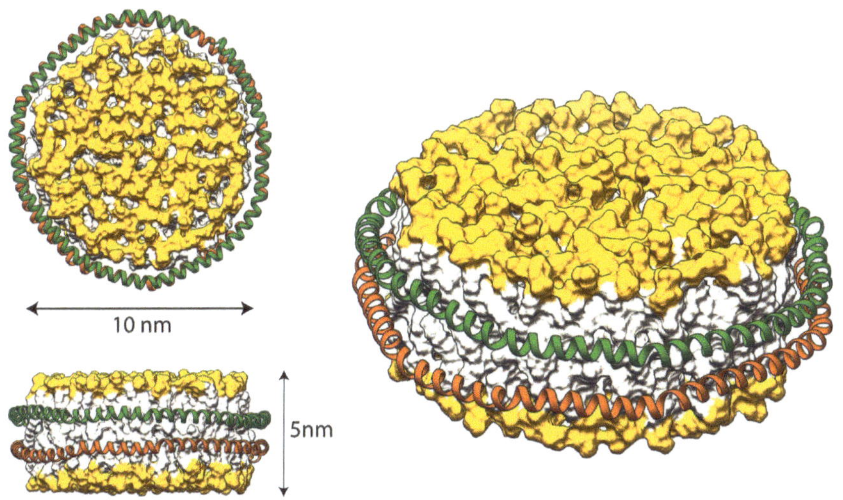

Fig. 1 Model of MSP1–POPC nanodisc structure. The two copies of MSP, surrounding the phospholipid bilayer, are shown in *green* and *brown*. Lipid headgroups are colored in *yellow*. The model was produced using the CHARMM-GUI server.

various numbers of amphipathic helices have been designed so that different sizes of nanodiscs can be obtained reaching diameters between ~10 and ~17 nm. This corresponds to ~100–600 lipids per nanodisc (Ritchie et al., 2009; Schuler et al., 2013). While MSP defines the size limits of the nanodiscs, as can be observed by negative stain electron microscopy (EM), the actual dimensions of the reconstituted discs vary significantly (Nasr et al., 2017). A recently reported method to generate covalently circularized nanodiscs using sortase produced significantly more homogeneous nanodiscs with diameters of up to 80 nm (Nasr et al., 2017).

Nanodiscs have shown to be potentially useful in studying membrane protein complexes, in particular fragile temporary signaling complexes, like those formed by GPCRs (Leitz, Bayburt, Barnakov, Springer, & Sligar, 2006). Detergents can disrupt the hydrophobic interactions involved in protein–protein and protein–lipid–protein interactions. For example, the β_2-adrenergic receptor, where mainly two palmitic acid and two cholesterol molecules mediate the interactions between two protomers of the natural dimer, binds only to G-proteins when reconstituted in nanodiscs but not when solubilized in detergents (Leitz et al., 2006). Other examples are the proton pump bacteriorhodopsin (Johnson et al., 2014) the ABC transporters MsbA (Kawai et al., 2011) and MalK (Bao & Duong, 2014). It has been shown that the functional properties of these proteins reconstituted in nanodiscs are similar to those in native membranes (Bao & Duong, 2014; Johnson et al., 2014; Kawai et al., 2011). Nanodiscs have been successfully used to determine high-resolution structures of membrane proteins using solution NMR (Hagn & Wagner, 2015) and single-particle cryo-EM (Efremov et al., 2015; Frauenfeld et al., 2011; Gao, Cao, Julius, & Cheng, 2016; Gatsogiannis et al., 2016; Shen et al., 2016).

Although nanodiscs are capable of mimicking the situation in a native lipid membrane, some biophysical properties of the lipids are slightly different and have to be taken into account when working with nanodiscs. For instance, the phase transition temperatures of lipids are higher and the thermal expansion coefficients of lipids are lower than in liposomes (Denisov, McLean, Shaw, Grinkova, & Sligar, 2005). Molecular dynamics simulations showed that the lipid layer that is in direct contact with MSP is perturbed due to the hydrophobic mismatch between MSP and lipids. However, 15–20 Å away from MSP, corresponding to two layers of lipid molecules, the properties of the lipids were similar to those in a planar bilayer (Schuler et al., 2013; Shih, Denisov, Phillips, Sligar, & Schulten, 2005).

Indeed, it has been shown for the ABC transporter MsbA that its activity is higher in larger nanodiscs containing more lipids (Kawai et al., 2011). EPR experiments with labeled lipids indicated that lipids within nanodiscs are generally more ordered than in liposomes and to some extent resemble a lipid bilayer in the presence of cholesterol (Stepien, Polit, & Wisniewska-Becker, 2015).

Examples of single-particle cryo-EM structures solved of membrane proteins reconstituted into lipid nanodiscs include the early work on the complex between the bacterial ribosome and the SecYE translocon (Frauenfeld et al., 2011), the muscular isoform of the rabbit ryanodine receptor (RyR1) (Efremov et al., 2015), more recent high-resolution studies of the Tc toxin from *Photorhabdus luminescens* (Gatsogiannis et al., 2016), the transient receptor potential channels TRPV1 (Gao et al., 2016) and PKD2 (Shen et al., 2016) (Fig. 2; Table 1). These examples demonstrate the feasibility of high-resolution 3-D structural analysis of membrane proteins embedded in nanodiscs, even with relatively small extracellular domains, as in the case of the transient receptor potential channels. In spite of the heterogeneous shape of the nanodisc, the transmembrane region of the proteins is one of the best-resolved parts of the structures.

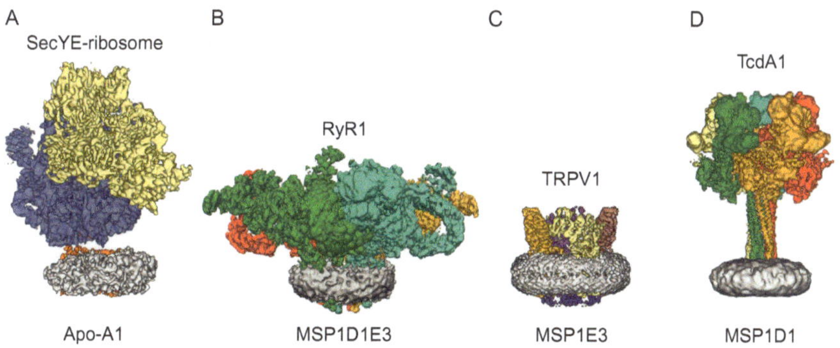

Fig. 2 Cryo-EM structures of proteins reconstituted in nanodiscs. (A–D) Examples of membrane proteins that have been incorporated in nanodiscs and successfully analyzed by cryo-EM include the bacterial ribosome-SecYE complex (7.1 Å, EMD-1858; Frauenfeld et al., 2011) (A), the rabbit ion channel RyR1 in the closed state (6.1 Å, EMD-2751; Efremov et al., 2015) (B), the human ion channel TRPV1 (3.0 Å, EMD-8117; Gao et al., 2016) (C), and the TcdA1 toxin subunit from *P. luminescens* in the pore state (3.5 Å, EMD-4068; Gatsogiannis et al., 2016) (D). The nanodisc density is shown in *gray*. The type of MSP protein used for reconstitution is indicated.

Table 1 Reconstitution conditions for membrane proteins in which medium- to high-resolution single-particle cryo-EM structures were obtained

Protein	Protein solubilizing detergent	MSP construct	Size of protein (total)	Lipids	Lipid/MSP/protein molar ratio	Lipid solubilizing detergent	Detergent removal	Reconstitution temperature (°C)	Resolution (Å)	References
SecYE-ribosome	DDM	Apo-A1	SecYEG 74 kDa bound to *E. coli* 70S ribosome (2.4 MDa)	*E. coli* phospholipids	185:11:1	Cholate	Bio-Beads	37	7.1	Frauenfeld et al. (2011)
RyR1	CHAPS	MSP1D1E3	2.3 MDa	POPC	140:2:1	CHAPS	Bio-Beads	4	6.1	Efremov et al. (2015)
TRPV1	DDM	MSP1E3, MSP2N2	330 kDa	Soybean polar lipids extract	600:4:1, 900:6:1, 400:4:1	DDM	Bio-Beads	4	3.2	Gao et al. (2016)
Tc toxin TcdA1	Tween-20	MSP1D1	1.4 MDa	POPC	440:8:1	CHAPS	Dialysis	4	3.5	Gatsogiannis et al. (2016)
PKD2 channel	DDM	MSP2N2	340 kDa	Soybean polar lipids	200:1:1 (or 0.25)	None	Bio-Beads	4	3.0	Shen et al. (2016)

Below, we first describe general considerations and approaches that need to be taken into account when reconstituting membrane proteins into lipid nanodiscs for single-particle cryo-EM. Next, we provide detailed protocols of reconstitution and cryo-EM sample preparation for two proteins, the rabbit ryanodine receptor RyR1 and the TcdA1 subunit of the Tc toxin complex from *P. luminescens*.

2. GENERAL CONSIDERATIONS

Reconstitution of membrane proteins into lipid nanodiscs is a self-assembling process. The most frequently applied method is to mix the purified membrane protein, lipids, and MSP in detergent solution and remove the detergent by dialysis or Bio-Beads (Fig. 3). This approach provides the best control over the reconstitution process and is the most reproducible. Each of the components, as well as the method for depleting the detergent from the solution, influences the assembly process (Ritchie et al., 2009). Therefore, the MSP construct, lipids, and detergents need to be chosen carefully. Single-particle EM studies require structurally homogeneous samples. Hence, the aim is to produce homogeneous nanodiscs with a single reconstituted protein, almost no empty nanodiscs, and no excess of proteins or lipids. To achieve this, the molar ratios between the membrane protein, MSP, and lipids during reconstitution should reflect their composition in the assembled discs.

Another way to reconstitute membrane proteins is to incorporate them into preformed empty nanodiscs. This method is especially suited for

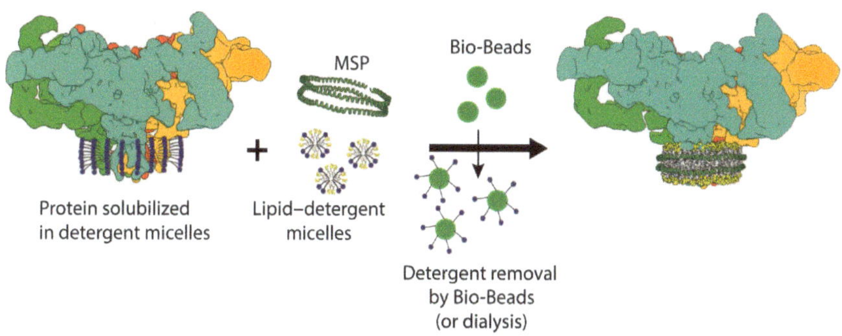

Fig. 3 Schematic of membrane protein reconstitution in nanodiscs. The detergent-solubilized membrane protein, lipids, and MSP are mixed and the detergent is removed by dialysis or adsorption to Bio-Beads. This results in the reconstitution of the membrane protein in nanodiscs.

adventitious membrane proteins, such as pore-forming toxins, that can directly penetrate membranes (Gatsogiannis et al., 2016).

Membrane proteins can also be reconstituted into nanodiscs directly from a detergent-solubilized membrane fraction before protein purification. This direct reconstitution approach has been used on bacterial (Civjan et al., 2003; Duan, Civjan, Sligar, & Schuler, 2004; Marty, Wilcox, Klein, & Sligar, 2013) and eukaryotic (Roy, Pondenis, Fan, & Das, 2015; Shirzad-Wasei et al., 2015) proteins. It can be of advantage for membrane proteins that are unstable in detergent or require weakly bound native lipids for protein function. When using the direct reconstitution approach, the lipid composition and concentration cannot be controlled. This may result in less homogeneous nanodiscs. Furthermore, the additional reconstitution of smaller membrane proteins together with the protein of interest in the same nanodisc cannot be prevented. Since many MSPs not only surround the protein of interest but also other membrane proteins, different concentrations and ratios are needed. In general, the requirements for this method differ from those of the standard approach.

When preparing nanodiscs, an optimal MSP construct that determines the size of the nanodisc should be first selected (see above). Optimally, the nanodisc should be just large enough to accommodate the protein, but keep at least two layers of lipids between protein and MSP to avoid unwanted interactions with the MSP and ensure a proper mimic of the lipid bilayer environment. Too large nanodiscs may result in less homogeneous dimensions of the discs and the "floating" of the protein within the disc (Fig. 5H and section describing TcA reconstitution). This can in turn interfere with particle alignment during image processing. In addition, the larger the nanodiscs, the higher the probability is to reconstitute more than one protein in a disc (Fig. 5F). A range of MSP constructs generated by the Sligar lab (Grinkova, Denisov, & Sligar, 2010) are available from Addgene (www.addgene.org). They allow assembling nanodiscs with diameters ranging from ~10 nm (MSP1D1 construct holding ~100 lipids) up to ~17 nm (MSP2N3 construct surrounding ~600 lipid molecules) (Ritchie et al., 2009; Schuler et al., 2013). The MSPs can be expressed in *Escherichia coli* and purified by affinity chromatography (see Ritchie et al., 2009 for a detailed protocol). MSPs as well as preformed nanodiscs are also commercially available.

The lipids should be chosen based on prior knowledge of the protein's native environment or on previous successful reconstitutions in liposomes or other bilayer systems. Besides total lipid extracts, for example, total *E. coli* or soybean lipid extracts (Gao et al., 2016; Hernández-Rocamora, García-Montañés, Rivas, & Llorca, 2012; Ritchie et al., 2009), mostly synthetic lipids have been successfully used. In particular, zwitterionic and charged phospholipids like 1,2-dimyristoyl-*sn*-glycero-3-phosphocholine (DMPC), 1-palmitoyl-2-oleoyl-*sn*-glycero-3-phosphocholine (POPC), 1,2-dipalmitoyl-*sn*-glycero-3-phosphocholine (DPPC), 1-palmitoyl-2-oleoyl-*sn*-glycero-3-phosphoserine (POPS), 1,2-dioleoyl-*sn*-glycero-3-phosphoserine (DOPS), 1,2-dimyristoyl-*sn*-glycero-3-phosphoglycerol (DMPG), and 1-palmitoyl-2-oleoyl-*sn*-glycero-3-phosphoglycerol (POPG) alone or in mixture have achieved good results (Bao, Dalal, Wang, Rouiller, & Duong, 2013; Nasr & Singh, 2014; Orlando et al., 2014; Ritchie et al., 2009; Wadsäter et al., 2013). The gel-to-liquid phase transition temperature of the lipid bilayer is an important parameter to consider when preparing nanodiscs. The lipid bilayer should be in the liquid phase under experimental conditions, while optimal reconstitution occurs at temperatures around the phase transition temperature.

Sodium cholate is the most commonly used detergent for producing nanodiscs (Ritchie et al., 2009). Even though the protein is solubilized in another detergent, it is normally used for solubilizing the lipids (Bayburt, Grinkova, & Sligar, 2006) and for reconstituting proteins in preformed nanodiscs (Gatsogiannis et al., 2016). The presence of sodium cholate, however, should not interfere with protein stability. Alternatively other detergents have been used to solubilize lipids, namely, *n*-dodecyl-β-D-maltoside (DDM) (Boldog et al., 2006), CHAPS (Efremov et al., 2015), *n*-octyl-β-D-glucopyranoside (OG) (Justesen et al., 2013), Triton X-100 (Mi et al., 2008), and digitonin (Eggensperger, Fisette, Parcej, et al., 2014).

For an optimal reconstitution, the protein:MSP:lipid ratio should exactly match the composition of the nanodiscs and the protein:MSP molar ratio should be kept at 1:2 unless an excess of empty nanodiscs is desired.

The optimal amount of lipids to reach a certain diameter of empty nanodiscs has been experimentally determined for some MSP constructs and common synthetic lipids (see table 11.2 in Ritchie et al., 2009). When a membrane protein is reconstituted, the number of lipids should be adjusted to account for the lipids displaced by the protein. To estimate the membrane cross-section of an α-helical membrane protein, one can assume an area of $\sim 140\,\text{Å}^2$ per transmembrane helix (Ritchie et al., 2009). Phospholipids occupy, respectively, an area of 52, 57, and 69 Å^2 per molecule for DPPC, DMPC, and POPC (Nagle & Tristram-Nagle, 2000).

Thus the number of lipid molecules per nanodisc can be estimated as such:

$$N_{\mathrm{lip}} = 2^{*}\left(S_{\mathrm{ND}} - S_{\mathrm{prot}}\right)/S_{\mathrm{lip}}$$

N_{lip} is the number of lipids per nanodisc, S_{ND} is the surface area of the lipid nanodisc, S_{prot} is the area of the protein's transmembrane section, and S_{lip} is the area occupied by one phospholipid molecule. Since membranes are lipid bilayers, a factor of two has to be introduced.

The surface area of the lipid nanodisc can be estimated from the number of amino acids, *M*, in MSP using the empirical formula (Denisov, Grinkova, Lazarides, & Sligar, 2004):

$$S_{\mathrm{ND}} = (0.423^{*}M - 9.75)^{2}\left(\mathring{A}^{2}\right)$$

The approximate membrane cross-section of a protein with N_{TM} transmembrane helices is

$$S_{\mathrm{prot}} = 140^{*}N_{\mathrm{TM}}\left(\mathring{A}^{2}\right)$$

The reconstitution process is initiated by removing the detergent using Bio-Beads or dialysis lasting from few hours up to several days at temperatures between 4°C and 37°C. Reconstitution is optimally performed close to the phase transition temperature of lipids (Ritchie et al., 2009). High glycerol concentrations in the buffer should be avoided since glycerol interferes with the reconstitution process at concentrations above 4% (Ritchie et al., 2009). It is often useful to analyze the reconstituted nanodisc by SEC. A clear peak corresponding to the reconstituted protein should be observed and if both protein and lipid concentrations are measured in the peak, they should closely correspond to the calculated protein:lipid ratio. If the reconstituted protein is to be further used for structural studies, the homogeneity of the nanodisc dimensions should be determined using, for example, negative stain EM and/or cryo-EM. The absence of aggregates, liposomes, and homogeneous dimensions of the protein particles indicates a successful reconstitution (Figs. 4A and 5D–F). In practice, however, it is difficult to obtain completely homogenous nanodiscs (Nasr et al., 2017) and a certain degree of inhomogeneity has to be taken into account during cryo-EM image processing. In the case of the magnesium channel CorA, for example, inhomogeneity of nanodiscs interfered with particle alignment and precluded a high-resolution 3-D reconstruction (Matthies et al., 2016).

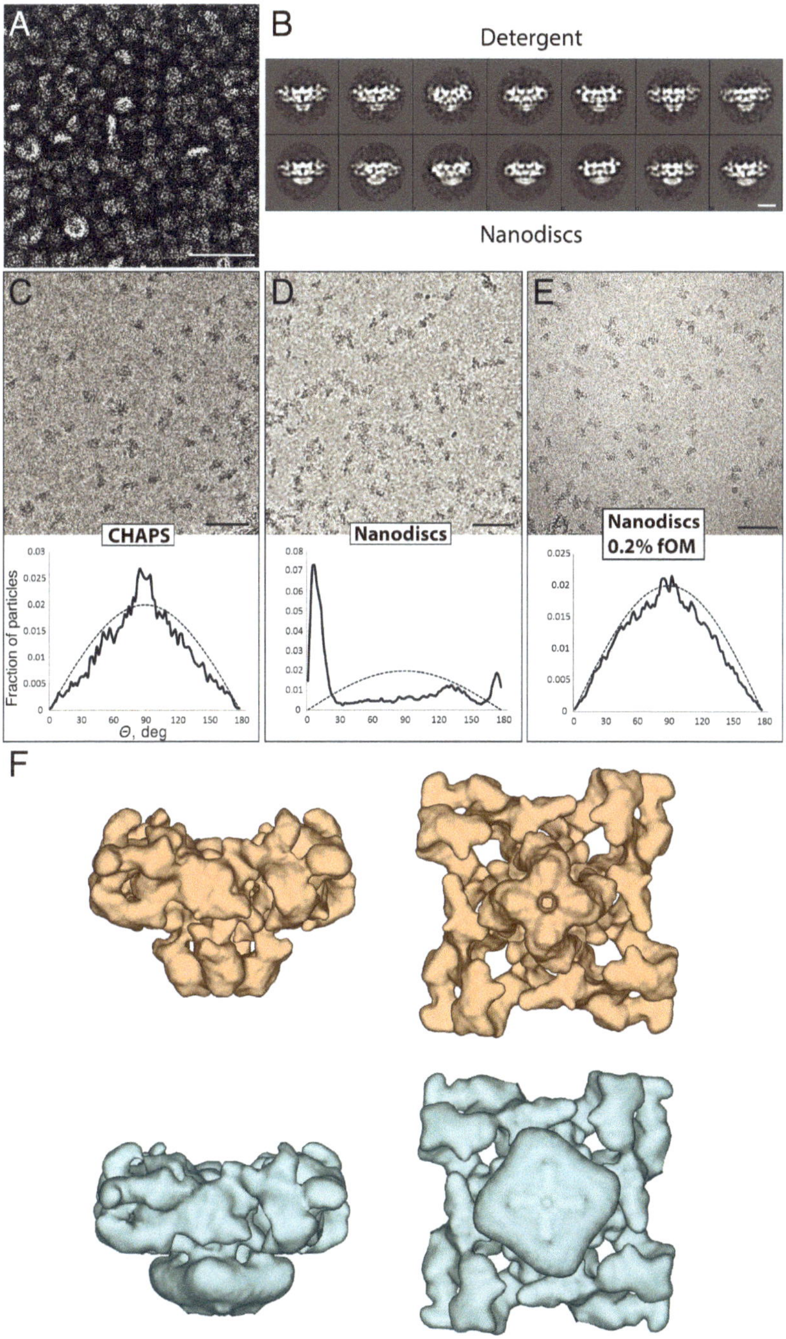

Fig. 4 Single-particle EM of RyR1 reconstituted into lipid nanodiscs. (A) An image of negatively stained RyR1 reconstituted into lipid nanodiscs. It shows an example of suboptimal reconstitution resulting in small liposome-like structures and large elongated *(Continued)*

If the reconstitution is not optimal, the following parameters can be varied: the amount and type of the lipids, the MSP construct, and the detergent. In addition, the temperature and rate at which the detergent is removed can be modified. The optimal protein:MSP:lipid ratio needs to be fine-tuned by scanning various ratios around the calculated one. An alternative approach is to perform the reconstitution under conditions that result in an excess of empty nanodiscs (Table 1) followed by affinity and/or SEC to remove empty nanodiscs (Fig. 5F).

3. SINGLE-PARTICLE CRYO-EM OF THE RYANODINE RECEPTOR RyR1 RECONSTITUTED INTO LIPID NANODISCS

In the following paragraph, we describe how we reconstituted the rabbit ryanodine receptor RyR1 into lipid nanodiscs and determined its structure using single-particle cryo-EM. Ryanodine receptors (RyRs) are tetrameric cation channels with high conductance (Yuchi & Van Petegem, 2016) that reside in the sarcoplasmic or endoplasmic reticulum (SR, ER). RyRs play a central role in muscular contraction, where the channels control calcium release from the SR, initiating muscle contraction. RyRs are gated by Ca^{2+} and through direct interaction with the voltage-gated calcium channel, $Ca_v1.1$ (Wu, Yan, et al., 2016), which resides in the cytoplasmic membrane. The gating of RyRs is tuned by interaction with multiple small molecules and proteins (Van Petegem, 2014). RyRs are the largest known ion channels with a molecular weight of 2.3 MDa for the

Fig 4—cont'd or discoid lipidic structures along with densely packed particles of RyR1. (B) 2-D class averages of vitrified RyR1 displaying side views of the channel solubilized in CHAPS or reconstituted into lipid nanodiscs. Additional density surrounding the transmembrane domain corresponds to nanodiscs. Scale bar, 10 nm. (C–E) Cryo-EM images and angular distribution of RyR1 solubilized in CHAPS (C), or reconstituted in nanodiscs (D), or reconstituted in nanodiscs and vitrified in the presence of 0.2% fluorinated octyl maltoside (E). Distribution of the angle between the fourfold symmetry axis of RyR1 and the normal of the sample plane is shown as a plot for each condition. The Θ distribution for particles with random orientations is described by sinΘ shown in the plots using a *dashed line*. Scale bars, 100 nm. (F) 3-D reconstructions of RyR1 solubilized in CHAPS (*brown*) and reconstituted in nanodiscs (*cyan*) show that the density corresponding to the nanodisc is *square with rounded corners*. The cryo-EM images were acquired on a JEOL JEM 3200 TEM, operated at 200 kV on a CMOS TVIPS F816 camera and processed in SPARX (Hohn et al., 2007).

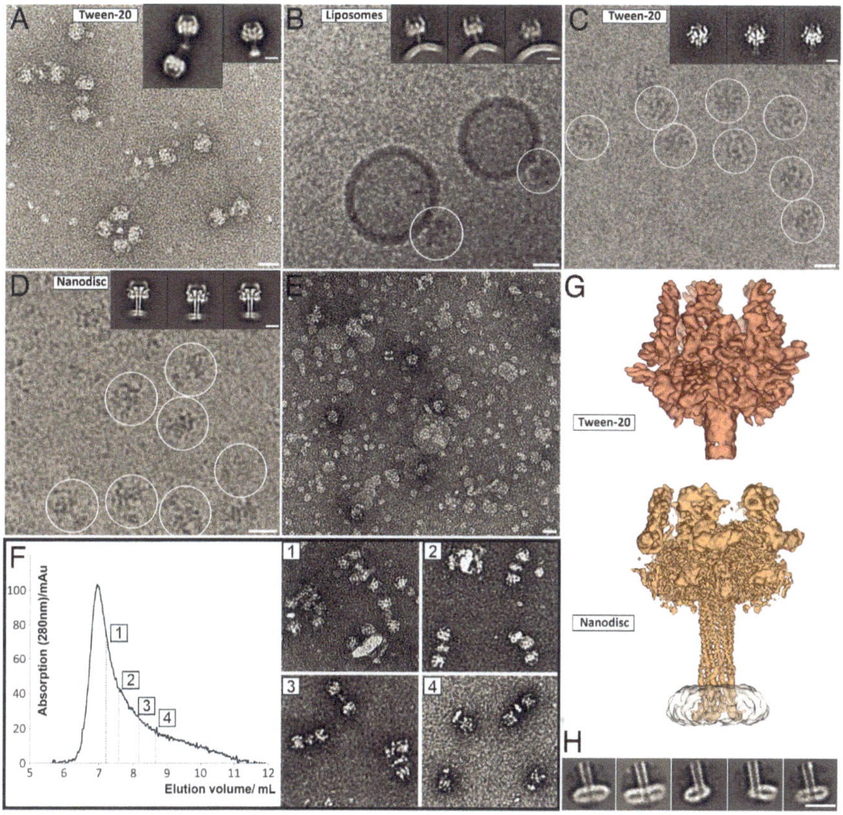

Fig. 5 Single-particle EM of the TcdA1 pore complex embedded in lipid nanodiscs. (A) Typical micrograph area of the negatively stained TcdA1 pore, induced at pH 11 in the presence 0.05% Tween-20. Note that complexes aggregate at the tip of the channel, which is the hydrophobic transmembrane region. The *insets* show representative class averages of one or two tip-to-tip-oriented TcdA1 complexes. (B–D) Typical subareas of cryo-EM images of TcdA1 pores embedded in liposomes (B), in the presence of Tween-20 (C), and reconstituted in lipid nanodiscs (D). Some particles are highlighted with *white circles*. *Insets* show representative class averages. (E) Typical micrograph area of negatively stained TcdA1 embedded into lipid nanodiscs, as an example of a sub-optimal protein:MSP ratio. Note the excess of empty nanodiscs and small liposome-like structures. (F) Gel filtration chromatogram of TcdA1 reconstituted in lipid nanodiscs. Fractions used for EM are marked by *dashed lines*, and typical micrograph areas are shown. (G) 3-D volumes of TcdA1 in Tween-20 (*pink*) and TcdA1 reconstituted in nanodiscs (*brown*). Note the absence of the transmembrane domain in the Tween-20 reconstruction of the pore state. Cryo-EM images of TcdA1 in Tween-20 and nanodiscs were taken on a JEOL JEM 3200 EM operated at 200kV on a CMOS TVPIS F816 camera and a FEI Titan Krios operated at 300kV on a Falcon II direct detector, respectively. (H) 2-D classification of TcdA1 reconstituted into lipid nanodiscs focused on the transmembrane domain. Note the variability in the diameter of the nanodisc and position of the transmembrane domain relative to the nanodisc. Scale bars for micrographs and class averages are, respectively, 20 and 10nm.

tetramer. The square-shaped cytoplasmic domain of the channel constitutes 80% of the protein and has an edge length of ~270 Å (Van Petegem, 2014).

For many years, the skeletal muscle isoform RyR1 from rabbit has been a model specimen for single-particle cryo-EM. The most common purification protocol of RyR1 uses a mixture of CHAPS with phospholipids to keep the protein in a stable and active form (Hawkes, Díaz-Muñoz, & Hamilton, 1989). For single-particle cryo-EM, the lipids were removed shortly before plunge-freezing to improve the image contrast (Samsó, Feng, Pessah, & Allen, 2009; Samsó, Wagenknecht, & Allen, 2005). Our primary motivation for reconstituting RyR1 into lipid nanodiscs for single-particle cryo-EM was to leave it in a lipid environment in order to ensure the active conformation of the protein. At the time, only low- to medium-resolution structures of RyR1 were available (Samsó et al., 2005). The number of transmembrane helices was not known and we could only estimate the diameter of the transmembrane region. Since its largest dimensions are between ~100 and ~110 Å, we chose the MSP1D1E3 construct, which forms nanodiscs with a diameter of ~12 nm (Denisov et al., 2004), for the reconstitution.

We decided to use the synthetic phospholipid POPC for the nanodiscs because it is a common lipid for reconstituting membrane proteins and its phase transition temperature of −2°C (https://avantilipids.com/tech-support/physical-properties/phase-transition-temps/) is close to the temperature at which we typically work with the protein.

We calculated the number of lipids per nanodisc as follows:

$$N_{\text{lip}} = \pi/4\left(D_{\text{ND}}{}^2 - D_{\text{RyR1}}{}^2\right)$$

Here, D_{ND} is the diameter of the nanodisc (120 Å), and D_{RyR1} is the average diameter of the membrane-embedded part of RyR1 (90 Å). This results in a membrane area of ~5000 Å^2 corresponding to ~145 POPC lipids per nanodisc. We have provided a detailed description of the reconstitution protocol:

- RyR1 was purified by solubilizing SR membranes in CHAPS in the presence of soy lecithin following SEC and sucrose gradient centrifugation. The excess of lipids was removed by an additional SEC in 0.8% CHAPS without lipids (see Efremov et al., 2015 for details).
- POPC, solubilized in chloroform, was purchased from Avanti Polar Lipids (USA). After evaporating the solvent under a stream of nitrogen, the phospholipid was rehydrated in 10 m*M* MOPS, pH 7.4 at a

concentration of 10 mg/mL. The lipid stock solution was frozen and stored in aliquots at −80°C.

- Prior to the reconstitution process, POPC was solubilized at a concentration of 3.8% (42 m*M*) in 10 m*M* MOPS, pH 7.4, 150 m*M* NaCl, and 8.5% CHAPS.
- MSP1E3D1 was expressed in *E. coli* using the pET expression system and purified using Ni-NTA affinity chromatography. A detailed protocol can be found elsewhere (Ritchie et al., 2009). Afterward, MSP was dialyzed overnight against 20 m*M* MOPS, pH 7.4, 100 m*M* NaCl at 4°C. The purified protein was then incubated overnight with TEV proteases at a ratio of 1:200 (w/w) MSP:TEV in the presence of 1 m*M* DTT in order to cleave the His-tag. Subsequently, the TEV proteases and cleaved peptide were removed by passage through a Ni-NTA column. Concentrated to ~10 mg/mL, MSP was aliquoted and stored at −80°C until used.
- Reconstitution was accomplished by mixing 400 n*M* (0.9 mg/mL) solubilized RyR1, 800 n*M* MSP (0.026 mg/mL) and 56 μ*M* (0.043 mg/mL) POPC in 20 m*M* MOPS, pH 7.4, 700 m*M* NaCl, 2 m*M* DTT, 1 m*M* EGTA, 10% sucrose, and 0.5% CHAPS. The mixture was incubated for 2 h at 4°C.

Next, the detergent was removed by adsorption to Bio-Beads (Bio-Rad) overnight at 4°C. The capacity of Bio-Beads to absorb different detergents has been tabulated by Rigaud et al. (Lévy et al., 1999). To ensure complete detergent absorption, we added two to three times more Bio-Beads than the theoretical calculated amount.

To ensure that all the detergent is removed by Bio-Beads, the concentration of the detergent in the protein solution needs to be measured beforehand by, for example, a colorimetric assay (Urbani & Warne, 2005) and the amount of Bio-Beads adjusted accordingly. This is particularly important for detergents with a low CMC. In this case, the protein-bound detergent constitutes the major part of the total amount of detergent (Crichton et al., 2013; Ilgü et al., 2014). Care should be taken that not too excessive amounts of Bio-Beads are used because they can absorb lipids along with the detergent (Lévy et al., 1999).

We first visualized reconstituted RyR1 by negative stain EM, which showed that the protein had the expected tetrameric shape, and aggregates or liposomes were absent (Fig. 4A). RyR1 in nanodiscs was then plunge-frozen using a cryoplunge CP3 (Gatan) following a standard protocol: 2 μL of RyR1 in nanodiscs at a concentration of 1–2 mg/mL were applied onto glow-discharged C-flat CF-1.2/1.3-4C carbon grids (Protochips), automatically blotted from both sides for 1 s and then plunge-frozen.

Cryo-EM images of RyR1 nanodiscs revealed a notable difference in particle distribution compared to the CHAPS-solubilized protein (Fig. 4C and D). In nanodiscs, RyR1 had the tendency to cluster nonspecifically. The most pronounced effect, however, was on the orientation of the particles. Upon reconstitution into nanodiscs, RyR1 was oriented with its symmetry axis perpendicular to the plane of the grid (Fig. 4D). This is detrimental for calculating a 3-D single-particle reconstruction. Surprisingly, the analysis of the particle distribution showed a single strong peak near zero degrees Θ angle (Fig. 4D), indicating that the particles stuck to only one air–water interface instead of being equally distributed in the vitreous ice. We assumed that in the absence of detergent hydrophobic interactions between the soluble regions of the RyR1 molecules were enhanced, resulting in the observed aggregation and strong interaction with the air–water interface. To solve this problem, we screened a number of small molecules that modulate hydrophobic interactions but, unlike detergents, preserve the lipid bilayer. We found that fluorinated octyl maltoside (fOM, Anatrace) resolved the problem when added at a concentration of 0.2%. The particles were then randomly oriented and did not aggregate (Fig. 4E). Due to the lipophobic property of their perfluorinated tails, fluorinated detergents do not solubilize lipid bilayers (Breyton, Chabaud, Chaudier, Pucci, & Popot, 2004). Using negative stain EM, we showed that even at a concentration of 1%, fOM does not solubilize nanodiscs.

During image processing, the nanodiscs became clearly visible in 2-D class averages as an additional density around the membrane region of RyR1 (Fig. 4B). In the case of CHAPS-solubilized RyR1, this density is absent, indicating that the size and therefore the electron scattering density of nanodiscs are significantly larger than that of a detergent micelle. Accordingly, in the 3-D reconstruction the density corresponding to nanodiscs was considerably larger than the one of CHAPS micelles. Unexpectedly, the nanodiscs had a square shape with rounded corners (Fig. 4F). This was also the case for tetrameric TRP ion channels in nanodiscs (Gao et al., 2016). We do not know whether the square shape of the nanodiscs is an artifact of image processing or whether the shape of the nanodiscs adapts to the shape of the reconstituted protein. It has been shown that larger empty nanodiscs can take a noncircular shape at certain lipid:MSP ratios (Nasr et al., 2017). The structures of more membrane proteins with different shapes and symmetries reconstituted into nanodiscs with different diameters will be needed to answer these open questions.

The horizontal density profile of the nanodisc also corresponds well with the expected density profile of a lipid bilayer. Two high density regions separated by a distance of ~30–35 Å at both surfaces of the bilayer corresponding to phospholipid headgroups and a lower density at the center of the bilayer corresponding to the hydrophobic tails (Fig. 4B).

Finally, in case of RyR1 its structure was determined under three conditions: solubilized in CHAPS (Zalk et al., 2015), solubilized in Tween-20 (Yan et al., 2015), and in nanodiscs (Efremov et al., 2015). The structure of the transmembrane domains under all three conditions is virtually identical when the structures of the closed state determined at resolutions between 3.8 and 6 Å are compared. The data set of RyR1 reconstituted into nanodiscs contained significantly less particles than those of detergent-solubilized proteins, which resulted in a lower resolution of the reconstruction and did not allow resolving protein side chains. The structure of TRPV1 reconstituted into nanodiscs and determined by single-particle cryo-EM (Gao et al., 2016) demonstrated that the high-resolution reconstruction of a transmembrane region embedded into nanodiscs is possible, even for proteins of relatively small dimensions in comparison to the nanodiscs (Fig. 2C). In this study not only protein side chains were well resolved inside the nanodiscs, but also specifically bound lipids and ligands were clearly visualized (Fig. 6), demonstrating the power of nanodiscs for studies of lipid protein interactions in a near-native lipid bilayer environment.

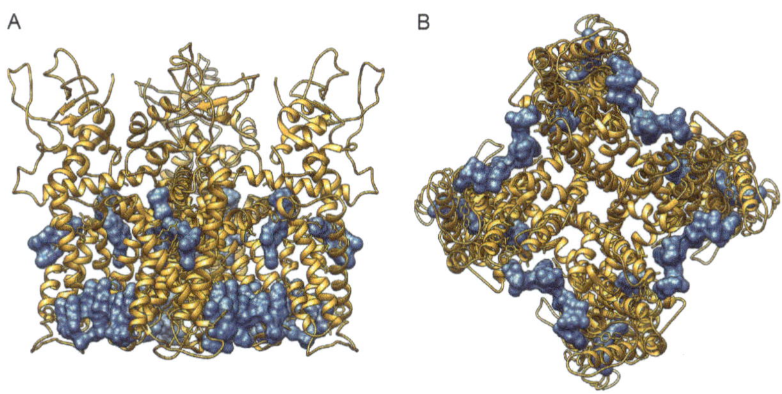

Fig. 6 Annular lipids bound to TrpV1. (A and B) Side (A) and top view (B) of the molecular model of apo TrpV1 depicted as ribbon diagram (*gold*) (PDB-ID: 5IRZ). Annular lipids specifically bound to the protein are shown in surface representation (*blue*).

4. EM ANALYSIS OF THE TcdA1 PORE COMPLEX EMBEDDED IN LIPID NANODISCS

In the following paragraph, we describe how we reconstituted the TcdA1 (TcA) component of the Tc toxin complex from *P. luminescens* into lipid nanodiscs and determined its structure using single-particle cryo-EM. Insecticidal Tc toxin complexes were first identified in *P. luminescens* (Bowen & Ensign, 1998; Bowen et al., 1998), but have later been found in other bacteria, including human pathogens. Tc toxin complexes are soluble virulence factors that perforate the host membrane similar to a syringe and translocate a toxic enzyme into the host cell. This damages and ultimately kills the host cell (Gatsogiannis et al., 2013; Lang et al., 2010). TcA (TcdA1 in *P. luminescens*), one of the three components of Tc complexes, forms a large pentameric structure with a molecular weight of 1.4 MDa. It is responsible for channel formation and the transfer of the biologically active enzyme into the host cell (Gatsogiannis et al., 2013; Meusch et al., 2014). Membrane insertion of TcA is triggered at low or high pH (Fig. 5A) (Gatsogiannis et al., 2013). Previously we solved the structure of TcdA1 from *P. luminescens* in the prepore state, i.e., before membrane penetration, by X-ray crystallography and cryo-EM (Gatsogiannis et al., 2013; Meusch et al., 2014). To obtain the structure of TcdA1 in the pore state, we shifted the pH level and simultaneously reconstituted the complex into liposomes. Single-particle cryo-EM analysis using a CMOS detector revealed the first structure of TcdA1 in a lipid bilayer (Fig. 5B) (Gatsogiannis et al., 2013). The resolution, however, of the resulting 3-D reconstruction was limited to 20 Å. This was probably due to the relatively thick ice layer resulting from the large diameter of the proteoliposomes, the strong signal of the membrane bilayer, and image drift.

In a later study (Meusch et al., 2014), we induced TcA pore formation in solution and stabilized the protein by detergent (0.05% Tween-20). Subsequently, we determined the structure of the complex at 9 Å by single-particle cryo-EM. However, already at the stage of 2-D classification, we noticed that the transmembrane domain appeared blurry, indicating that it was not sufficiently stabilized by the detergent (Fig. 5C). Consequently, the transmembrane domain was not resolved in the resulting 3-D volume (Fig. 5G).

To finally reveal the structure of the transmembrane region, we reconstituted the toxin complex into nanodiscs. Besides ensuring a native-like lipid environment, we anticipated that we could obtain more stable and homogeneous TcA pore complexes suitable for high-resolution cryo-EM. According to our low-resolution studies of the TcA pore and the crystal structure of the prepore state, the outer diameter of the transmembrane domain measures ~40 Å. Therefore, we chose a small MSP construct, namely, MSP1D1, that forms nanodiscs with a diameter of 95–100 Å (Denisov et al., 2004). To test the suitability of POPC lipids for TcA insertion, we first reconstituted the protein in POPC liposomes and analyzed the reconstitution efficiency by negative stain EM.

In contrast to a typical detergent-solubilized membrane protein, the TcA pore conformation must be induced by shifting the pH before the protein can be reconstituted into nanodiscs. Therefore, we coupled the reconstitution process directly with a shift to a high pH. Based on the available low-resolution structure, we estimated the TcA pore membrane section to be 1300 $Å^2$, and as a starting point we used 110 lipids per nanodisc for the reconstitution. We also mixed protein, MSP, and lipids at various molar ratios and monitored the reconstitution success by SEC and negative stain EM. The best result was achieved with a large excess of empty nanodiscs (1 TcdA1: 4 nanodiscs), as described below. A detailed description of the reconstitution protocol follows:

- The POPC lipids were prepared as described for RyR1.
- TcdA1 from *P. luminescens* was cloned, overexpressed, and purified as described previously (Gatsogiannis et al., 2016). The final purification step was performed on a Superose 6 10/300 GL column (GE Healthcare Life Sciences) in a buffer containing 50 m*M* Tris–HCl, pH 8.0, 100 m*M* NaCl, and 0.05% Tween-20.
- POPC was solubilized at a concentration of 1 mg/mL in 10 m*M* MOPS, pH 7.4, 150 m*M* NaCl, and 10% CHAPS in order to use in nanodisc reconstitution.
- MSP1D1 was purchased from Cube Biotech (Germany) and solubilized in 20 m*M* Tris–HCl, pH 8, 250 m*M* NaCl, and 1% sodium cholate.
- MSP1D1 and POPC were incubated for 30 min in buffer containing 20 m*M* Tris–HCl, pH 8, 250 m*M* NaCl, and 1% (w/v) sodium cholate at 4°C.
- Reconstitution was performed by mixing TcdA1, MSP1D1, and POPC in a molar ratio of 1:8:440 in 20 m*M* Tris–HCl, pH 8, 250 m*M* NaCl, and 1% sodium cholate. After incubation for 1 h, the sample was dialyzed against buffer at pH 11 without detergents (20 m*M* CAPS, pH 11, and 250 m*M* NaCl) overnight at 4°C to induce the pH-dependent conformational change of TcdA1 and reconstitution into nanodiscs. We also

tried to remove the detergent by Bio-Beads instead of dialysis, but this resulted in a loss of a large amount of protein.

- The quality of the reconstituted TcdA1 protein in nanodiscs was checked by negative stain EM. Before the sample was used for cryo-EM, the TcdA1–nanodiscs were further purified by SEC using a Superose 6 column in 20 m*M* CAPS, pH 11, 250 m*M* NaCl.

A negative stain EM analysis directly after reconstitution revealed that one to three TcdA1 molecules were embedded in the same nanodisc. A large fraction of particles showed two TcdA1 proteins embedded from opposite sides. To isolate the population of nanodiscs harboring only a single TcdA1 molecule, we performed an additional purification step by SEC. This step also removed the excess of empty nanodiscs. We screened fractions of the elution peak by negative stain EM and used the best one for single-particle cryo-EM (Fig. 5F). Before vitrification, we added 0.01% fOM (Anatrace) to the buffer to obtain a homogeneous thickness of the amorphous ice. Subsequently, 4 μL of the sample (0.04 mg/mL) was applied to freshly glow-discharged carbon grids (C-flat, 2/1, Protochips), covered with a thin film of continuous carbon and vitrified after double-side blotting at 90% humidity by plunging into liquid ethane using a Cp3 (Cryoplunge3, Gatan).

We collected 1,957 movies with a Falcon II detector on a Cs-corrected Titan Krios (FEI) and manually selected ~30,000 TcdA1–nanodisc particles (for more details, see Gatsogiannis et al., 2016). The 2-D class averages showed the typical mushroom-shaped TcdA1 pores and a density corresponding to the nanodisc. The transmembrane channel surrounded by the nanodisc was clearly resolved and in an obvious open conformation (Fig. 5D). In contrast to detergent the native-like environment of the nanodisc induced the opening of the pore and stabilized the transmembrane channel in its open conformation. Moreover, the 2-D class averages showed a high level of structural detail, indicating that the nanodiscs did not negatively influence the image alignment.

We finally solved the 3-D structure of the TcdA1 pore at an average resolution of 3.5 Å (Fig. 5G) (Gatsogiannis et al., 2016). However, in contrast to other membrane proteins (Bai et al., 2015; Gao et al., 2016), the transmembrane domain was less resolved than the other domains. In an attempt to improve the resolution in the transmembrane region, we reselected the particles with the box center shifted to the center of the nanodisc and reextracted the particles in smaller boxes. Then we performed focused 2-D and 3-D classifications and subsequently a local refinement. The resolution of the transmembrane helices could not be improved, but the analysis revealed that TcdA1 was positioned at different locations inside the nanodiscs. This indicates that the relatively large diameter of the nanodisc allows the protein to freely move inside the boundaries of the MSP (Fig. 5H). To find out why the resolution of the transmembrane helices was limited, we calculated MD simulations of this region in a model membrane. The simulations indicated that the transmembrane domain of the TcdA1 channel is very flexible and partially loses its fivefold symmetry. Thus, the lower local resolution in this area most likely does not stem from the "positional freedom" of this domain within the nanodisc, but is a result of an intrinsic flexibility of the domain. In addition, the MD simulations as well as the cryo-EM density indicated that, at the rim of the channel, lipids reach into the space between protomers, possibly stabilizing the open state. This could also explain the stabilizing effect of the nanodisc on the TcdA1 pore and their superiority over detergents for studying the pore conformation of Tc toxins (Fig. 5G).

5. CONCLUSIONS AND PERSPECTIVES

In this chapter, we have described advantages and practical methods for using lipid nanodiscs in structural studies of membrane proteins. While ensuring a native or a nearly native environment in which membrane proteins are fully functional, nanodiscs provide other possible advantages. Adding a molecular weight of more than 100 kDa, reconstitution into nanodiscs results in an increase of the overall size of the particle. This may be of particular advantage to single-particle cryo-EM of smaller membrane proteins, making it easier to image, select, and process them. However, the additional mass of the nanodisc will only be advantageous if their dimensional heterogeneity does not interfere with particle alignment. Unfortunately, nanodisc heterogeneity cannot be completely prevented at the moment especially if the nanodisc is chosen large enough to avoid unwanted interactions with the MSP. Therefore, it is challenging to study membrane proteins that have very small extramembrane domains. New image-processing algorithms and an improved nanodisc preparation will be needed to overcome this hurdle. In addition, increasing the contrast by taking images with a Volta phase plate could be beneficial. It will be interesting to explore the size limits of reconstituted proteins for high-resolution cryo-EM in the future.

The exposure of membrane proteins to detergents is reduced to a minimum when nanodiscs are used (Civjan et al., 2003; Duan et al., 2004; Marty et al., 2013; Roy et al., 2015; Shirzad-Wasei et al., 2015). This is especially beneficial for the structural analysis of highly unstable membrane proteins or multisubunit membrane protein complexes that interact weakly or only transiently. The interaction of several membrane proteins reconstituted in the same nanodisc presents an exciting opportunity for structural studies (Boldog et al., 2006). Such a system would closely mimic a real biological membrane environment with 2-D protein diffusion under crowded conditions and would be ideal to study at high-resolution processes, such as signal transduction, involving formation of transient complexes.

It remains to be seen how well other lipid-containing bilayer mimetics, in particular those based on SMA instead of MSP and liposomes, are suited for high-resolution single-particle cryo-EM. If both of these alternatives work, they would beautifully complement the MSP-based nanodisc system and together equip structural biologists with powerful tools for studying membrane proteins in a near to native environment.

ACKNOWLEDGMENTS

This work was funded by Vlaams Instituut voor Biotechnologie (to R.E.), Humboldt Foundation (to R.E.), Fonds voor Wetenschappelijk Onderzoek Vlaanderen (grant number G.0266.15N to R.E.), Max Planck Society (to S.R.), and the European Council under the European Union's Seventh Framework Programme (FP7/2007–2013) (grant no. 615984 to S.R.).

REFERENCES

Allegretti, M., Klusch, N., Mills, D. J., Vonck, J., Kühlbrandt, W., & Davies, K. M. (2015). Horizontal membrane-intrinsic α-helices in the stator *a*-subunit of an F-type ATP synthase. *Nature*, *521*, 237–240. http://dx.doi.org/10.1038/nature14185.

Bai, X.-C., Yan, Z., Wu, J., Li, Z., & Yan, N. (2016). The central domain of RyR1 is the transducer for long-range allosteric gating of channel opening. *Cell Research*, *26*, 995–1006. http://dx.doi.org/10.1038/cr.2016.89.

Bai, X.-C., Yan, C., Yang, G., Lu, P., Ma, D., Sun, L., et al. (2015). An atomic structure of human γ-secretase. *Nature*, *525*, 212–217. http://dx.doi.org/10.1038/nature14892.

Bao, H., Dalal, K., Wang, V., Rouiller, I., & Duong, F. (2013). The maltose ABC transporter: Action of membrane lipids on the transporter stability, coupling and ATPase activity. *Biochimica Et Biophysica Acta (BBA)-Biomembranes*, *1828*, 1723–1730. http://dx.doi.org/10.1016/j.bbamem.2013.03.024.

Bao, H., & Duong, F. (2014). Nucleotide-free MalK drives the transition of the maltose transporter to the inward-facing conformation. *Journal of Biological Chemistry*, *289*, 9844–9851. http://dx.doi.org/10.1074/jbc.M113.545525.

Bayburt, T. H., Grinkova, Y. V., & Sligar, S. G. (2006). Assembly of single bacteriorhodopsin trimers in bilayer nanodiscs. *Archives of Biochemistry and Biophysics*, *450*, 215–222. http://dx.doi.org/10.1016/j.abb.2006.03.013.

Bayburt, T. H., & Sligar, S. G. (2003). Self-assembly of single integral membrane proteins into soluble nanoscale phospholipid bilayers. *Protein Science*, *12*, 2476–2481. http://dx.doi.org/10.1110/ps.03267503.

Boldog, T., Boldog, T., Grimme, S., Grimme, S., Li, M., Li, M., et al. (2006). Nanodiscs separate chemoreceptor oligomeric states and reveal their signaling properties. *Proceedings of the National Academy of Sciences of the United States of America*, *103*, 11509–11514. http://dx.doi.org/10.1073/pnas.0604988103.

Bowen, D. J., & Ensign, J. C. (1998). Purification and characterization of a high-molecular-weight insecticidal protein complex produced by the entomopathogenic bacterium *Photorhabdus luminescens*. *Applied and Environmental Microbiology*, *64*, 3029–3035.

Bowen, D., Rocheleau, T. A., Blackburn, M., Andreev, O., Golubeva, E., Bhartia, R., et al. (1998). Insecticidal toxins from the bacterium *Photorhabdus luminescens*. *Science*, *280*, 2129–2132.

Breyton, C., Chabaud, E., Chaudier, Y., Pucci, B., & Popot, J.-L. (2004). Hemifluorinated surfactants: A non-dissociating environment for handling membrane proteins in aqueous solutions? *FEBS Letters*, *564*, 312–318. http://dx.doi.org/10.1016/S0014-5793(04)00227-3.

Brilot, A. F., Chen, J. Z., Cheng, A., Pan, J., Harrison, S. C., Potter, C. S., et al. (2012). Beam-induced motion of vitrified specimen on holey carbon film. *Journal of Structural Biology*, *177*, 630–637. http://dx.doi.org/10.1016/j.jsb.2012.02.003.

Civjan, N. R., Bayburt, T. H., Schuler, M. A., & Sligar, S. G. (2003). Direct solubilization of heterologously expressed membrane proteins by incorporation into nanoscale lipid bilayers. *BioTechniques*, *35*, 556–563.

Crichton, P. G., Crichton, P. G., Harding, M., Harding, M., Ruprecht, J. J., Ruprecht, J. J., et al. (2013). Lipid, detergent, and Coomassie blue G-250 affect the migration of small membrane proteins in blue native gels: Mitochondrial carriers migrate as monomers not dimers. *The Journal of Biological Chemistry*, *288*, 22163–22173. http://dx.doi.org/10.1074/jbc.M113.484329.

Cross, T. A., Murray, D. T., & Watts, A. (2013). Helical membrane protein conformations and their environment. *European Biophysics Journal*, *42*, 731–755. http://dx.doi.org/10.1007/s00249-013-0925-x.

Das, A., & Sligar, S. G. (2009). Modulation of the cytochrome P450 reductase redox potential by the phospholipid bilayer. *Biochemistry*, *48*, 12104–12112. http://dx.doi.org/10.1021/bi9011435.

Deisenhofer, J., Epp, O., Miki, K., Huber, R., & Michel, H. (1985). Structure of the protein subunits in the photosynthetic reaction centre of *Rhodopseudomonas viridis* at 3Å resolution. *Nature*, *318*, 618–624.

Denisov, I. G., Grinkova, Y. V., Lazarides, A. A., & Sligar, S. G. (2004). Directed self-assembly of monodisperse phospholipid bilayer nanodiscs with controlled size. *Journal of the American Chemical Society*, *126*, 3477–3487. http://dx.doi.org/10.1021/ja0393574.

Denisov, I. G., McLean, M. A., Shaw, A. W., Grinkova, Y. V., & Sligar, S. G. (2005). Thermotropic phase transition in soluble nanoscale lipid bilayers. *The Journal of Physical Chemistry. B*, *109*, 15580–15588. http://dx.doi.org/10.1021/jp051385g.

Dörr, J. M., Koorengevel, M. C., Schäfer, M., Prokofyev, A. V., Scheidelaar, S., van der Cruijsen, E. A. W., et al. (2014). Detergent-free isolation, characterization, and functional reconstitution of a tetrameric K^+ channel: The power of native nanodiscs. *Proceedings of the National Academy of Sciences of the United States of America*, *111*, 18607–18612. http://dx.doi.org/10.1073/pnas.1416205112.

Dörr, J. M., Scheidelaar, S., Koorengevel, M. C., Dominguez, J. J., Schäfer, M., van Walree, C. A., et al. (2016). The styrene-maleic acid copolymer: A versatile tool in membrane research. *European Biophysics Journal*, *45*, 3–21. http://dx.doi.org/10.1007/s00249-015-1093-y.

Duan, H., Civjan, N. R., Sligar, S. G., & Schuler, M. A. (2004). Co-incorporation of heterologously expressed Arabidopsis cytochrome P450 and P450 reductase into soluble nanoscale lipid bilayers. *Archives of Biochemistry and Biophysics*, *424*, 141–153. http://dx.doi.org/10.1016/j.abb.2004.02.010.

Efremov, R. G., Leitner, A., Aebersold, R., & Raunser, S. (2015). Architecture and conformational switch mechanism of the ryanodine receptor. *Nature*, *517*, 39–43. http://dx.doi.org/10.1038/nature13916.

Eggensperger, S., Fisette, O., Parcej, D., et al. (2014). An annular lipid belt is essential for allosteric coupling and viral inhibition of the antigen translocation complex TAP (transporter associated with antigen processing). *The Journal of Biological Chemistry*, *289*, 33098–33108. http://dx.doi.org/10.1074/jbc.M114.592832.

Fiedorczuk, K., Letts, J. A., Degliesposti, G., Kaszuba, K., Skehel, M., & Sazanov, L. A. (2016). Atomic structure of the entire mammalian mitochondrial complex I. *Nature*, *538*, 406–410. http://dx.doi.org/10.1038/nature19794.

Frauenfeld, J., Gumbart, J., Sluis, E. O. V. D., Funes, S., Gartmann, M., Beatrix, B., et al. (2011). Cryo-EM structure of the ribosome-SecYE complex in the membrane environment. *Nature Structural & Molecular Biology*, *18*, 614–621. http://dx.doi.org/10.1038/nsmb.2026.

Frauenfeld, J., Löving, R., Armache, J.-P., Sonnen, A. F.-P., Guettou, F., Moberg, P., et al. (2016). A saposin-lipoprotein nanoparticle system for membrane proteins. *Nature Methods*, *13*, 345–351. http://dx.doi.org/10.1038/nmeth.3801.

Gao, Y., Cao, E., Julius, D., & Cheng, Y. (2016). TRPV1 structures in nanodiscs reveal mechanisms of ligand and lipid action. *Nature*, *534*, 347–351. http://dx.doi.org/10.1038/nature17964.

Gatsogiannis, C., Lang, A. E., Meusch, D., Pfaumann, V., Hofnagel, O., Benz, R., et al. (2013). A syringe-like injection mechanism in *Photorhabdus luminescens* toxins. *Nature*, *495*, 520–523. http://dx.doi.org/10.1038/nature11987.

Gatsogiannis, C., Merino, F., Prumbaum, D., Roderer, D., Leidreiter, F., Meusch, D., et al. (2016). Membrane insertion of a Tc toxin in near-atomic detail. *Nature Structural & Molecular Biology*, *23*, 884–890. http://dx.doi.org/10.1038/nsmb.3281.

Grinkova, Y. V., Denisov, I. G., & Sligar, S. G. (2010). Engineering extended membrane scaffold proteins for self-assembly of soluble nanoscale lipid bilayers. *Protein Engineering Design and Selection*, *23*, 843–848. http://dx.doi.org/10.1093/protein/gzq060.

Gupta, K., Donlan, J. A. C., Hopper, J. T. S., Uzdavinys, P., Landreh, M., Struwe, W. B., et al. (2017). The role of interfacial lipids in stabilizing membrane protein oligomers. *Nature*, *541*, 421–424. http://dx.doi.org/10.1038/nature20820.

Hagn, F., & Wagner, G. (2015). Structure refinement and membrane positioning of selectively labeled OmpX in phospholipid nanodiscs. *Journal of Biomolecular NMR*, *61*, 249–260. http://dx.doi.org/10.1007/s10858-014-9883-6.

Hawkes, M. J., Díaz-Muñoz, M., & Hamilton, S. L. (1989). A procedure for purification of the ryanodine receptor from skeletal muscle. *Membrane Biochemistry*, *8*, 133–145. http://eutils.ncbi.nlm.nih.gov/entrez/eutils/elink.fcgi?dbfrom=pubmed&id=2641949&retmode=ref&cmd=prlinks.

Hernández-Rocamora, V. M., García-Montañés, C., Rivas, G., & Llorca, O. (2012). Reconstitution of the *Escherichia coli* cell division ZipA–FtsZ complexes in nanodiscs as revealed by electron microscopy. *Journal of Structural Biology*, *180*, 531–538. http://dx.doi.org/10.1016/j.jsb.2012.08.013.

Hiller, S., & Wagner, G. (2009). The role of solution NMR in the structure determinations of VDAC-1 and other membrane proteins. *Current Opinion in Structural Biology*, *19*, 396–401. http://dx.doi.org/10.1016/j.sbi.2009.07.013.

Hohn, M., Tang, G., Goodyear, G., Baldwin, P. R., Huang, Z., Penczek, P. A., et al. (2007). SPARX, a new environment for cryo-EM image processing. *Journal of Structural Biology*, *157*, 47–55. http://dx.doi.org/10.1016/j.jsb.2006.07.003.

Ilgü, H., Jeckelmann, J.-M., Gachet, M. S., Boggavarapu, R., Ucurum, Z., Gertsch, J., et al. (2014). Variation of the detergent-binding capacity and phospholipid content of membrane proteins when purified in different detergents. *Biophysical Journal*, *106*, 1660–1670. http://dx.doi.org/10.1016/j.bpj.2014.02.024.

Johnson, P. J. M., Halpin, A., Morizumi, T., Brown, L. S., Prokhorenko, V. I., Ernst, O. P., et al. (2014). The photocycle and ultrafast vibrational dynamics of bacteriorhodopsin in lipid nanodiscs. *Physical Chemistry Chemical Physics*, *16*, 21310–21320. http://dx.doi.org/10.1039/c4cp01826e.

Justesen, B. H., Justesen, B. H., Hansen, R. W., Hansen, R. W., Martens, H. J., Martens, H. J., et al. (2013). Active plasma membrane P-type H^+-ATPase reconstituted into nanodiscs is a monomer. *The Journal of Biological Chemistry*, *288*, 26419–26429. http://dx.doi.org/10.1074/jbc.M112.446948.

Kawai, T., Caaveiro, J. M. M., Abe, R., Katagiri, T., & Tsumoto, K. (2011). Catalytic activity of MsbA reconstituted in nanodisc particles is modulated by remote interactions with the bilayer. *FEBS Letters*, *585*, 3533–3537. http://dx.doi.org/10.1016/j.febslet.2011.10.015.

Kim, J., Wu, S., Tomasiak, T. M., Mergel, C., Winter, M. B., Stiller, S. B., et al. (2015). Subnanometre-resolution electron cryomicroscopy structure of a heterodimeric ABC exporter. *Nature*, *517*, 396–400. http://dx.doi.org/10.1038/nature13872.

Kühlbrandt, W. (2014). The resolution revolution. *Science*, *343*, 1443–1444. http://dx.doi.org/10.1126/science.1251652.

Lam, E., & Packer, L. (1983). Nonionic detergent effects on spectroscopic characteristics and the photocycle of bacteriorhodopsin in purple membranes. *Archives of Biochemistry and Biophysics*, *221*, 557–564.

Lang, A. E., Schmidt, G., Schlosser, A., Hey, T. D., Larrinua, I. M., Sheets, J. J., et al. (2010). *Photorhabdus luminescens* toxins ADP-ribosylate actin and RhoA to force actin clustering. *Science*, *327*, 1139–1142. http://dx.doi.org/10.1126/science.1184557.

Lee, A. (2004). How lipids affect the activities of integral membrane proteins. *Biochimica Et Biophysica Acta (BBA)-Biomembranes*, *1666*, 62–87. http://dx.doi.org/10.1016/j.bbamem.2004.05.012.

Lee, T.-Y., Yeh, V., Chuang, J., Chung Chan, J. C., Chu, L.-K., & Yu, T.-Y. (2015). Tuning the photocycle kinetics of bacteriorhodopsin in lipid nanodiscs. *Biophysical Journal*, *109*, 1899–1906. http://dx.doi.org/10.1016/j.bpj.2015.09.012.

Leitz, A. J., Bayburt, T. H., Barnakov, A. N., Springer, B. A., & Sligar, S. G. (2006). Functional reconstitution of β_2-adrenergic receptors utilizing self-assembling nanodisc technology. *BioTechniques*, *40*, 601–612. http://dx.doi.org/10.2144/000112169.

Letts, J. A., Fiedorczuk, K., & Sazanov, L. A. (2016). The architecture of respiratory supercomplexes. *Nature*, *537*, 644–648. http://dx.doi.org/10.1038/nature19774.

Lévy, D., Rigaud, J. L., Lévy, D., Mosser, G., Lambert, O., Mosser, G., et al. (1999). Two-dimensional crystallization on lipid layer: A successful approach for membrane proteins. *Journal of Structural Biology*, *127*, 44–52. http://dx.doi.org/10.1006/jsbi.1999.4155.

Li, X., Mooney, P., Zheng, S., Booth, C. R., Braunfeld, M. B., Gubbens, S., et al. (2013). Electron counting and beam-induced motion correction enable near-atomic-resolution single-particle cryo-EM. *Nature Methods*, *10*, 584–590. http://dx.doi.org/10.1038/nmeth.2472.

Liao, M., Cao, E., Julius, D., & Cheng, Y. (2013). Structure of the TRPV1 ion channel determined by electron cryo-microscopy. *Nature*, *504*, 107–112. http://dx.doi.org/10.1038/nature12822.

Marty, M. T., Wilcox, K. C., Klein, W. L., & Sligar, S. G. (2013). Nanodisc-solubilized membrane protein library reflects the membrane proteome. *Analytical and Bioanalytical Chemistry*, *405*, 4009–4016. http://dx.doi.org/10.1007/s00216-013-6790-8.

Matthies, D., Dalmas, O., Borgnia, M. J., Dominik, P. K., Merk, A., Rao, P., et al. (2016). Cryo-EM structures of the magnesium channel CorA reveal symmetry break upon gating. *Cell*, *164*, 747–756. http://dx.doi.org/10.1016/j.cell.2015.12.055.

McMullan, G., Faruqi, A. R., Clare, D., & Henderson, R. (2014). Comparison of optimal performance at 300keV of three direct electron detectors for use in low dose electron microscopy. *Ultramicroscopy*, *147*, 156–163. http://dx.doi.org/10.1016/j.ultramic.2014.08.002.

Meusch, D., Gatsogiannis, C., Efremov, R. G., Lang, A. E., Hofnagel, O., Vetter, I. R., et al. (2014). Mechanism of Tc toxin action revealed in molecular detail. *Nature*, *508*, 61–65. http://dx.doi.org/10.1038/nature13015.

Mi, L.-Z., Grey, M. J., Nishida, N., Walz, T., Lu, C., & Springer, T. A. (2008). Functional and structural stability of the epidermal growth factor receptor in detergent micelles and phospholipid nanodiscs. *Biochemistry*, *47*, 10314–10323. http://dx.doi.org/10.1021/bi801006s.

Nagle, J. F., & Tristram-Nagle, S. (2000). Structure of lipid bilayers. *Biochimica et Biophysica Acta*, *1469*, 159–195. http://eutils.ncbi.nlm.nih.gov/entrez/eutils/elink.fcgi?dbfrom=pubmed&id=11063882&retmode=ref&cmd=prlinks.

Nasr, M. L., Baptista, D., Strauss, M., Sun, Z.-Y. J., Grigoriu, S., Huser, S., et al. (2017). Covalently circularized nanodiscs for studying membrane proteins and viral entry. *Nature Methods*, *14*, 49–52. http://dx.doi.org/10.1038/nmeth.4079.

Nasr, M. L., & Singh, S. K. (2014). Radioligand binding to nanodisc-reconstituted membrane transporters assessed by the scintillation proximity assay. *Biochemistry*, *53*, 4–6. http://dx.doi.org/10.1021/bi401412e.

Orlando, B. J., McDougle, D. R., Lucido, M. J., Eng, E. T., Graham, L. A., Schneider, C., et al. (2014). Cyclooxygenase-2 catalysis and inhibition in lipid bilayer nanodiscs. *Archives of Biochemistry and Biophysics*, *546*, 33–40. http://dx.doi.org/10.1016/j.abb.2014.01.026.

Overington, J. P., Al-Lazikani, B., & Hopkins, A. L. (2006). How many drug targets are there? *Nature Reviews. Drug Discovery*, *5*, 993–996. http://dx.doi.org/10.1038/nrd2199.

Phillips, R., Ursell, T., Wiggins, P., & Sens, P. (2009). Emerging roles for lipids in shaping membrane-protein function. *Nature*, *459*, 379–385. http://dx.doi.org/10.1038/nature08147.

Popot, J.-L., Althoff, T., Bagnard, D., Banères, J.-L., Bazzacco, P., Billon-Denis, E., et al. (2011). Amphipols from A to Z. *Annual Review of Biophysics*, *40*, 379–408. http://dx.doi.org/10.1146/annurev-biophys-042910-155219.

Raunser, S., & Walz, T. (2009). Electron crystallography as a technique to study the structure on membrane proteins in a lipidic environment. *Annual Review of Biophysics*, *38*, 89–105. http://dx.doi.org/10.1146/annurev.biophys.050708.133649.

Ritchie, T. K., Grinkova, Y. V., Bayburt, T. H., Denisov, I. G., Zolnerciks, J. K., Atkins, W. M., et al. (2009). Chapter 11—Reconstitution of membrane proteins in phospholipid bilayer nanodiscs. *Methods in Enzymology*, *464*, 211–231. http://dx.doi.org/10.1016/S0076-6879(09)64011-8.

Roy, J., Pondenis, H., Fan, T. M., & Das, A. (2015). Direct capture of functional proteins from mammalian plasma membranes into nanodiscs. *Biochemistry*, *54*, 6299–6302. http://dx.doi.org/10.1021/acs.biochem.5b00954.

Samsó, M., Feng, W., Pessah, I. N., & Allen, P. D. (2009). Coordinated movement of cytoplasmic and transmembrane domains of RyR1 upon gating. *PLoS Biology*, *7*, e85. http://dx.doi.org/10.1371/journal.pbio.1000085.sv002.

Samsó, M., Wagenknecht, T., & Allen, P. D. (2005). Internal structure and visualization of transmembrane domains of the RyR1 calcium release channel by cryo-EM. *Nature Structural & Molecular Biology*, *12*, 539–544. http://dx.doi.org/10.1038/nsmb938.

Scheidelaar, S., Koorengevel, M. C., Pardo, J. D., Meeldijk, J. D., Breukink, E., & Killian, J. A. (2015). Molecular model for the solubilization of membranes into nanodisks by styrene maleic acid copolymers. *Biophysical Journal*, *108*, 279–290. http://dx.doi.org/10.1016/j.bpj.2014.11.3464.

Schmidt-Krey, I., & Rubinstein, J. L. (2011). Electron cryomicroscopy of membrane proteins: Specimen preparation for two-dimensional crystals and single particles. *Micron, 42*, 107–116. http://dx.doi.org/10.1016/j.micron.2010.07.004.

Schuler, M. A., Denisov, I. G., & Sligar, S. G. (2013). Nanodiscs as a new tool to examine lipid-protein interactions. *Methods in Molecular Biology, 974*, 415–433. http://dx.doi.org/10.1007/978-1-62703-275-9_18.

Shen, P. S., Yang, X., DeCaen, P. G., Liu, X., Bulkley, D., Clapham, D. E., et al. (2016). The structure of the polycystic kidney disease channel PKD2 in lipid nanodiscs. *Cell, 167*. 763–773.e11. http://dx.doi.org/10.1016/j.cell.2016.09.048.

Shih, A. Y., Denisov, I. G., Phillips, J. C., Sligar, S. G., & Schulten, K. (2005). Molecular dynamics simulations of discoidal bilayers assembled from truncated human lipoproteins. *Biophysical Journal, 88*, 548–556. http://dx.doi.org/10.1529/biophysj.104.046896.

Shirzad-Wasei, N., van Oostrum, J., Bovee-Geurts, P. H. M., Kusters, L. J. A., Bosman, G. J. C. G. M., & DeGrip, W. J. (2015). Rapid transfer of overexpressed integral membrane protein from the host membrane into soluble lipid nanodiscs without previous purification. *Biological Chemistry, 396*, 1–13. http://dx.doi.org/10.1515/hsz-2015-0100.

Stepien, P., Polit, A., & Wisniewska-Becker, A. (2015). Comparative EPR studies on lipid bilayer properties in nanodiscs and liposomes. *Biochimica et Biophysica Acta, 1848*, 60–66. http://dx.doi.org/10.1016/j.bbamem.2014.10.004.

Uhlén, M., Fagerberg, L., Hallström, B. M., Lindskog, C., Oksvold, P., Mardinoglu, A., et al. (2015). Tissue-based map of the human proteome. *Science, 347*, 1260419. http://dx.doi.org/10.1126/science.1260419.

Urbani, A., & Warne, T. (2005). A colorimetric determination for glycosidic and bile salt-based detergents: Applications in membrane protein research. *Analytical Biochemistry, 336*, 117–124. http://dx.doi.org/10.1016/j.ab.2004.09.040.

Van Petegem, F. (2014). Ryanodine receptors: Allosteric ion channel giants. *Journal of Molecular Biology*, 1–23. http://dx.doi.org/10.1016/j.jmb.2014.08.004.

Vinothkumar, K. R., & Henderson, R. (2010). Structures of membrane proteins. *Quarterly Reviews of Biophysics, 43*, 65–158. http://dx.doi.org/10.1017/S0033583510000041.

Wadsäter, M., Maric, S., Simonsen, J. B., Mortensen, K., Bendall, D. S., & Cárdenas, M. (2013). The effect of using binary mixtures of zwitterionic and charged lipids on nanodisc formation and stability. *Soft Matter, 9*, 2329. http://dx.doi.org/10.1039/c2sm27000e.

Wang, L., & Sigworth, F. J. (2010). Liposomes on a streptavidin crystal: A system to study membrane proteins by cryo-EM. *Methods in Enzymology, 481*, 147–164. http://dx.doi.org/10.1016/S0076-6879(10)81007-9.

Wu, M., Gu, J., Guo, R., Huang, Y., & Yang, M. (2016). Structure of mammalian respiratory supercomplex I1III2IV1. *Cell, 167*. 1598–1609.e10. http://dx.doi.org/10.1016/j.cell.2016.11.012.

Wu, J., Yan, Z., Li, Z., Qian, X., Lu, S., Dong, M., et al. (2016). Structure of the voltage-gated calcium channel Cav1.1 at 3.6 Å resolution. *Nature, 537*, 191–196. http://dx.doi.org/10.1038/nature19321.

Yan, Z., Bai, X.-C., Yan, C., Wu, J., Li, Z., Xie, T., et al. (2015). Structure of the rabbit ryanodine receptor RyR1 at near-atomic resolution. *Nature, 517*, 50–55. http://dx.doi.org/10.1038/nature14063.

Yeagle, P. L. (2014). Non-covalent binding of membrane lipids to membrane proteins. *Biochimica et Biophysica Acta, 1838*, 1548–1559. http://dx.doi.org/10.1016/j.bbamem.2013.11.009.

Yuchi, Z., & Van Petegem, F. (2016). Ryanodine receptors under the magnifying lens: Insights and limitations of cryo-electron microscopy and X-ray crystallography studies. *Cell Calcium, 59*, 209–227. http://dx.doi.org/10.1016/j.ceca.2016.04.003.

Zalk, R., Clarke, O. B., Georges, A. d., Grassucci, R. A., Reiken, S., Mancia, F., et al. (2015). Structure of a mammalian ryanodine receptor. *Nature*, *517*, 44–49. http://dx.doi.org/10.1038/nature13950.

Zhou, H.-X., & Cross, T. A. (2013). Influences of membrane mimetic environments on membrane protein structures. *Annual Review of Biophysics*, *42*, 361–392. http://dx.doi.org/10.1146/annurev-biophys-083012-130326.

Zhu, J., Vinothkumar, K. R., & Hirst, J. (2016). Structure of mammalian respiratory complex I. *Nature*, *536*, 354–358. http://dx.doi.org/10.1038/nature19095.

CHAPTER TWO

SSM-Based Electrophysiology for Transporter Research

Andre Bazzone*,†, Maria Barthmes†, Klaus Fendler*,1
*Max Planck Institute of Biophysics, Frankfurt/Main, Germany
†Nanion Technologies GmbH, Munich, Germany
[1]Corresponding author: e-mail address: fendler@biophys.mpg.de

Contents

Methods in Enzymology, Volume 594
ISSN 0076-6879
http://dx.doi.org/10.1016/bs.mie.2017.05.008

Abstract

Functional characterization of transport proteins using conventional electrophysiology can be challenging, especially for low turnover transporters or transporters from bacteria and intracellular compartments. Solid-supported membrane (SSM)-based electrophysiology is a sensitive and cell-free assay technique for the characterization of electrogenic membrane proteins. Purified proteins reconstituted into proteoliposomes or membrane vesicles from cell culture or native tissues are adsorbed to the sensor holding an SSM. A substrate or a ligand is applied via rapid solution exchange. The electrogenic transporter activity charges the sensor, which is recorded as a transient current. The high stability of the SSM allows cumulative measurements on the same sensor using different experimental conditions. This allows the determination of kinetic properties including EC_{50}, IC_{50}, K_m, K_D, and rate constants of electrogenic reactions. About 100 different transporters have been measured so far using this technique, among them symporters, exchangers, uniporters, ATP-, redox-, and light-driven ion pumps, as well as receptors and ion channels. Different instruments apply this technique: the laboratory setups use a closed flow-through arrangement, while the commercially available SURFE^{2}R N1 resembles a pipetting robot. For drug screening purposes high-throughput systems, such as the SURFE^{2}R 96SE enable the simultaneous measurement of up to 96 sensors.

1. BACKGROUND

1.1 Scope

Solid-supported membrane (SSM)-based electrophysiology is a technique developed for the measurement of transporters which are difficult to investigate using conventional electrophysiology (Schulz, Garcia-Celma, & Fendler, 2008; Tadini-Buoninsegni & Fendler, 2015). The method is based on the black lipid membrane technique, but exhibits much higher membrane stability and therefore allows for fast solution exchange and measurements under various conditions while using the same sensor.

The method allows the use of a wide variety of samples. Essentially every membrane preparation containing the protein of interest can be used. This includes purified proteins reconstituted into proteoliposomes, membrane vesicles from cell lines overexpressing the protein of interest or purified from native tissues like bovine heart, or rabbit skeletal muscle. Even intracellular membranes from organelles can be used.

Another important aspect is the large sensor size of up to 3 mm diameter which can be coated with membrane preparations containing the transporter of interest. The signal amplitude, therefore, can be several orders of magnitude larger compared to whole-cell patch clamp measurements, where the cell surface limits the measured current (Geibel, Flores-Herr, Licher, & Vollert, 2006). The lactose/H^+ symporter LacY with a turnover of $30\,s^{-1}$ at optimal conditions generates current amplitudes in the range of 1 nA using proteoliposomes with a lipid-to-protein ratio (LPR) of 5 (Garcia-Celma, Smirnova, Kaback, & Fendler, 2009). With a detection limit of 20 pA (Fig. 3A), we can estimate that the lowest turnover (assuming one transported elementary charge/turnover) detectable with SSM-based electrophysiology is $\sim 1\,s^{-1}$.

On the other hand, the noise and time resolution of SSM-based electrophysiology and whole-cell patch clamp using a standard perfusion system are similar. This is why the signal-to-noise ratio and sensitivity in SSM-based electrophysiology are usually higher compared to patch clamp measurements. In addition, SSM-based electrophysiology allows the use of purified samples with an extremely high protein density which further enhances sensitivity. Therefore SSM-based electrophysiology makes transporters with low turnover accessible for electrophysiological functional characterization. In addition, the fast solution exchange ensures that fast presteady-state kinetics, e.g., electrogenic binding or conformational transitions with low electrogenicity can be measured. Even single turnover events like substrate binding or presteady-state conformational transitions with low electrogenicity can be detected (Bazzone, Madej, Kaback, & Fendler, 2016; Ganea, Pourcher, Leblanc, & Fendler, 2001; Tadini-Buoninsegni, Bartolommei, Moncelli, Guidelli, & Inesi, 2006).

Most transporters studied using SSM-based electrophysiology are symporters (Garcia-Celma et al., 2009; Krause, Watzke, Kelety, Dorner, & Fendler, 2009; Zhou et al., 2004) and exchangers (Rycovska, Hatahet, Fendler, & Michel, 2012; Schulz, Werner, Stauber, Henriksen, & Fendler, 2010; Zuber et al., 2005), followed by ATP-driven ion pumps (Pintschovius, Fendler, & Bamberg, 1999; Tadini-Buoninsegni et al., 2006). In addition, redox- (Watzke, Diekert, & Obrdlik, 2010) and light-driven (Seifert, Fendler, & Bamberg, 1993) ion pumps, as well as receptors and ion channels (Balannik et al., 2010; Niessen et al., 2016; Schulz, Dueck, Mourot, Hatahet, & Fendler, 2009) have been characterized so far. The method is suitable for the detection of any kind of reaction associated with a charge displacement or with the change in water accessibility close to a charge. Mostly this is the steady-state transport of charged compounds across the membrane. But also other reactions can be electrogenic and observed as presteady-state components of the transient current, for example, the movement of charged residues or electrical dipoles (α-helices) inside the protein or the shielding of charged residues upon substrate binding. Even overall electroneutral transporters are accessible, e.g., the electroneutral sodium proton antiporter NhaP (Calinescu, Paulino, Kuhlbrandt, & Fendler, 2014) by isolating an electrogenic partial reaction.

Due to its mechanical robustness, SSM-based electrophysiology has a high potential for screening applications. For this reason, high-throughput systems (HTS) and setups capable of performing automated measurements have been developed. So far it is possible to analyze 96 sensors in parallel allowing determination of the dose dependence of 100 compounds in less than 30 min.

1.2 Measuring Principle

The basis of this technique is a gold-coated sensor chip with an alkylmercaptane monolayer on top (Fig. 1A). The SSM is formed by adding lipid and rinsing the electrode with buffer. The SSM on the sensor chip is the measuring electrode which is coated with the membrane preparation containing the protein of interest. The sensor is placed inside a Faraday cage and connected to a fluidic system which allows fast solution exchange.

A typical single solution exchange experiment consists of the sequential flow of nonactivating (NA), activating (A), and nonactivating (NA) solution (written as NA-A-NA). The activating solution contains the substrate of the transporter immobilized on the SSM. Therefore, the NA-A solution exchange generates a substrate concentration gradient which drives the electrogenic transport reaction leading to an increase of the potential inside the adsorbed proteoliposomes (Fig. 1B). Note that the figure shows proteoliposomes. However, open membranes adsorbed to the SSM are electrically equivalent to adsorbed proteoliposomes (Tadini-Buoninsegni & Fendler, 2015). Due to capacitive coupling, this leads to a charging current in the external measuring circuit (on-signal) (Fig. 1D), which is recorded by the amplifier. The subsequent switch to nonactivating solution generates the opposite substrate concentration gradient and therefore restores initial conditions. A transient current with opposite sign is recorded (off-signal, not shown).

A double solution exchange (DSE) is applied to establish an ion or substrate concentration gradient. Hereafter the NA-A-NA sequence a resting solution (R) flows followed by an incubation time in R to adjust the intraliposomal solution to the resting solution R. In this flow protocol,

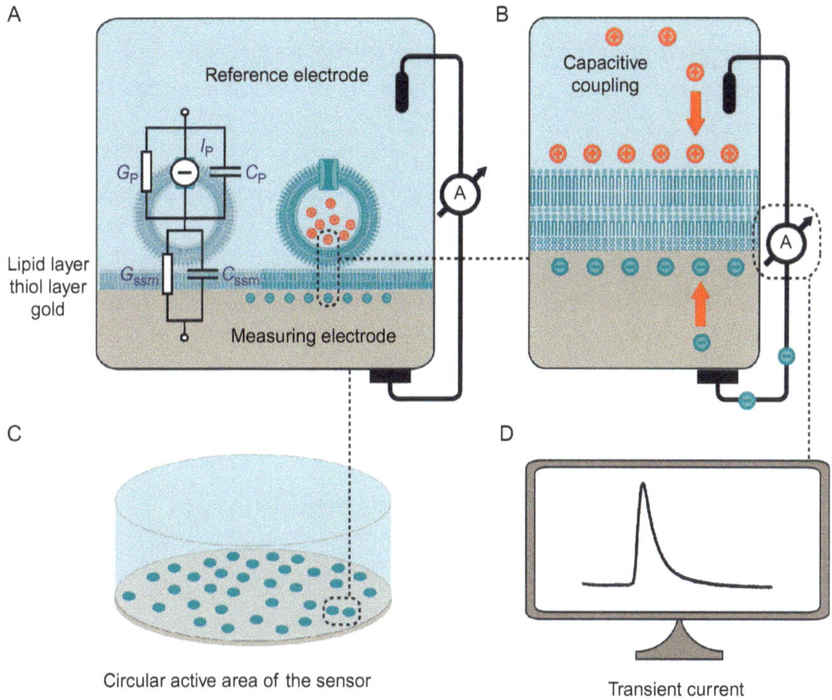

Fig. 1 Measurement principles of SSM-based electrophysiology. (A) Adsorbed proteoliposomes on the SSM. The compound membrane (=SSM + proteoliposomes) forms a capacitively coupled system. Both membranes proteoliposomes (P) and SSM can be described using their capacitance *C* and conductance *G*. In addition, the current of the membrane transporter can be included in the electrical circuit as I_P. Electrogenic transport charges the proteoliposomal membrane which is accompanied by the charging of the measuring electrode. (B) Expansion of the interface of the SSM and the proteoliposomal membrane. Charging of the membranes leads to a current flow in the external measuring circuit which is recorded by the amplifier (A). (C) Millions of proteoliposomes adsorbed to the SSM contribute to the measured current. (D) Currents measured at the SSM are in general transient currents (see text).

the R-NA exchange establishes the gradient, while the NA-A exchange activates the transporter at gradient conditions. Examples for these protocols are given in Section 4.

Due to the capacitive coupling principle only transient currents are obtained in SSM-based electrophysiology. This is true also for experiments where steady-state transport would be observed under short circuit conditions. Therefore, we refer to these currents as "steady-state current components" although they represent strictly speaking not a steady-state and decline with time constants of 50–500 ms due to the rising membrane potential. However, multiple turnovers contribute to these currents, and the peak current amplitude is proportional to the turnover rate of the transporter, since there is no significant membrane potential at the beginning of the transport reaction. In general, the peak current of the on-signal is used for analysis. But also the time constants for the current rise and decline can yield information about transporter kinetics (rate constants), especially when observing presteady-state currents.

One measurement only takes a few seconds. Due to the high sensor stability sequential measurements under different conditions are possible. This allows kinetic data, such as half-saturation constants (EC_{50}, IC_{50} K_m, K_D), rate constants (k_{obs}), substrate specificities (relative V_{max}), or pKs for pH-dependent transporters to be derived.

1.3 Instruments for SSM-Based Electrophysiology

SSM-based electrophysiology was developed in the 1990s (Pintschovius & Fendler, 1999; Pintschovius et al., 1999; Seifert et al., 1993). Since then several specialized systems and technical improvements have been established.

Custom-built laboratory setups (briefly "laboratory setups" in the following) enable continuous solution flow in a closed system. Pressurized solution containers, tubing, valves, and a cuvette in which the sensor chip and the reference electrode are mounted are placed inside a Faraday cage (Figs. 2A–D and 4). Valve opening/closing and data acquisition are controlled by a computer. Rapid sample exchange is not possible in this system.

Further development of laboratory setups aiming at higher time resolution use specialized cuvettes with two solution inlets (Fig. 2B). This allows investigation of fast electrogenic reactions with low electrogenicity, such as binding reactions or conformational transitions.

Commercial semiautomatic and automatic systems like the SURFE^{2}R product line (Nanion Technologies, Munich, Germany) combine a liquid handling approach based on pipetting robots and an open cuvette geometry. The SURFE^{2}R N1 incorporates a 53 positions autosampler on top of the device. The solutions are loaded and injected into an open sensor well in a fully automated process using the so-called IonJet (Fig. 2G). This allows rapid sample exchange. Also, fully automated high-throughput pipetting systems have been introduced. The SURFE^{2}R 96SE is capable of measuring 96 sensors in a fully parallel mode using a 96-well plate-based sensor.

In academia two systems are currently used for basic research purposes: the laboratory setup and the commercially available SURFE^{2}R N1. These

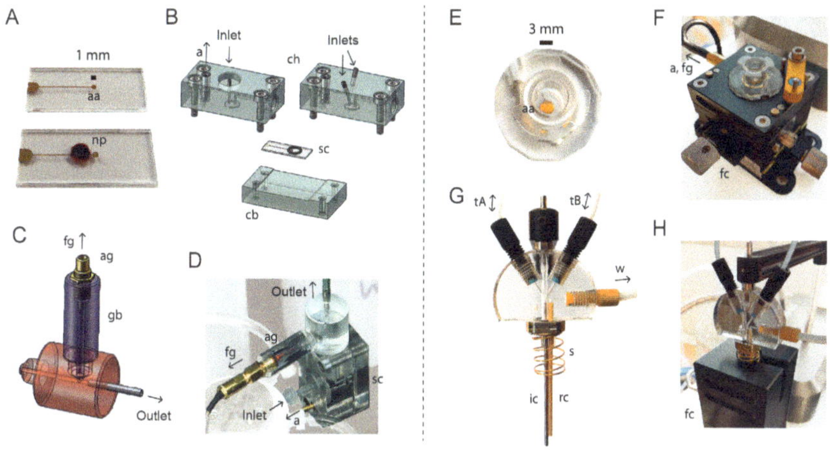

Fig. 2 Sensors and reference electrodes in the laboratory setup (A–D) and the SURFE^{2}R N1 (E–H). (A) Sensor chip with gold-coated surface. The circular active area (aa) has a diameter of 1 mm. During sensor preparation the contact strip needs to be isolated using nail polish (np). (B) The cuvette consists of the cuvette base (cb) and the cuvette head (ch) which contains the solution flow pathway. There are two different cuvette configurations: The I-cuvette (*left*) has one inlet and is used in the low time resolution (LTR) and double solution exchange (DSE) setups. The Y-cuvette (*right*) has two inlets and is used in the high time resolution (HTR) setups. The sensor chip (sc) is placed between the cuvette base (cb) and the cuvette head (ch). The contact strip of the sensor is contacted by a spring contact pin inside the cuvette head which connects to the amplifier (a). (C) The reference electrode assembly of the cuvette is screwed on top of the main part of the cuvette. The reference electrode is an Ag/AgCl wire (ag) which is connected to the function generator (fg) also operating as ground. The reference electrode is isolated from the main solution flow pathway by a gel bridge (gb). (D) The complete cuvette assembly consists of the main cuvette and the reference electrode assembly. The sensor chip (sc) contacts the amplifier (a) and the reference electrode (ag) contacts the function generator (fg). (E) Sensor well with gold-coated surface. The circular active area (aa) has a diameter of 3 mm. (F) Sensor well mounted on the Faraday cage (fc) contacted to amplifier (a) and function generator (fg) inside the instrument. (G) IonJet unit of the instrument. The injecting stainless steel capillary (ic) controls the loading and injection of solutions to and from tubing A (tA) and tubing B (tB). The removing PEEK capillary (rc) controls the removal of solutions from the sensor well to the waste container (w). The injecting stainless steel capillary (ic) simultaneously operates as reference electrode. The spring (s) makes contact between the reference electrode and the Faraday cage (ground). (H) The IonJet unit is positioned on top of the Faraday cage with a mounted sensor well inside.

two systems primarily differ in the liquid handling principle and the potential for customization and automatization. The laboratory setups are designed for easily changing components and developing new applications, like integrated optical fibers for analyzing the kinetics of light-driven pumps (unpublished). In contrast, the SURFE^{2}R N1 is an all-in-one device capable of semiautomated measurements. The high-throughput SURFE^{2}R 96SE addresses the requirements of the pharmaceutical industry. This instrument is especially suitable for drug screening purposes.

2. TECHNICAL IMPLICATIONS

The decision of which SSM-based electrophysiological system is the most suitable for each lab depends on the exact application. The following chapter focuses on the hardware specifications to highlight the differences between the setups.

2.1 Electrodes and Electrophysiological Hardware

The sensor chips are coated with a 100-nm gold layer on top of a 10-nm adhesive layer. The structured coating consists of an active circular area where the SSM is formed and a connection strip for the contact to the amplifier. Two sizes of the active area are commonly used: one with 1 mm and one with 3 mm diameter. The ~10 × higher surface area of the 3-mm sensor results in ~10 × larger signals. In the laboratory setup, the sensor chip is a Borofloat® glass plate (Fig. 2A) which is mounted inside a cuvette (Fig. 2B) connecting the sensor to the fluidic system. The SURFE^{2}R N1 sensor chip consists of polycarbonate well with a volume of about 250 μL (Fig. 2E) and a 3-mm diameter gold sensor area on the bottom. The sensor can be mounted directly inside the Faraday cage (Fig. 2F) and buffer can be injected and removed by a specialized pipette head, the IonJet (Fig. 2G and H).

The solution above the sensor is connected to the ground using a reference electrode. In the laboratory setup, the reference electrode is a silver chloride wire which is separated from the main fluid pathway by a gel bridge saturated with 100 m*M* KCl (Fig. 2C). In the SURFE^{2}R N1 the steel capillary of the injecting pipette works as the reference electrode (Fig. 2G). It is connected to the Faraday cage (ground) using a contact spring (Fig. 2H).

While the laboratory setup is a modular system which in principle can be used with any existing electrophysiological equipment, the SURFE^{2}R N1 is a preconfigured all-in-one instrument.

In the laboratory setup, all fluidic components including buffer containers and the cuvette with the sensor chip are placed inside a Faraday cage which can be opened to change the buffers and mount the cuvette (Fig. 2D). Amplifier, oscilloscope, voltage generator, and a computer are individual and exchangeable devices outside the Faraday cage and connected to the electrodes via shielded cables. A computer controls the valves and records the electrical signals at the amplifier output via a USB interface. The SURFE^{2}R One Software controls both valve switching and current recording.

The SURFE^{2}R N1 has only a small Faraday cage, which encloses the sensor well on top of the device, with an opening for the injecting pipette (Fig. 2H). The sensor is connected through the Faraday cage to the amplifier and the voltage generator inside the instrument. Also fluid flow control hardware and the computer are placed inside the instrument. Both fluidic components and signal recording are controlled by the software SURFE^{2}R N1 Control (Nanion Technologies).

In both systems, a voltage function generator is used for measuring capacitance and conductance of the membrane allowing quality control of the sensor before starting a measurement. Table 1 shows a summary of electrophysiological components of the SURFE^{2}R N1 and a typical laboratory setup.

2.2 Data Quality: Signal Amplitudes, Noise, Artifacts, and Time Resolution

Signal amplitudes, noise, solution exchange artifacts, and time resolution of the system are important characteristics which determine if a specific transporter is measurable or not. These parameters are influenced by the solution exchange geometry and the sensor size.

The time resolution of the SSM-based electrophysiology instrument is limited by the solution exchange at the surface of the SSM. Because nonactivating and activating solutions are unstirred layers before reaching the sensor, a fast solution exchange is difficult. The time resolution has a large influence on the shape of the transient current including the signal amplitude. For the systems described here, the time resolution ranges from ~4 to ~40 ms (Fig. 3A). To measure the rise time of the substrate concentration, advantage is taken of the strong interaction of ions with the lipid headgroups of the SSM (Garcia-Celma, Hatahet, Kunz, & Fendler, 2007). The displaced charge can be used as a relative measure of the local ion concentration at the surface of the SSM. We have determined the concentration rise time for different instruments using varying configurations as shown in Fig. 3 by performing a $KCl/KClO_4$ solution exchange.

The consequence of time resolution variations is different for slow declining steady-state currents and fast presteady-state currents. In steady-state currents the rise time and slightly also the peak current could be affected,

Table 1 Comparison of Electric Components

		Laboratory Setup	**SURFE^{2}R N1**
Electrodes	Measuring electrode	Reusable gold-coated sensor chip, Ø 1 mm	Reusable gold-coated sensor well, Ø 3 mm
	Reference electrode	Chlorinated silver wire	Steel capillary of IonJet
Amplifier and A/D converter	Amplifier	Keithley 428 (external), typical configuration	Custom built (integrated)
	Gain	10^9 V/A	10^8–10^{11} V/A
	Rise time (10%–90%)	3–10 ms	0.7 ms
	Low pass	—	480 Hz
	A/D converter	NI USB-6210 (National Instruments)	USB-6009 OEM (National Instruments)
	Resolution	—	−10 V to +10 V (1.2 mV steps)
	RMS noise	—	0.5–22 pA
IT interface	Valve controller	NI USB-6210	Custom built
	Computer	Variable	Integrated, Win10
	Software	SURFE^{2}R One Software	SURFE^{2}R N1 Control (Nanion Technologies)
Function generator		Custom built: 100 mV ⊓, 50 mV ΛΛ	Custom-built: 230 mV ⊓, 62 mV ΛΛ
Oscilloscope		Variable	—

but generally the effects are negligible at time resolutions <20 ms. Steady-state currents often decline with time constants between 50 and 200 ms. In contrast, fast presteady-state currents can easily reach the limit of the time resolution and, therefore, are affected by the time resolution of the system. Here, the time resolution of the solution exchange limits the decline of the signal which is used to determine the rate constant. A good time resolution consequently enables a better resolution for fast electrogenic reactions with low electrogenicity: it allows higher current amplitudes (higher S/N) to be

A

	Laboratory setup ∅1 mm sensor	SURFE²R N1 ∅3 mm sensor	SURFE²R 96SE ∅3 mm sensor
Typical signal amplitudes	0.1–2 nA	0.1–20 nA	0.1–20 nA
Detection limit	~20 pA	~50 pA	~30 pA
Noise	~2 pA	~20 pA	~10 pA
Time resolution	4 ms (HTR setup) 16 ms (LTR/DSE setup)	38 ms 6 ms (∅1 mm sensor)	7 ms

B

$$\tau = \frac{\pi l d^2}{8\dot{V}}$$

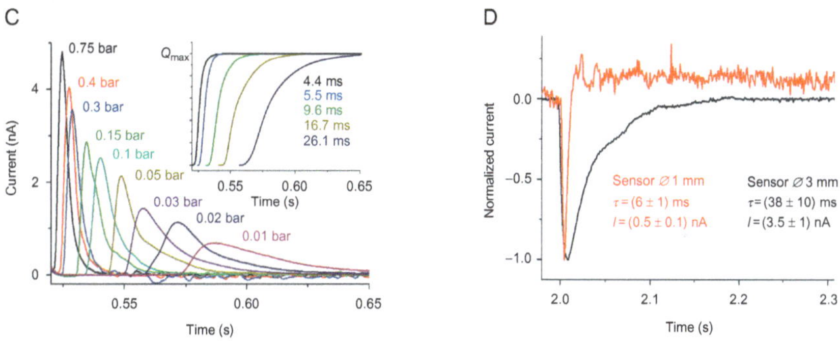

Fig. 3 Time resolution and signal amplitudes in SSM-based electrophysiology instruments. (A) Comparison of important characteristics for typical results using the laboratory setups, the SURFE²R N1 and the high-throughput system SURFE²R 96SE. (B) The time resolution depends on the length of the fluidic pathway from the merging point of the solutions to the sensor (*l*), the diameter of the fluidic pathway (*d*), and the volume speed ($\dot{V}$, mL/s). The given equation is derived from the law of Hagen–Poiseuille (Garcia-Celma et al., 2008). (C) Transient currents generated by a 100-m*M* $KClO_4$/KCl solution exchange (both solutions contain 100 m*M* KPi, pH 7.6) in the laboratory setup in HTR configuration. Different pressures were used as indicated to vary the speed of solution exchange. The lower the pressure, the poorer the time resolution. The valve switching occurs at $\tau = 0.5$ s. The activating solution reaches the sensor 25–75 ms after valve switching. The inset shows selected peak integrals which were fitted using an empirical equation (Garcia-Celma et al., 2008) to determine the time resolution indicated in the figure. (D) Normalized transient currents following a 20-m*M* $KClO_4$/KCl solution exchange using different sensor sizes (both solutions contain 100 m*M* KPi, pH 7.6) on the SURFE²R N1. Time resolutions and current amplitudes are indicated. With 1 mm diameter, the time resolution is six times better compared to 3 mm diameter sensors. On the other hand the current amplitude decreases depending on sensor area and time resolution. For both experiments, a solution speed of 200 μL/s and a 0.8-mm capillary was used.

measured, as well as a better estimation of the rate constant of the reaction. On the other hand ion–lipid interactions generate fast transient currents, so-called solution exchange artifacts, which are more prominent at better time resolution and complicate signal interpretation. Strategies for reducing artifacts are discussed in Section 4.

There are three factors which influence the time resolution of the system: (1) the fluidic geometry, (2) the sensor size, and (3) the rise time of the amplifier. Since the rise time of the amplifier can easily be set below the fluidic time resolution and the sensor size can be adjusted, the solution exchange geometry is often the limiting factor. The relevant process is fluid dispersion described by the law of Hagen–Poiseuille (Garcia-Celma et al., 2008) (Fig. 3B). This leads to mixing of nonactivating and activating solutions in the fluidic pathway between the merging point of nonactivating and activating solutions and the sensor. The result is a concentration gradient of activating solution, making an abrupt switch from nonactivating to activating solution impossible. The duration of the gradient (τ) on the sensor surface depends on the volume speed ($\dot{V}$), the length (l), and the thickness (d) of the capillary or tube between the merging point of nonactivating and activating solutions and the sensor (Garcia-Celma et al., 2008). Better time resolutions can, therefore, be reached by increasing the flow velocity (Fig. 3C) or decreasing the length and the thickness of the solution path.

Equation in Fig. 3B is only valid for a system where the sensor has the same size as the fluidic tube and is located close to the tube ending. Also, ideal mixing behavior has to be assumed on the sensor surface. When the sensor size is larger than the tubing diameter, the time resolution will be poorer since the solution takes more time to reach the outer edge of the sensor. The diameter of the solution flow pathway in the described systems varies between 0.5 and 1 mm. Therefore, sensors with 1 mm diameter show 5–10 times higher time constants compared to 3 mm sensors (Fig. 3D).

The sensor size does not only affect the time resolution but also the transferred charge by increasing the number of adsorbed transporters. In the case of steady-state currents, the sensor size is proportional to the amplitude, because the signal shape is not or only slightly affected by the change in time resolution. Therefore, 3 mm sensors should be preferred for slow transport currents to yield larger signal amplitudes. For fast transient currents limited by the time resolution of the instrument, the increased sensor size would not necessarily yield higher signal amplitudes. The result is a broader signal shape due to the poorer time resolution, meaning a loss of information. Therefore 1 mm sensors are preferred when investigating fast electrogenic reactions.

With larger sensor surface also the noise increases, since the capacitance at the amplifier affects the noise. Therefore, S/N is not necessarily proportional to the sensor size. The noise correlates also with the rise time of the amplifier. The lower the rise time, the higher the noise. Therefore, the rise time should be set only a few ms below the time resolution of the solution exchange. This maximizes the time resolution, while minimizing the noise.

Beside solution exchange artifacts (see Section 5), there are artifacts originating from mechanical or electrical components of the instruments (compare Fig. 7). In the laboratory system, the mechanical switching of the valves lead to large artifacts, also during the NA-A solution exchange. Vibrations can lead to mechanical artifacts interfering with the transporter signal, especially if the time for the activating solution to reach the sensor is low. This could be an issue in the HTR setup (Fig. 2B right), where this time can be as low as 30 ms. In the SURFE^{2}R N1, the valve is isolated from the sensor and induces only minor artifacts. Mechanical artifacts are observed when the solution flow starts with nonactivating solution.

2.3 Liquid Handling in the Laboratory Setup

The closed solution flow pathway for the laboratory setup consists of pressurized solution storage containers connected through tubes and rapidly switching valves with a cuvette containing the sensor chip (Figs. 3A–D and 4A and B).

This instrument usually has two to three places for solution storage containers which are used for single solution exchange and double solution exchange protocols, respectively. The solution containers are standard 50–100 mL polyethylene bottles which are pressurized by compressed air or nitrogen gas. Pressures of 0.3–1.0 bar are typically used in these experiments. A manometer controls the pressure, which has to be adjusted to yield a defined solution flow speed of ~0.5–2 mL/s. The valves are switched using computer-controlled valve drivers. Solution flow can be triggered by manually switching the valves, e.g., for cleaning procedures, or starting a flow sequence under software control. When changing experimental conditions, the solution containers are exchanged manually: pressure is released, followed by unscrewing the containers from the tubing system. The containers can be closed and stored for subsequent experiments.

There are two types of cuvettes requiring different kinds of valve/tubing configurations. These two different systems are used for different applications. One is well suited for measurements with high time resolutions (HTR setup), whereas the other is used for measurements with lower time resolutions (LTR setup) and can also be used for double solution exchange

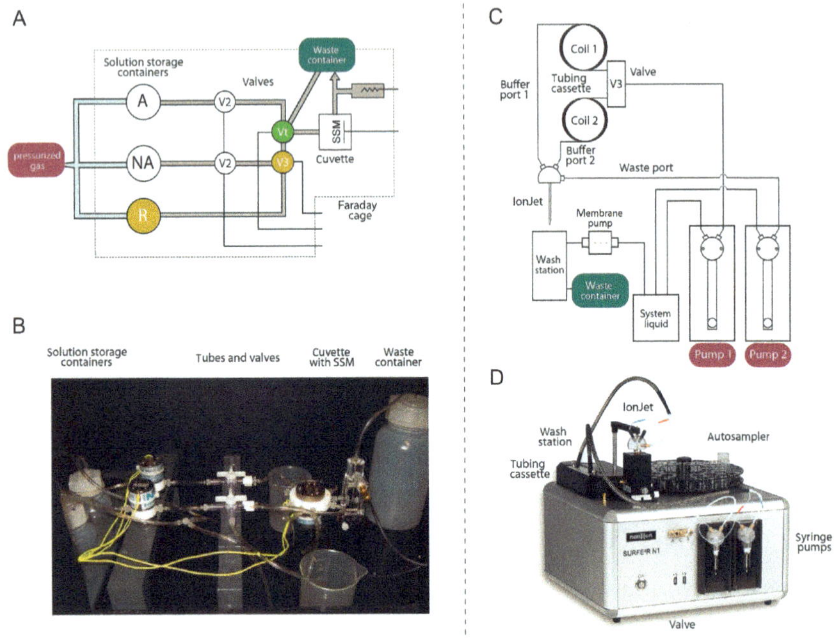

Fig. 4 Solution flow and liquid handling in the (A, C) laboratory setup and the (B, D) SURFE^{2}R N1. (A) Schematics of the liquid handling components of the laboratory setup in DSE configuration. Three solution storage containers for activating (A), nonactivating (NA), and resting (R) solutions which are pressurized by gas. Flow speed is adjusted via a manometer. Flow control: Both V2 valves control A and NA solutions, V3 switches from NA to R solution, and Vt switches from NA/R to A solution, while redirecting the other solution to a waste container. The LTR configuration can be obtained by removing R and V3 (*yellow/light gray*) from the system. In this configuration, V2 controlling NA is directly connected to Vt. The HTR setup only contains the two V2 which are directly connected to the two inlets of the cuvette; in addition to R and V3, Vt (*green/dark gray*) is not required in this configuration. (B) Picture of the fluidics inside the Faraday cage of the laboratory setup in LTR configuration. (C) Schematics of the liquid handling components of the SURFE^{2}R N1. The IonJet unit takes A and NA solutions from up to 53 buffers stored in the autosampler of the instrument. The pipetting speed is controlled by two syringe pumps which can be set by software. Depending on the activation status of the valve V3, the solution is directed into coil 1 or coil 2; therefore, A and NA are loaded into separate tubings. After loading the solutions, the IonJet moves to the sensor and starts injecting the solutions controlled by pump 1; V3 controls the solution exchange by switching from coil 1 to coil 2. At the same time, the PEEK capillary of the IonJet removes the solution from the sensor well which is controlled by pump 2. This enables a continuous solution flow. The liquid handling components behind the coils, including V3 and the syringe pumps are always filled with system liquid to achieve higher pipetting accuracy. (D) Picture of the SURFE^{2}R N1 instrument. The syringe pumps are placed on the front, the autosampler is located on top of the system. The IonJet is mounted on a moving arm; the coils are hidden inside the shielded tubing cassette.

experiments (DSE setup). The valves from NResearch (West Caldwell, USA) are driven with a voltage of 17 V, which is slightly higher than the specified value (12 V) to allow fast switching.

The valve configuration for the LTR setup consists of two 2-way valves V2 (NR225T011) controlling the nonactivating and activating solution, respectively, and a terminal mechanically coupled 3 × 3-way valve Vt (NR360T331) switching between the two solutions. Recently, the 3 × 3-way valve has been replaced by a simple 3-way valve (NR225T031) with practically identical results. This reduces valve costs and fluid consumption, and should therefore be preferred. The two 2-way valves are switched simultaneously to start the experiment. The terminal valve ensures that during the entire experiment both A and NA solutions are in motion. In ground state it directs the NA solution to the sensor and the A solution to a waste container. When activated by the software or manually it directs the A solution to the sensor and the NA solution to a waste container. The terminal valve directly connects to the cuvette holding the sensor chip. Therefore, the distance between sensor and merging point of NA and A solutions inside the terminal valve is as short as possible, yielding a time resolution of about 16 ms with a solution flow of 1.6 mL/s (Schulz et al., 2008).

The DSE setup contains an additional 3-way valve V3 (NR225T031) switching between nonactivating and resting solution to enable a DSE. The lower time resolution of NA-R exchange (100–200 ms) is due to the large distance between the merging point of NA and R solutions and the sensor, but this is not critical for the measurements.

The HTR setup uses a specialized cuvette with two inlets (Fig. 2B, right), one for the A and one for the NA solution. The two pathways merge shortly before the sensor chip inside the cuvette, resembling a Y-geometry. Therefore, this cuvette is called the Y-cuvette, in contrast to the I-cuvette used in the LTR and DSE setups which only has one inlet. The shorter distance between the sensor and the merging point of NA and A solutions allows for a higher time resolution. The HTR setup only contains two 2-way valves which, similar to the LTR/DSE setups, control the flow of NA and A solutions. In contrast to the LTR/DSE setups the two valves are switched sequentially to control the solution exchange: A few milliseconds after valve 1 is turned off to stop the flow of the NA solution, valve 2 opens to start the flow of the A solution. The delay between switching valve 1 off and valve 2 on has to be optimized for each configuration. The time resolution of solution exchange in this configuration can be as low as 4 ms (Fig. 3A).

In both cuvettes, the merged pathway of NA and A solutions is centered directly above the sensor and leads to the cylindrical cuvette volume of about 17 μL surrounded by an O-ring. Inlet and outlet bores have a diameter of 1 mm which equals the diameter of the sensor. The outlet of the cuvette is connected to a waste container. In addition, for the removal of air bubbles introduced when solution containers are exchanged, manual valves are integrated (Fig. 4B).

2.4 Liquid Handling in the SURFE^{2}R N1

In contrast to the laboratory setups, the SURFE^{2}R N1 resembles a pipetting robot. On top of the device is an autosampler with space for up to 53 open solution containers, each holding 10 mL (Fig. 4D). The robot can be programmed to sequentially supply all solutions during one experiment using the same sensor. The sensor is placed in an open well allowing injection of the solutions from above. The solution flow is controlled by two pumps, one for injecting solutions (P1) and one for removing solutions (P2) from the sensor well (Fig. 4C and D). P1 is connected to one fast switching 3-way valve for fast solution exchange. The speed of the solution flow can be adjusted in the software and is preset to 200 μL/s.

The pipette arm carries the IonJet (Fig. 2G), which controls the solution exchange by its Y-fluidic geometry, similar to the cuvette of the HTR laboratory setup (Fig. 2B, right). A steel capillary is used for loading solutions from the solution containers. The steel capillary merges in the Y-fluidic geometry of the IonJet which is connected to two different tubings (coil 1 and coil 2) inside a tubing cassette (Fig. 4C). Depending on the status of a 3-way valve (V3), P1 controls the loading of A solution into coil 1 or NA solution into coil 2.

After loading solutions, the pipettor moves to the sensor well and positions the steel capillary ~0.2 mm above the sensor for optimal solution exchange. P1 starts injecting NA solution and the fast switching valve controls the solution exchange. For a typical configuration (3 mm sensor, 200 μL/s solution flow, steel capillary of 0.8 mm in diameter), the time resolution of solution exchange is 38 ms. When using smaller sensors of 1 mm in diameter, the time resolution of solution exchange under the same conditions is 6 ms. Beside the steel capillary working as the injecting pipette there is a PEEK capillary for removing the solutions remote from the sensor surface. It is 1 cm shorter than the injecting pipette to remove the solution from above, while simultaneously adding solution directly to the sensor. Since the well volume of 250 μL is much larger compared to the cuvette volume of 17 μL in the laboratory setups, the second flow of the NA solution often is followed by an additional rinse in NA solution to completely remove the A solution. Similar to the laboratory setups, the SURFE^{2}R N1 also allows DSE experiments. Here, the rinse in NA solution is replaced by a rinse and incubation time in R solution.

While in the laboratory setup theoretically up to 100 mL of each solution can be used for solution exchange experiments without waiting times between the measurements, the pump, and the tubing of the SURFE^{2}R N1 can only load up to 2.5 mL of solution, making fast repetitive measurements and long solution flow times difficult. On the other hand the solution consumption for the same flow time is 5–10 times lower compared to the laboratory setup.

At the beginning of the experiment, pumps and tubings are filled with system liquid (water) leaving only a small air gap to the A as well as NA solutions inside the coils. The system liquid is stored in a 2-L storage container and connected with separate tubings to both pumps and the washing station for the IonJet. The pipette controlling the removal of solution from the sensor and the washing station are connected to a waste container (Fig. 4C). For a comparison of the liquid-handling components of the above mentioned setups see Table 2.

2.5 Throughput and Automatization

Another important factor of the instrument is throughput and automatization. These factors are mainly limited by the software, the space for solutions, and the sensors, as well as the process of sensor preparations.

Throughput depends on the time required for both sensor preparation and the electrophysiological measurement itself. Here, the possibility to prepare and measure a number of sensors simultaneously tremendously increases the throughput. For simultaneous measurements of more than one sensor specialized HTS are used, e.g., the SURFE^{2}R 96SE (96 sensors in parallel) as well as the outdated SURFE^{2}R N96 (8 sensors in parallel) (Tadini-Buoninsegni & Fendler, 2015). In the SURFE^{2}R N1 and the laboratory setup only one sensor can be measured at a time. For basic research, higher throughput by means of parallel measurements is mostly not required. A direct comparison of the workflow shows that the SURFE^{2}R N1 and the laboratory setup still significantly differ in their throughput.

In the laboratory setup, the sensor has to be mounted inside the cuvette which itself has to be mounted inside the fluidic system before the SSM is formed and the membranes containing the protein of interest are adsorbed.

Table 2 Comparison of Liquid-Handling Components

		Laboratory Setup	SURFE^{2}R N1
Characteristics of liquid handling	Liquid-handling components	Closed system	Open system
	Flow control	Pressurized air	Two syringe pumps
	Valve configuration	Multiple valve/tubing configurations	One 3-way valve
	Valve control	Manual or software controlled	Software controlled
	Solution exchange	HTR system: Y-fluidic in cuvette, LTR/DSE systems: Terminal 3 × 3-way valve or simple 3-way valve	Y-fluidic in IonJet
	Sensor volume	Cuvette with 17 μL	Sensor well with 250 μL
Properties of solution flow	Solution consumption per substrate	0.5–4 mL	0.2–0.5 mL
	Solution consumption per inhibitor	1.5–12 mL	0.4–1.3 mL
	Max volume per measurement	100 mL (container volume)	2.5 mL (syringe pump volume)
	Solution flow speed	0.5–2 mL/s (0.3–1 bar)	200 μL/s (10–300 μL/s)
	Time resolution of solution exchange	4–16 ms (HTR vs LTR setup)	6–38 ms (1 mm vs 3 mm sensor)
	Continuous flow time per measurement	1.5–6 s	1.5–6 s

Table 2 Comparison of Liquid-Handling Components—cont'd

		Laboratory Setup	$SURFE^2R$ N1
Solution containers	Container positions	2–3 pressurized containers	Autosampler with 53 positions
	Container volume	50–100 mL each	10 mL each
	Container change	Manual change of buffer containers	Automated

Consequently only one sensor can be prepared at a time which takes about 2 h. The $SURFE^2R$ N1 use sensor wells which can be prepared simultaneously and even stored at −80°C for several weeks before use, increasing the throughput by a factor of up to 10, depending on the measurements per sensor and the number of sensors used for one experiment.

The number of data points per hour using different sensors compared to using the same sensor is different in both systems. Switching between sensors is much faster in the $SURFE^2R$ N1, since they can be prepared in batches and manually exchanged within seconds. When it comes to measurements with the same sensor, the two instruments require a similar time per data point: In the laboratory setup the manual change of solutions takes about 2–3 min which defines the time between measurements under different conditions. The $SURFE^2R$ N1 requires time between the measurements for loading the solutions and washing the tubing which also takes about 2–3 min. In contrast to the $SURFE^2R$ N1, the measurement itself using the laboratory setup takes as long as the solution flows, i.e., 1–2 s. Repetitive measurements using the same solutions without incubation times in-between are, therefore, much faster in the laboratory setup.

Table 3 shows the number of data points per day of different systems under different experimental settings. Using the $SURFE^2R$ N1 system during an 8-h day up to 150 data points using 20 sensors prepared batchwise on the same day can be obtained. Using the laboratory setup a similar amount of data points can be generated each day, even in triplicates. But the generated data points are limited to one or two sensors each day. Where the success rate of sensor preparation may be the same in both instruments, the case of one sensor failure has a greater impact on the workflow for the laboratory setup. If one sensor fails in the commercial system it can be immediately replaced by the next sensor. In the laboratory setup a new sensor has to be prepared which takes about 2 h.

Table 3 Comparison of the Throughput Under Different Experimental Settings for Different Instruments

Setup	1: Time for One Measurement	2: Number of Measurements Per Day (Without Sensor Preparation)	3: Substrate Screening for 20 Different Compounds With Higher Statistics, e.g., Using 10 Sensors (Including Sensor Preparation)	Involvement of Researcher
Laboratory setup	10 s ($n=1$) Triplicate measurement using 1 sensor 2–3 min ($n=1$) Including buffer change	>150 ($n=1$) Using 1 sensor, triplicate measurements	5 days ($n=10$) Using 2 sensors per day, each 20 triplicate measurements. Sensor preparation takes about 2 h/sensor	Exchanging solutions, preparing new sensors one by one
SURFE^{2}R N1	2–3 min ($n=1$) Single measurement using 1 sensor 6–9 min ($n=1$) Triplicate measurement	>150 ($n=1$) Using 1 sensor, single measurements	1 day ($n=10$) Using 10 sensors, each 20 single measurements. Sensor preparation takes about 1 h in total	Exchanging sensors
SURFE^{2}R 96SE	5 min ($n=96$) Using 1 sensor plate	>10,000 ($n\geq96$) E.g., 10 measurements of 10 sensor plates each with $n=96$	160 min ($n=96$) 20 measurements of one sensor plate equal ~2000 data points. Sensor preparation takes about 60 min	Exchanging sensor plates where required

The number (n) represents the number of sensors used over the mentioned time or for generating the respective amount of data points. One measurement or data point equals one transient current for one experimental condition, e.g., one substrate or inhibitor concentration. The differences in sensor preparation time between the systems were considered in column 3. 1 day is 8 h.

Automatization denotes the time period the setup can operate without human intervention. The laboratory setup has the least potential for automatization. The solutions need to be changed manually. The software controls the valve switching, but only within one solution exchange measurement which only lasts seconds. Therefore, the researcher has to attend the whole experiment. The SURFE^{2}R N1 contains a liquid handling system based on a fully programmable pipetting robot with 53 solution storing positions. Multiple measurements can be done automatically with the same sensor well; only switching between sensors must be done manually. Therefore, depending on the experiment, the device can operate for 3min–2h unattended and the operator only needs to be present to switch sensors, a process which takes mere seconds. The SURFE^{2}R 96SE also encompasses a pipetting robot allowing the fully automated measurement of 96 sensors using 7 positions for 96-well plates with different nonactivating and activating solutions. Similar to the SURFE^{2}R N1 the operator needs to change the sensor well plate every 3min to 2h, depending on the number of experiments for one sensor. But in this period 96 datasets are gathered instead of one.

2.6 96 Sensors in Fully Parallel Mode: The SURFE^{2}R 96SE

The increasing interest in membrane transporters as drug transporters, as possible pathways for drug distribution or the cause of drug–drug interactions has triggered the development of SSM-based electrophysiological measuring systems with much higher throughput than the systems discussed so far. In the SURFE^{2}R 96SE this goal is achieved mostly due to high parallelization.

The SURFE^{2}R 96SE employs the pipetting platform Cybio Felix (Analytik Jena) for all liquid handling steps (Fig. 5A). A measuring chamber (Fig. 5B) is incorporated onto the robot deck. It includes Faraday shielding and establishes electrical connection between the sensor plate (Fig. 5C), the amplifier, and the ground contact. The amplifier and necessary command electronics are placed below the robot. The measurement traces are displayed using specialized recording software (Nanion Technologies), allowing the management of large data volumes (Fig. 5D, left). For parallel data analysis and compound management, a further software tool is provided (Fig. 5D, right).

The system records currents simultaneously from 96 individual SSM sensors, which are accommodated in a standard 96-well plate (Fig. 5C). The geometry of the sensor is identical to the previously described systems, with a diameter of 3mm. In contrast to the other systems, the SURFE^{2}R 96SE records with 96 individual gold-coated ground electrodes, which dip into each well. The SURFE^{2}R 96SE uses the same principle as the SURFE^{2}R N1:

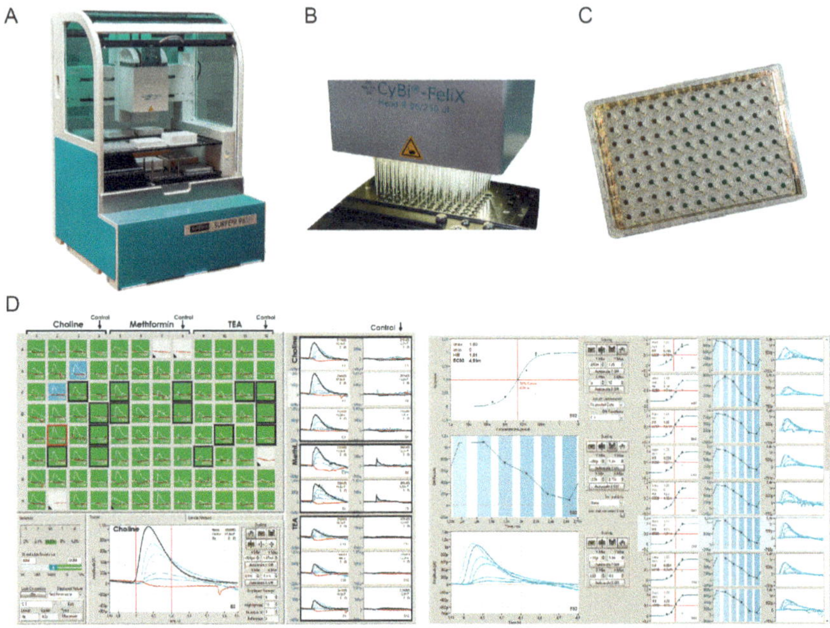

Fig. 5 The SURFE^{2}R 96SE, a commercial device applying SSM-based electrophysiology in a high-throughput manner. (A) The device encompasses the liquid-handling platform CyBio® FeliX and integrates a SURFE^{2}R 96SE module consisting of the measuring chamber, a low-noise 96-channel amplifier, and a control unit. (B) Pipette head of the instrument with 96 pipettes injecting solutions into a sensor plate inside the measuring chamber. The instrument measures 96 sensors fully parallel. (C) The sensor plate is based on the standard 96-well plate format. It contains in each well a gold-coated sensor with 3 mm diameter. (D) Screenshots of the recording (*left*) and data analysis (*right*) software for the SURFE^{2}R 96SE. It shows the results of an assay for characterization of the human organic cation transporter OCT2. The sensor plate was loaded with membranes from CHO cells (columns 1–3, 5–7, and 9–11) and negative control membranes (columns 4, 8, and 12). Activation took place by concentration jumps of different substrates: choline (columns 1–4), metformin (columns 5–8), and TEA (columns 9–12). Each substrate was used in different concentrations cumulatively in each well to determine EC_{50} values. The statistical analysis is an automated process. The EC_{50} curves of each well and the averaged curves can be viewed separately and exported for processing using third party software.

The fluidic system is separated from the SSM sensor, which allows batchwise preparation of the sensors independently from the experiment. The preparation of the 96 SSM sensors in one well plate is performed by the instrument itself or another liquid handling platform.

In contrast to the systems discussed so far, the SURFE^{2}R 96SE operates without continuous perfusion. The pipetting head is used either to add or to remove solution using 96 pipette tips, each with 250 μL capacity (Fig. 5B). For the activation of a transporter, two to three buffers are stacked in layers in the pipette tips, which results in the sequential addition to the sensor. The solution consumption is the lowest among SSM-based electrophysiological instruments: only 10–30 μL of the activating solution is required per sensor (compare Table 3). The rinsing of the sensor for the following measurement is performed by removing and refilling cycles which may be repeated as necessary. Although in principle this simplified fluidic system could increase mixing of the assay solutions, the observed resulting time resolution of 7 ms is even higher compared to other systems using a 3-mm electrode. Also, signal amplitude and S/N are comparable to other systems. One limitation of the perfusion system is the DSE experiment: A fast DSE could be problematic in cases where a long NA solution flow is required, due to the limited volume of 250 μL for solution stacking.

The SURFE^{2}R 96SE is a fully automated system, which runs unattended. It generates 96 data points every 5 min. User input is only required to exchange the sensor plate and solution or compound vials. For further automatization and increased unattended runtime, the system can be included into a robotic environment.

3. DESCRIPTION OF THE WORKFLOW

For the laboratory setup, a detailed protocol for setting up the system and performing experiments was published recently (Bazzone et al., 2013). This section focuses on the general procedures as well as the differences between the laboratory setup and the SURFE^{2}R N1.

3.1 Sample Preparation

Each membrane containing the protein of interest is suitable for electrophysiological measurements using SSM-based electrophysiology. Nevertheless sample preparation can be the key to successful measurements, since a high protein density and purity can compensate for low turnover, low electrogenicity, and large artifacts.

For sample preparation, different source materials are possible. Fig. 6A gives an overview of the sample preparation procedure. Mostly the protein of interest is either recombinantly overexpressed or used from native tissue such as heart, kidney, liver, brain, skeletal muscle, or gastric mucosa from different organisms including pig, rabbit, or mice. The nAChR (nicotinic acetylcholine receptor) channel was characterized using preparations from

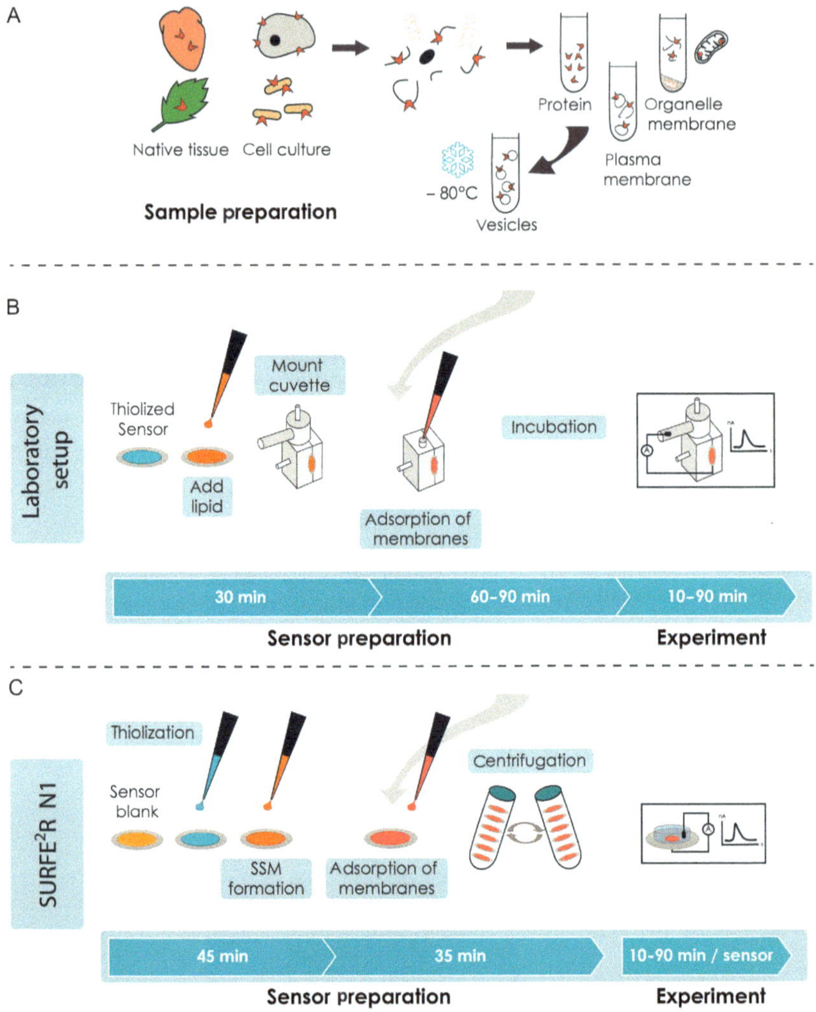

Fig. 6 Timeline of the experimental workflow including sample preparation, sensor preparation, and electrophysiological experiment. Comparison of the SURFE^{2}R N1 and the laboratory setup. (A) Summary of the sample preparation. Using native tissue or eukaryotic/bacterial cell culture, the cells first have to be ruptured. Then either proteins (protein purification) or membranes (sucrose gradient centrifugation) are isolated. Proteins need to be reconstituted into vesicles; membranes are used directly for adsorption on the sensor. The ready-to-use sample is stored at −80°C. (B) Timeline for working with the laboratory setup. Sensor preparation occurs partly inside the cuvette. Therefore only one sensor can be prepared at a time. (C) Timeline for working with the SURFE^{2}R N1. Sensor preparation and electrophysiological experiments are independent workflows. Sensors can be prepared in parallel and even stored at −80°C for weeks before use, depending on the transporter.

eel electric organ (Niessen et al., 2016). Commonly used expression systems range from bacteria to eukaryotic cell lines such as CHO, HEK, and COS-1. Also cell-free expression systems have been used to assay transporter function with SSM-based electrophysiology and it was recently shown that even nanodiscs containing the protein of interest adsorb to the SSM (unpublished observations).

Using ChR2 (channelrhodopsin 2)- and PEPT1 (peptide transporter 1)-expressing CHO cells, we recently observed that whole cells can be adsorbed to the sensor and used directly for measurements of transporter activity. However, typically several purification steps are performed before starting with the electrophysiological experiment. Sometimes membrane vesicles are used for SSM-based electrophysiology directly after cell disruption. In most cases the membranes are purified after cell disruption using a sucrose density gradient centrifugation step. This is especially appropriate when membranes from distinct organelles like mitochondria (Siebels & Drose, 2016; Watzke et al., 2010), endoplasmatic reticulum (Tadini-Buoninsegni et al., 2006, 2015), synaptic vesicles (Obrdlik et al., 2010), microsomes (Tadini-Buoninsegni et al., 2010; Tadini-Buoninsegni & Smeazzetto, 2017), or lysosomes are required. Such purification step is also used for enrichment of plasma membrane fragments and mostly yields a significant signal enhancement compared to nonpurified samples.

The ideal sample regarding signal amplitudes is prepared by protein purification followed by reconstitution into liposomes at high protein densities. Usually liposomes with $\sim$100 nm in diameter are used for the experiments, but also giant unilamellar vesicles can be used (Barthmes, Liao, Jiang, Bruggemann, & Wahl-Schott, 2016). Transporters with turnover rates below $100\,s^{-1}$ should be characterized using proteoliposomes with low LPR of 5–10. LPR of successfully tested proteoliposomes range from 5 (LacY with turnover rates of $30\,s^{-1}$, Garcia-Celma et al., 2009) to 500 (NhaA with turnover rates of $1000\,s^{-1}$; Zuber et al., 2005). The use of proteoliposomes is also the right choice when the native membrane contains with similar substrate specificity as the protein of interest and therefore could produce unwanted signals. Another advantage of reconstituted samples is the possibility to vary the membrane composition of the sample which can affect the protein function or ion gradient stabilities. Higher turnover transporters usually also work with membrane preparations, e.g., membrane vesicles from plasma membranes, but show lower signal to noise. When using reconstituted samples, it is critical to ensure that no residual detergent remains in the membrane preparation after reconstitution. Different procedures are described for detergent removal, e.g., rapid dilution or incubation with Bio-Beads® (Bazzone et al., 2016). If detergent remains in the protein sample, the sensor can be destroyed during the adsorption process.

Before starting experiments, the lipid concentration for the adsorption of the membranes to the sensor has to be optimized. Too high concentrations as well as too low concentrations can reduce the signal amplitude. For adjusting the concentrations, the sample can be diluted in the respective nonactivating assay buffer. Usually lipid concentrations from 0.2 to 5 mg/mL are ideal, which has been tested for different reconstituted samples with known lipid concentrations.

In the case of membrane preparations with unknown lipid concentration, the total protein concentration can be used as a benchmark. Typical samples are prepared with 2–10 mg/mL total protein concentration. For adsorption to the SSM total protein concentrations between 0.1 and 1 mg/mL are used. Since the concentration of the transporter protein cannot be adjusted directly, the output mainly depends on the expression efficiency which, therefore, also has to be optimized.

In all cases the sample should be stored at −80°C in high concentration and diluted directly before the measurement.

3.2 Preparation of Thiol Solution, Lipid, and Membranes

The commercial and the laboratory setup make use of the same solutions for sensor preparation and the same procedure for sample preparation, both described in the following.

Preparation of thiol solution

Prepare a 0.5-m*M* solution of 1-octadecanethiol in isopropanol or ethanol. Use a glass bottle and slightly heat and stir the solution until the thiol is dissolved completely. Avoid evaporation of the solvent. Store the solution in the dark at room temperature and check for precipitates before use. If necessary, reheat carefully to dissolve precipitates.

Preparation of lipid solution

Prepare 7.5 μg/μL 1,2-diphytanoyl-*sn*-glycero-3-phosphocholine in *n*-decane. When the lipid is dissolved in chloroform, remove the solvent using a rotary evaporator. 25 mg of the lipid is dissolved in 3.33 mL *n*-decane; the lipid dissolves immediately. Transfer the solution in a glass bottle with a Teflon lid, and store at −80°C or −20°C. Always allow the bottle to warm up to room temperature before opening to avoid water condensation forming on the inside of the bottle.

Preparation of the membranes

Rapidly thaw an aliquot of the membrane preparation. Prepare a homogenous membrane suspension by diluting the membranes with nonactivating buffer to yield the optimized lipid or total protein concentration for the adsorption to the sensor. The dilution factor has to be optimized empirically for each membrane preparation. Sonicate the membrane suspension in a 1.5-mL Eppendorf tube. For proteoliposomes use gentle sonication with a bath sonicator (Sonorex RK 52 H, Bandelin). Use three to five sonication cycles, each lasting 10 s and put the membranes on ice in-between to prevent damage by overheating. For membrane vesicles use a tip sonicator (ultrasonic processor UP50H, Dr. Hielscher GmbH, Germany, equipped with MS 1 tip) by applying 10 bursts with an amplitude of 20% and a cycle time ratio of 0.5. After sonication, immediately add the membranes to the sensor.

3.3 Sensor Preparation

In the case of the SURFE^{2}R N1, sensor preparation (addition of lipid and membrane preparation) can be done in batches, e.g., 10 sensors at the same time. The measurements can be repeated on the same day with different sensors for better statistics. The SURFE^{2}R sensors also can be stored after preparation at 4°C for 1–3 days without significant loss of activity, depending on the sensitivity of the protein. If kept at −80°C, they may be ready-to-use for weeks or even months. This is mainly limited by the stability of the transporter protein. In the case of the laboratory setup, only one sensor can be prepared at a time. The sensor is destroyed when removed from the cuvette due to drying.

The sensor preparation procedure is described separately for the laboratory setup and the commercial SURFE^{2}R N1. Fig. 6B and C gives an overview of the differences of both systems.

Laboratory setup

For details of the different manipulations, see Bazzone et al. (2013).

(1) Thiolization of the sensor surface

- The sensor is stored in thiol solution. Therefore no incubation time is required. The sensor is ready to use when stored in thiol solution for at least 1 h.

- Remove the sensor from the thiol solution and rinse with isopropanol or pure ethanol.
- Dry the sensor under pressurized air or nitrogen gas.

(2) Assembly of the lipid layer

- Isolate the contact strip of the sensor with nail polish to ensure that the SSM is only formed at the circular sensor surface (active area) (Fig. 2A).
- Apply 1.5 μL of the lipid to the surface of the thiolated sensor without touching the gold surface with the pipette tip.
- Immediately mount the sensor inside the cuvette. After cuvette assembly mount it inside the fluidic system.
- Flush the system with nonactivating buffer to spontaneously form the SSM. The nonactivating buffer differs depending on the assay conditions.

(3) Check for sensor quality

- Before adding the membrane sample, the quality of the membrane can be checked by measuring capacitance and conductance by applying a voltage using the function generator.
- Calculate the parameters (Bazzone et al., 2013) and compare the output with the ideal values: For the sensors with 1 mm diameter capacitance should be 1.5–3 nF and conductance should be below 0.5 nS.
- When the parameters do not fulfill the requirements, the sensor should be discarded and a new sensor has to be prepared.

(4) Application of membranes to the sensor

- Apply 30 μL of the membrane suspension directly to the outlet of the cuvette until the cuvette volume above the sensor is filled.
- Incubate for about 1 h to adsorb the membranes to the SSM.

SURFE^{2}R N1

(1) Thiolization of the sensor surface

- Fill the sensor well with 50 μL of 1-octadecanethiol solved in isopropanol.
- Incubate for 30 min at room temperature in a closed Petri dish.
- Remove the solution by tapping the sensor upside-down on a tissue.
- Rinse the sensor three times with isopropanol or pure ethanol.
- Rinse the sensor three times with deionized water.
- Dry the sensor thoroughly by tapping it on a tissue.

(2) Assembly of the lipid layer
- Apply 1.5 μL of the lipid to the surface of the thiolated sensor without touching the gold surface with the pipette tip.
- Immediately fill the sensor well carefully with 50 μL of nonactivating buffer to spontaneously form the SSM. The nonactivating buffer differs depending on the assay conditions.

(3) Application of membranes to the sensor
- Apply 5–10 μL of the diluted and sonicated membrane/proteoliposome suspension by submerging the pipette tip in the solution covering the sensor. Slowly dispense the suspension onto the sensor surface. Do not touch the surface of the sensor with the pipette tip.
- Centrifuge the sensor for 30 min at 2000–3000 × *g*. Use a plate centrifuge with suitable adapter or put the sensors in 50 mL falcon tubes.
- You can use the sensor immediately, incubate the sensor at 4°C for several hours or overnight or freeze the sensor at −20°C or −80°C for long-term storage. Make sure that the sensor does not dry out during incubation. The sensor stability mainly depends on the stability of the transporter protein.

(4) Check for sensor quality
- Position the socket of the sensor on the contact pin located on top of the device. Move the locking to the left and shield the sensor using the Faraday cage.
- Before starting the measurement, check for sensor quality by measuring capacitance and conductance of the compound membrane by choosing the respective protocol in the software.
- Compare the output with the ideal values: For the commonly used sensors with 3 mm diameter capacitance should be 15–30 nF and conductance should be below 5 nS.
- When the parameters do not fulfill the requirements, the sensor should be discarded.

3.4 Electrophysiological Measurement

After the sensor is prepared and the membrane preparation containing the protein of interest is adsorbed to the SSM, the electrophysiological measurement can begin. The procedure depends on the performed assay, but there are typical steps which differ between the setups and are summarized later.

Laboratory setup

(1) Prepare the device

- Start the computer and the software.
- Set the saving path for your workflows and data.
- Check that the waste containers are empty.
- Clean the system by flushing with 50 mL 30% ethanol. Manually switch the solution containers, set the system pressure to 0.5 bar, and open the valves using the controller. Repeat the cleaning procedure with 100 mL ddH_2O.

(2) Add the buffers

- Change the solutions to the buffers used for the first experiment. Usually these are the buffers for which the highest signal amplitudes are expected to test for adsorption efficiency and signal stability.
- Repeat the cleaning with the respective nonactivating, activating, and resting buffers. End the cleaning with resting or nonactivating buffer.

(3) Add the sensor

- After the setup is ready-to-use, the sensor can be prepared, involving the mounting of the cuvette, the formation of the SSM, checking for the membrane quality, and the adsorption of the membrane preparation (see Section 3.3).
- Sensor preparation cannot be performed in advance or in batches.

(4) Define the workflow

- The software can be used to define the valve configuration and assign the three solutions to a combination of on/off states of the respective valves.
- The workflow defines the time of each solution flow and the sequence of the solutions, e.g., single or DSE.
- The time of solution flow is only restricted by the volume of the solution containers (100 mL) and the volume speed which depends on the pressure used.

(5) Run the experiment

- The setup is semiautomatic, it performs one workflow with a defined set of three buffers independently. This can involve multiple solution exchange experiments, e.g., triplicate measurements. One measurement takes as long as the solution exchange.
- When measurements with different buffer compositions are performed, the buffers need to be changed manually between the

experiments. The pressure has to be released, the solution containers unscrewed, and replaced by another set of containers holding the respective buffers. When replacing containers, the tubing inside the container has to be dried using tissues in order to not contaminate the next buffer. Then air bubbles introduced by container exchange are removed using the manual valves.

- Depending on the measurement sequence it could be necessary to include incubation times between the measurements to adjust the intraliposomal composition to a new buffer.
- After finishing the experiment the cuvette can be unmounted and cleaned and another sensor can be prepared to perform another measurement, e.g., for statistical analysis or for testing a negative control under the same conditions.

SURFE^{2}R N1

(1) Prepare the device
- Turn on the SURFE^{2}R N1 device and start the software. Wait until initialization has finished and the software is ready.
- Set the saving path for your workflows and data.
- Check that system water is available and the waste container is empty.
- Start the workflow "initialize fluidics" to fill the tubings and syringe pumps with system liquid. This takes about 3 min.

(2) Define the workflow
- The "workflow" contains all the required information for the instrument to perform one experiment, either one single measurement or a whole sequence of measurements with up to 53 different solutions.
- Select one of the preconfigured workflows from the drop-down menu or design a custom workflow in the workflow editor.
- The workflow editor enables the researcher to define a specific experiment. The settings include the duration, speed, and volume of solution flow during the experiment, the number and sequence of different buffers, the number of repetitions per measurement, the incubation times between experiments, the volume used for rinsing the sensor after an experiment, and the choice to rinse and incubate with resting solution instead of nonactivating solution to perform DSE experiments.

- There are a few limitations which have to be considered: It is not possible to take up more than 2 mL of each buffer. The aspirated volume has to be at least 100 μL greater than the dispensed volume. If you use repetitive measurements, ensure that you do not run out of buffer. If air is pipetted on the sensor, it will be ruined.

(3) Add the buffers

- Position all the buffers in the autosampler according to the assignment in the chosen workflow. Usually position 1 contains activating buffer, position 2 nonactivating buffer, and position 3 washing buffer (nonactivating buffer or resting buffer). This can be repeated up to position 53 when measuring under different conditions.

(4) Add the sensor

- The sensors are prepared in batches beforehand. Add one of the sensors and check for quality (see Section 3.3).

(5) Run the experiment

- Start the measurement by clicking on the play button. During the measurement the indicator will flash yellow. You can pause and stop the run by clicking on the corresponding buttons.
- The measurement itself generally consists of loading and injecting the solutions as well as washing the pipette and tubing after each sequence. Everything is fully automated and takes about 3 min for each measurement. Depending on the number of sequential measurements the robot will work independently for several minutes or hours.
- Depending on the measurement sequence it may be necessary to include incubation times between the measurements to adjust the intraliposomal composition to a new buffer. This can be set within the workflow editor.
- After finishing the workflow, you can switch to another prepared sensor and repeat quality control and measurement for statistical analysis, comparison among different mutants or comparison with a negative control membrane.

3.5 Cleaning and Maintenance

After the experiment, the sensor has to be regenerated. Moreover regular cleaning of the tubing is important, since it prevents the formation of a biofilm. In the case of the laboratory setup, there are parts which need to be maintained on a regular basis.

Laboratory setup

(1) Daily cleaning procedure
- Wash the tubing sequentially with water and 30% isopropanol or 30% ethanol by replacing the solution storage containers and switching the valves manually.
- Leave the tubing in 30% ethanol. Release the pressure from the system, turn off all instruments and unmount the cuvette. The cuvette can be dried with pressurized gas.

(2) Regeneration of sensor
- Place the sensor chip in a glass beaker with pure ethanol and sonicate for 1–2 min using a bath sonicator.
- Dry the sensor chip under nitrogen gas.
- Store the sensor chip in a 10-m*M* octadecanethiol solution.

(3) Other maintenance procedures
- There are parts of the instrument which needs to be exchanged or regenerated occasionally. The gel bridge has to be renewed every month and the silver chloride electrode needs to be chlorinated on a regular basis. Moreover the O-rings of the cuvette need to be checked for mechanical abrasion.
- The valves can clog after intensive use over a long period of time when not using filtered solutions or due to tissue fibers. This should be checked by regularly comparing the solution flow speed using a defined pressure. When the speed decreases significantly, the valve is clogged and has to be exchanged. To prevent valve clogging use only filtered solutions and clean the tubing inside the solution containers with low-lint tissues, e.g., kimtech precision wipes.
- When the tubing shows a biofilm, either an intense cleaning protocol needs to be used or the tubing must be replaced.

SURFE^{2}R N1

(1) Daily cleaning procedure
- After finishing all experiments for the day, select the respective workflow for daily maintenance.
- Follow the instructions on the screen: The system will be sequentially flushed with 30% isopropanol or 30% ethanol and water. Afterward the tubing is emptied.

- Close the software and shut down all windows. The system will turn off automatically.

(2) Regeneration of the sensor
- The sensors have a limited lifetime, but can be reused multiple times using the following protocol. Be aware that mechanical contact with the sensor surface produces scratches and reduces its lifetime. Due to glued parts, sonication of the sensor is not possible.
- Rinse the sensor with 10 mL water using a pipette and dry it by tapping the sensor on a tissue or using pressurized gas.
- Fill the sensor with 100 μL isopropanol or ethanol and incubate for about 10 min.
- Presoak a cotton swab with isopropanol or ethanol and gently rotate the cotton swab on the sensor surface without applying pressure to the sensor.
- Rinse the sensor sequentially with 5 mL pure isopropanol or ethanol and 10 mL water.
- The sensor should dry overnight to prevent corrosion. Optically check the surface for scratches and cleanliness the next day.
- Store the sensor under inert gas for a longer lifetime.

(3) Other maintenance procedures
- When the tubing shows a biofilm, a more intense cleaning protocol is available in the software interface. This includes flushing the system with neodisher.
- No parts need to be exchanged or maintained on a regular basis.

4. ASSAY DEVELOPMENT

The assay conditions depend on the transport protein as well as the information the researcher wants to obtain. Nevertheless, there are a few aspects to consider for optimizing the experiment, e.g., for reducing solution exchange artifacts, establishing required gradients, controlling transporter stability during the assay, or using the accurate negative control.

4.1 Single Solution Exchange Measurements

The starting point for developing an assay is the single solution exchange experiment with a known substrate (Fig. 7A). Here, two different solutions are sequentially conducted through the cuvette: The nonactivating solution (NA) followed by the activating solution (A). The nonactivating solution contains all components required for transporter function apart from the

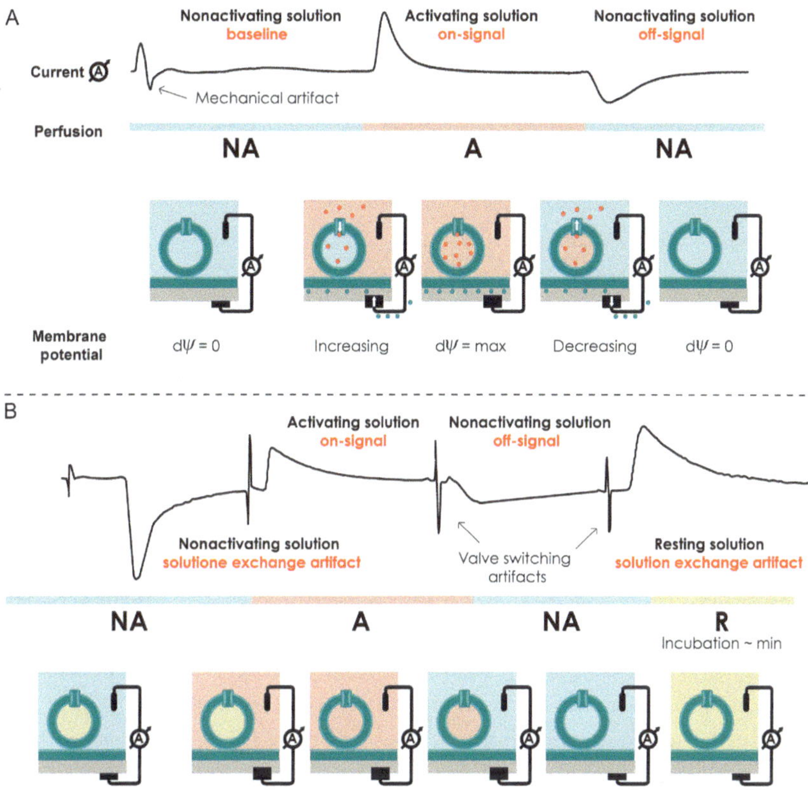

Fig. 7 Single and double solution exchange protocols and the resulting transient currents. (A) In the single solution exchange protocol nonactivating (NA) solution is replaced by activating (A) solution followed by the flow of NA solution to restore initial conditions. During the first NA flow the membrane potential is zero and a baseline is recorded. When switching to A solution the substrate gradient activates the transporter, electrogenic transport leads to an increase in membrane potential which itself leads to a decrease in transport and a decline in the measured charging current. When the membrane potential has reached its maximum, no net flux remains and the measured current is zero. The on-signal reflects the transporter activity. When switching back to NA solution, an opposite substrate gradient is applied leading to substrate efflux. This is accompanied by an off-signal with the opposite sign compared to the on-signal. Since the membrane is charged and the intraliposomal substrate concentration may be less than the concentration in the A solution, only the on-signal is used for analysis. The current was recorded using the SURFE^{2}R N1. Mechanical artifacts are observed at the beginning of the solution flow. (B) The double solution exchange protocol ends with the flow of the resting (R) solution which differs from NA and A solutions in at least one component. During incubation in R solution, the intraliposomal composition adjusts to the composition of the R solution. When starting the next experiment with the NA flow, a gradient (NA → R) is established. This is usually accompanied by a large solution exchange artifact depending on the solution exchange due to the different

(Continued)

component(s) used for activation of the transporter immobilized on the SSM. The latter is only added in the activating solution to produce a concentration jump at the SSM and to initiate transport and the concomitant charge displacement (on-signal). To restore initial conditions the A solution is replaced by NA solution at the end of the measurement. This is accompanied by the detection of a transient current of the opposite sign (off-signal). Usually only the on-signal is used for analysis. This protocol ensures that before and in-between experiments, the SSM is incubated with NA solution.

Different types of transporters have been characterized using SSM-based electrophysiology, uniporters, symporters, antiporters, ion pumps, and ion channels. Each type requires a well-adapted assay. So far only a few uniporters have been characterized, e.g., the organic cation transporter OCT2 (Gaiko et al., 2011), the high-affinity choline transporter CHT (Choudhary et al., 2017), and the ammonium transporter Amt (Wacker, Garcia-Celma, Lewe, & Andrade, 2014). For transporters with more than one substrate, there are different possibilities for single solution exchange experiments: In the Ca^{2+} ATPase either Ca^{2+} or ATP is used in all solutions (we refer to this as Ca^{2+} or ATP background), while the other is used for activation of the transport reaction (Tadini-Buoninsegni et al., 2006). In this case information about different reactions can be obtained: The concentration jump of ATP with Ca^{2+} background reflects the ATP-induced release of prebound calcium. The concentration jump of Ca^{2+} with ATP background shows calcium-induced transport. Other ATPases have been characterized using ATP concentration jumps, e.g., NaK-ATPase (Pintschovius & Fendler, 1999; Pintschovius et al., 1999), HK-ATPase (Kelety et al., 2006), V-ATPase (Obrdlik et al., 2010), fof1-ATPase (Burzik, Kaim, Dimroth, Bamberg, & Fendler, 2003), and a Cu-ATPase (Mattle et al., 2015; Tadini-Buoninsegni et al., 2010; Tadini-Buoninsegni & Smeazzetto, 2017).

Fig. 7—Cont'd compositions of the buffers. When switching to A solution, the transporter is activated while the established gradient affects its function. The assumption of this protocol is that the compound can permeate the liposomal membrane during the incubation time in R solution of several minutes, while not permeating during the first NA flow of about 1 s. The required times for both R incubation and NA flow have to be optimized according to the desired gradient and the sample membrane. The current was recorded using the laboratory setup in DSE configuration. Mechanical artifacts are observed every time when a valve switches.

For proton-coupled transporters, pH jumps are not advisable since they produce large artifacts interfering with possible transporter signals. Instead concentration jumps of the cosubstrate, e.g., a sugar, should be used. Comparison of different pH can be achieved by varying the background pH. By doing so, the pH dependencies of different proton-coupled transporters have been obtained, e.g., H^+/sugar symporters (Bazzone et al., 2016; Garcia-Celma, Ploch, Smirnova, Kaback, & Fendler, 2010), the Na^+/H^+ exchanger NhaA (Mager, Rimon, Padan, & Fendler, 2011) and the H^+/Cl^- exchanger Clc (Garcia-Celma, Szydelko, & Dutzler, 2013). For sodium-coupled transporters, sodium concentration jumps can be performed as shown for the melibiose permease (Ganea et al., 2001), a Na^+/melibiose symporter; the Na^+/Ca^{2+} exchanger NCX (Barthmes et al., 2016; Kelety et al., 2006); the Na^+/H^+ exchangers NhaA, NhaP, and NhaP1 (Calinescu, Danner, Bohm, Hunte, & Fendler, 2014; Calinescu et al., 2016; Calinescu, Paulino, et al., 2014; Mager et al., 2011; Zuber et al., 2005); and the amino acid symporters PutP (Zhou et al., 2004), GltP (Raunser et al., 2006), and EAAC1 (Krause et al., 2009). Sodium-induced artifacts are several orders of magnitude lower than proton-induced artifacts, possibly due to the lower interaction of sodium with the lipid headgroups of the SSM.

4.2 Double Solution Exchange Measurements

Transporter function may require a cosubstrate gradient, e.g., pH gradient, Na-gradient, or even a membrane potential. A DSE on the SSM allows establishing concentration gradients or even a membrane potential to modulate transporter activity.

The DSE experiment requires an additional resting (R) solution, which is conducted through the cuvette after the NA-A-NA solution exchange (Fig. 7B). Thus, in-between experiments, the SSM is incubated in resting solution to adjust the intraliposomal composition to the composition of R. At the beginning of an experiment, a gradient (NA → R) is generated and with the flow of A the transporter activity is measured under gradient conditions.

A DSE could be required for many different reasons. In every case, control experiments without the gradient should be used to quantify the effect of the gradient on transporter activity.

- For a ligand-gated ion channel, the resting solution is free of permeant ions while the nonactivating and the activating solutions contain the permeant ions. In addition, the activating solution contains the ligand (Schulz et al., 2009).

- The glutamate transporter EAAC1 requires opposite potassium and sodium gradients for high activity (Krause et al., 2009).
- A DSE is used to test the influence of intraliposomal and extraliposomal pH on transporter activity separately (Bazzone et al., 2016; Mager et al., 2011). The two p*K* values for alkaline and acidic downregulation could be assigned to the extraliposomal protonation and the intraliposomal deprotonation of the transporter, respectively.
- In the sodium calcium exchanger NCX this technique was used to preload the vesicles either with sodium or calcium followed by a concentration jump of the other ion to produce exchange activity in opposite directions (Barthmes et al., 2016).
- The sodium-coupled betaine transporter BetP needs to be activated by osmotic shock before the substrate is provided to measure transporter activity. This was also achieved with the DSE by exchanging a low osmolarity solution (R) by a high osmolarity solution (NA-A-NA) (Khafizov et al., 2012).
- We could show recently that a membrane potential accelerates the turnover of transporters adsorbed to the SSM by a factor of 2–10, depending on the transporter. This was done with the DSE by exchanging a potassium containing solution (R) with a potassium-free solution (NA-A-NA) on a sensor incubated with valinomycin. During the R-NA exchange potassium exits the proteoliposome and generates an internal negative potential to drive the influx of positive charge. This was shown to drive proton-coupled sugar transport in XylE as well as proton-coupled peptide transport in PEPT1.

There are two restrictions to the DSE protocol: First the incubation time in R depends on the permeability of the compound or ion for which the gradient should be established. For pH gradients incubation time is between 3 and 20 min, depending on proton concentration (Garcia-Celma et al., 2013). Also for Na^+ an incubation period of ~20 min was observed (Meyer-Lipp, Ganea, Pourcher, Leblanc, & Fendler, 2004). Some ions may not be permeable on a reasonable time scale. You can increase the permeability for an ion by using electroneutral carriers, e.g., monensin. On the other hand, when the transporter itself transports the ion, the gradient may be established very fast. The second restriction relates to the flow time of NA solution when starting an experiment. For compounds showing a high permeability, the gradient could decay before the transporter is activated.

The minimal time of NA solution flow depends on the size and shape of the artifact following the R-NA solution exchange. The minimal flow time has to be determined before starting the measurements, but usually is in the range of 1 s.

4.3 Sequence of Measurements and Stability of Transporter Activity

The high stability of a sensor allows measurements of different experimental conditions using the same sensor. It is possible to derive half-saturation constants like EC_{50}, IC_{50}, K_m, and K_i. Again, if there is more than one substrate, there are more possibilities to measure a concentration dependence: In the case of the Ca^{2+}-ATPase either by increasing the Ca^{2+} background concentration or by varying the Ca^{2+} concentration in the A solution (Tadini-Buoninsegni et al., 2006). The respective EC_{50} values have to be interpreted differently.

It is important to mention that current amplitudes can only be compared directly when obtained from the same sensor. All data points for the determination of EC_{50} or IC_{50} should be generated in one measurement sequence with different concentrations using the same sensor. The reason is the variability of the vesicle-SSM adsorption efficiency which also affects the current amplitude.

The number of measurements using the same sensor without loss of transporter activity depends on the transporter and membrane preparation as well as the experimental conditions. Usually at least 5–10 different conditions can be checked on the same sensor. In many cases one sensor lasts for 50–100 measurements. Before starting an experiment, it is advisable to check for the stability of the transporter signals by sequential measurements under identical conditions.

Even if the signal is stable over a longer time period, it is recommended to include a rundown control in each experiment. Start and end the experiment under conditions where high signal amplitudes are expected. When the signals have the same amplitude at the beginning and the end of the measurement sequence, the transporter activity was stable over the time period of the experiment. This is essential for measurements using extreme conditions like acidic or alkaline pH to ensure that the signal change is not due to irreversible transporter inactivation.

An important issue when performing multiple measurements under different conditions is the time used for rinsing and incubating the sensor after changing the conditions. When measuring with different substrate concentrations in the activating solution, in general, no additional incubation time is required. When changing the background concentration in all solutions, longer incubation periods may be necessary. For instance, in the case of a pH dependence, after changing the pH an incubation time of at least 3 min is required before starting the measurement with the new pH to adjust the intraliposomal pH. When adding an amphiphilic inhibitor in the background, incubate for at least 20 min before the measurement to equilibrate the compound concentration inside the membrane.

4.4 How to Avoid Artifacts: Guidelines for Preparing Buffers

Special attention is needed for the preparation of buffers for SSM-based electrophysiology. The interactions between the compound membrane and the buffer components can generate large solution exchange artifacts. This may be due to the interaction of the ions with lipid headgroups (Garcia-Celma et al., 2007) or with other components (membrane proteins) of the adsorbed membranes in the case of an unpurified sample.

In both cases, unwanted transient currents (artifacts) are observed which can overlay the transporter signal. Artifacts due to other membrane proteins can only be reduced by optimizing the sample preparation and using samples with higher purity (see above). Alternatively, a proper negative control membrane can be used for subtracting the artifacts from the transporter signals as described in Section 4.5.

The artifacts due to lipid–ion interactions may be reduced by keeping nonactivating and activating solutions as similar as possible. Any differences between activating and nonactivating solutions, e.g., ionic concentration, pH, and temperature may generate transient currents easily of the magnitude of protein-induced signals. Therefore, it is a good practice to prepare nonactivating and activating solutions in one batch and adjust pH. Afterward the buffer is divided into two volumes and the activating compound is added to the activating solution.

Even the activating compound (the substrate) may produce solution exchange artifacts which are not related to transporter function. For this reason use an inert compensation for the activating compound in the nonactivating solution. For instance, if transport by the Na^+/H^+ exchanger NhaA is initiated with 100 m*M* NaCl in the activating solution, use 100 m*M* KCl in the nonactivating solution as compensation. In special cases, it may be advantageous to choose different concentrations in the A and NA solutions for a better compensation. Moreover, high background salt reduces artifacts, e.g., a 200 m*M* KCl background concentration is used in both solutions to reduce possible Na^+ artifacts in assays with NhaA (Zuber et al., 2005). Also 10–30 m*M* $MgCl_2$ in the background can be helpful. As an example for a successful artifact compensation strategy see the ammonium transporter Amt, where the measurement was performed with different concentration ratios $NH_4Cl/NaCl$ and at a constant total osmolarity in the activating and nonactivating solutions (Wacker et al., 2014).

Low-affinity transporters which need high substrate concentration jumps generally require more attention concerning artifact controls and setting up the buffer systems. In this case, it is helpful to include a negative control membrane without the transporter of interest. It is worth determining the EC_{50}, since measurements should be performed using concentrations close to the EC_{50} for two reasons: First to measure at concentrations high enough to induce large current amplitudes with high S/N; and second to measure at concentrations low enough to reduce the artifact fraction of the measured current. A compound membrane with higher conductance generally produces larger artifacts. Be sure to only use compound membranes which pass the quality control (conductance and capacitance determination, see above). Moreover some compounds produce larger artifacts than others:

- Noncharged water-soluble organic substrates do not produce artifacts, even in higher concentrations, e.g., 30 m*M* glucose yields artifacts below 50 pA. They are, therefore, well suited for concentration jumps in sodium- or proton-coupled transporters.
- Experience shows that concentration jumps of monovalent cations like Na^+ or K^+ above 1–10 m*M* produce artifacts. Below there is often no compensation necessary.
- For highly charged ions such as Mg^{2+} or Ni^{2+} the half-saturation constants for lipid–ion interaction is in the micromolar range (Garcia-Celma et al., 2007). Therefore, metal transporters with substrates with valences larger than 1 are difficult to measure. Even at low concentrations, large artifacts can interfere with the transporter signal.
- Protons produce large artifacts of several nA depending on the proton concentration used. pH jumps can, therefore, only be used for high turnover transporters. In other cases, the pH dependence should be determined by variation of background pH.
- Preferentially use activating compounds which are well soluble in water. Lipophilic and amphiphilic compounds may generate large artifacts if used in high concentrations.

4.5 The Ideal Negative Control

To estimate the solution exchange artifact for a specific experiment, a negative control membrane should be used. Ideally, the only difference between the membrane containing the protein of interest and the negative control membrane would be the presence of the protein of interest. In the case of proteoliposomes, the reconstitution should take place ideally with the same lipid and a negative control protein preparation. The latter should be prepared using the same protocol and the same cells as used for the protein sample but without induction of protein expression in the negative control. If such control is not available, empty liposomes are also suitable for quick artifact testing. A bare SSM without adsorbed membranes is no reliable control measurement since artifacts can behave differently. It may, however, be used for a rough estimation of expected artifact current amplitudes.

The negative control can be used in assay development to reduce artifacts before starting the experiment with the transporter of interest. Different compensating compounds in different concentrations should be tested in the nonactivating solution to reduce the artifact before proceeding to measure the preparation under these "reduced artifact" conditions. If artifacts cannot be completely avoided, a duplicate measurement series with sample and the negative control may be performed and artifacts subtracted.

4.6 Control Measurements Using the Same Sensor

Artifact testing can also be performed with the membrane preparation containing the protein of interest. There are different possibilities, depending on the transporter.

- One quick test could involve solution exchange experiments using compounds with similar physicochemical properties as the substrate, e.g., for the lactose transporter LacY sucrose was used as a nonsubstrate (Garcia-Celma et al., 2009). Be sure that this compound does not interact with your protein of interest.
- If specific inhibitors are available, use them in the background of all solutions and repeat the solution exchange experiment. The remaining current amplitude should represent the artifact fraction of the signal.
- For ATPase or other transporters requiring at least two compounds for their activity, a good control is repetition of the solution exchange measurement without one required component in the background. In the case of the Ca^{2+} ATPase this could be an ATP jump without Ca^{2+} in the solution (Tadini-Buoninsegni et al., 2006). This only works because

ATP does not bind to the transporter and therefore does not induce an electrogenic reaction in absence of Ca^{2+}.

5. DATA INTERPRETATION AND ANALYSIS

Since SSM-based electrophysiology can be used for obtaining different kinds of kinetic information, it is important to know how to analyze and interpret the data. In the last section of this chapter, we will, therefore, offer some specific advice for analysis and interpretation of the transient currents obtained in SSM-based electrophysiology.

While the experiment itself can be performed very fast and easy, the interpretation of the transient current is not trivial. When the activating compound induces electrogenic transport activity, a transient current is recorded. This current is characterized by its rise time, its peak current amplitude, its integrated charge, and its decay time.

The interpretation of each parameter depends on the origin of the electrogenic reaction. When the signal is a result of continuous turnover (steady-state current component, see earlier), the decay time constant is due to the potential rise inside the liposomes/membranes inhibiting further transport. In this case, the relevant signal component is the signal amplitude which is proportional to the turnover rate of the transporter. In the case of a presteady-state reaction, e.g., the electrogenic binding of a substrate, the reciprocal decay time constant may ideally represent the rate constant of this reaction. In this case, it is important to know the time resolution of the system for a correct estimation of the upper limit of determinable rate constants. Also the integrated charge may be used for analysis. This is meaningful in cases where binding of a substrate needs to be quantified because the charge is proportional to the bound substrate (Zhou et al., 2004).

5.1 Transporter Characterization Using the Peak Current Amplitude

In most cases, the peak value of the measured transient current is used to quantify the transport activity of the investigated protein. The peak current is an initial current amplitude recorded just when the substrate reaches the binding site and transport starts. At this point the membrane has not yet built up a significant voltage to affect the kinetic properties of the enzyme. Experimental support for this conclusion comes from a comparison of proteoliposomes with an LPR of 500 and 10 (Zuber et al., 2005). Even at extremely high pump densities, the kinetic properties of the transporter were found to be unaltered.

Analyzing the peak current at different experimental conditions allows a detailed functional characterization of the transporter of interest. Here, we suggest sequences of measurements for revealing basic information about a transporter.

- Measurements with different substrate concentrations yield half-saturation constants (EC_{50}). For transport or binding reactions they correspond to K_m or K_D values (Fig. 8A), respectively. For the proton-coupled sugar transporter MelB, both EC_{50} for binding and transport have been obtained using SSM-based electrophysiology (Ganea et al., 2001).
- For proton-coupled transporters measurements at a fixed substrate concentration and different pH yields apparent pK values for protonation and deprotonation (Fig. 8B), as observed, for example, for H^+/sugar symporters (Bazzone et al., 2016; Garcia-Celma et al., 2010), the Na^+/H^+ exchanger NhaA (Garcia-Celma et al., 2009), and H^+/Cl^- exchanger Clc (Garcia-Celma et al., 2013).
- When EC_{50} values for different cosubstrate concentrations (in proton-coupled transporters for different pH values) are determined, information about cooperativity can be derived. For the sodium-coupled proline transporter, PutP it was shown that the presence of proline does not affect the apparent affinity for Na^+, while on the other hand the presence of Na^+ increased the apparent proline affinity 16-fold (Zhou et al., 2004).
- Substrate concentration jumps with different inhibitor concentrations in all solution yield IC_{50} values (Fig. 8C). Inhibition of different inhibitors can also be compared by the percentage of signal amplitude after inhibition. Inhibition experiments using SSM-based electrophysiology were performed for human transporters like PEPT1(Kelety et al., 2006), Clc7 (Schulz et al., 2010), the NaK-ATPase (Gramigni et al., 2009), and ion channels like the rat P2X2 and the electric ray nAChR receptors (Schulz et al., 2009).
- A DSE can be used to vary cosubstrate gradients. This can yield information about the coupling mechanism or driving force.

In general, parameters such as EC_{50} or IC_{50} have to be interpreted as apparent values. They depend on the experimental conditions. With sufficient information about the transporter, the real constants may be derived by kinetic analysis using a kinetic model, as shown for NhaA (Mager et al., 2013, 2011), LacY (Garcia-Celma et al., 2010), and EAAC1 (Krause et al., 2009).

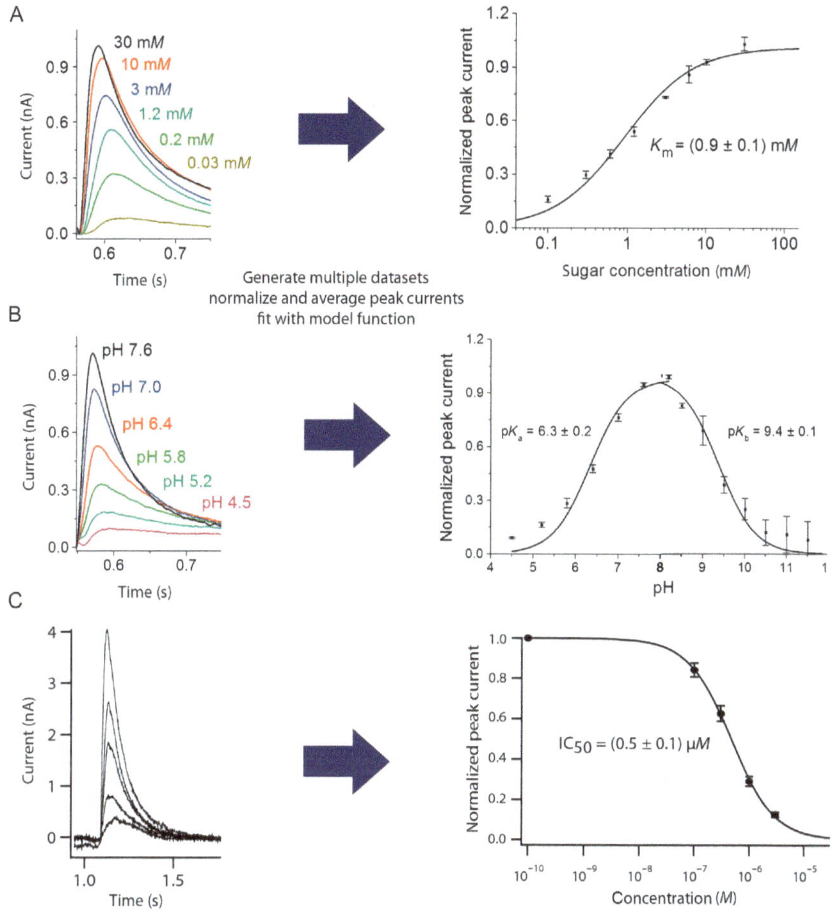

Fig. 8 Examples of datasets only using the current amplitude. (A) Currents for the proton-coupled sugar transporter FucP from *Escherichia coli* obtained for different L-fucose concentration jumps as indicated. All buffers contained 100 m*M* KP_i at pH 7.6. The current amplitudes were used for the calculation of the K_m value by fitting with a Michaelis–Menten equation. The dataset was generated with the laboratory setup in LTR configuration. (B) Currents for FucP obtained for 30 m*M* L-fucose concentration jumps at different pH values as indicated. All buffers contained 100 m*M* KP_i equilibrated to different pH. The current amplitudes were used for the calculation of apparent p*K*s for proton binding and release by fitting with titration equations (Bazzone et al., 2016). The dataset was generated with the laboratory setup in LTR configuration. (C) Currents for the human organic cation transporter OCT2 obtained for a choline concentration jump and different background concentrations of the inhibitor quinidine. The current amplitudes were used for the calculation of IC_{50} by dose–response fitting. The dataset was generated with the SURFE^{2}R N1.

5.2 Steady-State Currents and Presteady-State Signals

A persistent steady state does not exist in SSM-based electrophysiology. Because of capacitive coupling, continuous electrogenic turnover is gradually slowed down to zero within a few 100 ms due to the buildup of a membrane potential in the compound membrane (see Fig. 1). Therefore, we refer to a current or current component which would lead to a steady-state current in a noncapacitively coupled measurement (or a DC-coupled measurement), a steady-state signal (Fig. 9A, left). Since the peak current of this signal is recorded immediately after activation of transport when the membrane potential is not yet built up, the peak current represents steady-state (or stationary) transport of the membrane transporter.

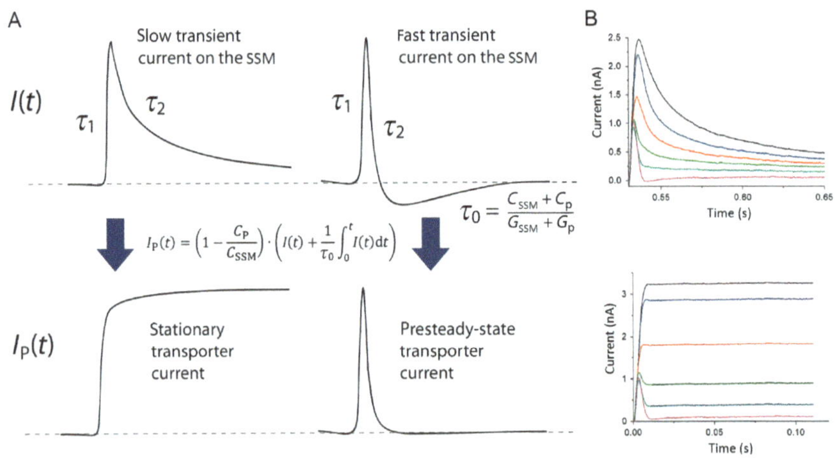

Fig. 9 Differences between steady-state and presteady-state currents on the SSM and reconstruction of transporter currents. (A) steady-state currents (*left*) and presteady-state currents (*right*) on the SSM (*top*) and after reconstruction of the transporter current (*bottom*). The reconstruction procedure using the equation as indicated yields the currents as if recorded in a DC-coupled electrophysiological experiment, and therefore, information about steady-state transport. Fast transient currents usually show a component with opposite polarity after the current decay. This is due to discharging of the compound membrane and can be described by the system time constant as indicated. No kinetic information about the transporter is comprised in this component. (B) Currents for the proton-coupled sugar transport XylE from *E. coli* after 30 m*M* D-xylose concentration jumps at different pH values (*top*). All buffers contained 100 m*M* KP_i adjusted to different pH values. At neutral pH steady-state currents are observed, while at acidic pH only a fast transient remains representing sugar binding (Bazzone et al., 2016). Reconstruction allows distinguishing stationary currents from presteady-state currents (*bottom*).

In contrast presteady-state signals originate from a single partial reaction in the transport cycle (Fig. 9A, right). A typical presteady-state signal is the sugar-induced conformational transition in proton-coupled sugar transporters which becomes visible at acidic pH, when proton release and turnover is inhibited (Bazzone et al., 2016). Here, forward and reverse rate constants have been determined using SSM-based electrophysiology (Garcia-Celma et al., 2010). The initial conditions also determine if a presteady-state signal is observed or not. In the Ca^{2+}-ATPase an ATP jump with Ca^{2+} background yields no presteady-state signal for Ca^{2+} binding, since Ca^{2+} is already bound. Instead only Ca^{2+} transport is observed. On the other hand Ca^{2+} concentration jumps without ATP background yields presteady-state signals representing Ca^{2+} binding (Tadini-Buoninsegni et al., 2006).

Presteady-state reactions may be hidden when steady-state currents dominate (Fig. 9B): Usually steady-state signals transfer a much larger charge than presteady-state reactions. Charges from steady-state signals usually represent multiple elementary charges, while charges from presteady-state signals may be only a fraction of an elementary charge per transporter. It is also possible that two presteady-state reactions transfer opposite charges and cancel each other out in the steady state as, e.g., electrogenic Na^+ and H^+ translocation in electroneutral Na^+/H^+ exchangers (Calinescu et al., 2016; Calinescu, Paulino, et al., 2014). At suitable experimental conditions steady state may be slowed down making one of these presteady-state reactions (Na^+ translocation) accessible for electrophysiological analysis (Calinescu et al., 2016; Calinescu, Paulino, et al., 2014).

Assignment of electrophysiological signals to steady-state or presteady-state reactions is important for correct interpretation of the data. To distinguish between steady-state signals and presteady-state signals, measurements with different protein densities or LPR can be used (Zuber et al., 2005). In the steady-state signal, the current declines due to the buildup of potential which in turn depends on transporter density: the higher the protein density, the higher the peak current, and the faster the current decline. In steady-state signals, therefore, the decay time constant depends on LPR. For presteady-state currents, the current decline generally is independent of LPR since it represents the reaction rate of the reaction itself. In this case only the peak current, but not the current decay varies with LPR.

Also current signals with biphasic time dependencies are observed (Fig. 9B). This is the case when a rapid electrogenic presteady-state reaction occurs early in the reaction cycle followed by a slower rate-limiting reaction which determines the steady-state turnover rate. In this case, a rapidly decaying presteady-state signal is followed by a slower decay of the steady-state component. An example where both reactions are electrogenic is represented by the proton-coupled sugar transporters: here fast electrogenic sugar binding can be distinguished from the electrogenic translocation of the proton-binding site (Bazzone et al., 2016). In the case of biphasic transient currents, the reconstruction of the transporter current can be useful for analyzing both components separately (Fig. 9B). This procedure involving numerical processing based on circuit analysis allows separation of the currents resulting from presteady-state and steady-state reactions (Tadini-Buoninsegni & Fendler, 2015).

If the presteady-state transient current decays sufficiently fast (that is, the related reaction is sufficiently fast), the transient currents usually display an additional component with opposite polarity (Fig. 9A, right). This component represents discharging of the compound membrane (Gropp, Cornelius, & Fendler, 1998) and does not contain information about the transporter. Its time constant is given by the system time constant reflecting the properties of the electrical circuit of the compound membrane (Schulz et al., 2008).

5.3 SSM and Transporter Orientation

An important parameter for the interpretation of the measured transport activities is the orientation of the protein in the adsorbed membranes. In proteoliposomes, the protein may be incorporated at random. But also native membranes may be randomly adsorbed to the surface of the SSM. If both populations of transporters are then analyzed on the SSM, the kinetic properties determined in the experiments represent an average of the characteristics of both transport directions. Exceptions are transporters which are exclusively activated in a defined orientation, for example, ATPases: an ATP concentration jump would only generate a signal for the fraction of inside-out-oriented transporters with the ATP-binding domain accessible from the outside. Also transporters may behave symmetrically (Mager et al., 2011). In this case, both populations show the same kinetics and, therefore, transporter orientation is no issue.

It is possible to obtain information about transporter symmetry by using samples with defined transporter orientation. Right-side-out as well as inside-out orientation in proteoliposomes have been found depending on the protein and/or the reconstitution technique (Mager et al., 2011; Meyer-Lipp et al., 2006; Rigaud, Pitard, & Levy, 1995). But also natural membranes can be prepared to yield preferentially inside-out or right-side-out vesicles (Rosen, 1986). For the sodium proton exchanger NhaA proteoliposomes with right-side-out orientation and native vesicles showing an inside-out orientation were used to compare forward and reverse transport modes (Mager et al., 2011).

5.4 Guidelines for Statistical Analysis and Comparison of Different Datasets

It is important to know how to set up the experiment to get reliable information. Especially when it comes to comparing data recorded on different sensors, there are a few things to take into account. Here, we offer guidance for statistically reliable data generation and processing.

Always include a rundown control to investigate transporter stability over the time period of the experiment. If the rundown exceeds 20%, the measurement should be discarded. A redesign of the experiment using less repetitions or solutions can reduce rundown. When the transporter always shows large rundown, it is possible to use a correction assuming a linear time dependence of the signal loss. However, it is usually worth spending time getting a good preparation and the right solution composition for stable protein signals.

When the transporter stability and solution consumption allows, always perform measurements in triplicates at the same conditions and the same sensor. Repeat the same sequence of conditions with at least three different sensors. The peak current amplitude not only depends on the rate of the transporter but also depends on the amount of active transporters on the sensor. Due to differences in the vesicle-SSM adsorption efficiency the peak amplitude can vary from sensor to sensor. The different datasets obtained using different sensors have to be normalized before averaging, e.g., to the data point with highest amplitude or to a fit parameter-like V_{max}. Thereby the variation in adsorption efficiency between different sensors is excluded from statistical analysis while the concentration dependence is preserved leading to more reliable values for IC_{50} or K_m.

Repeat the same sequence of measurements with a negative control membrane preparation to determine the artifact current. If the artifact exceeds 20% of the transporter signal, the averaged artifact peak currents should be subtracted from the peak current of the membrane preparation containing the transporter protein. This was done, for example, with the nitrate transporter NirC (Rycovska et al., 2012), the sodium-coupled proline transporter PutP (Zhou et al., 2004), and the chloride transporter Clc-7 (Schulz et al., 2010). If possible, complete the experiment by adding a specific inhibitor of the transporter protein and repeat the measurement with the rundown control solution. This is a specificity control for your measurement.

When amplitudes of measurements from different sensors under different conditions, but with the same sample need to be compared, all datasets require at least one reference measurement performed under the same conditions. All datasets are then normalized to this value to exclude an influence of the variability of the adsorption efficiency. On the other hand, when using different samples, a direct comparison of signal amplitudes is associated with a large standard deviation. Such comparison could be required for comparative analysis of different mutants, homologs, or isoforms. In addition to the variation of adsorption efficiency the protein density can vary in the different preparations due to differences in the expression, purification, and reconstitution steps, even when using a reconstituted sample with the same LPR. In these cases, comparisons yield only a rough estimate and a high number of experiments on different sensors are required.

It is important to note that SSM-based electrophysiology provides no information about absolute turnover rates since the exact amount of transporters contributing to the signal is not known. However, some transporters with known turnover rates were analyzed with SSM-based electrophysiology using samples with known protein densities. By comparison this information can be used to estimate an approximate turnover rate of a transporter with unknown turnover rate (Wacker et al., 2014).

REFERENCES

Balannik, V., Obrdlik, P., Inayat, S., Steensen, C., Wang, J., Rausch, J. M., et al. (2010). Solid-supported membrane technology for the investigation of the influenza A virus M2 channel activity. *Pflugers Archiv: European Journal of Physiology*, *459*(4), 593–605. http://dx.doi.org/10.1007/s00424-009-0760-1.

Barthmes, M., Liao, J., Jiang, Y., Bruggemann, A., & Wahl-Schott, C. (2016). Electrophysiological characterization of the archaeal transporter NCX_Mj using solid supported membrane technology. *The Journal of General Physiology*, *147*(6), 485–496. http://dx.doi.org/10.1085/jgp.201611587.

Bazzone, A., Costa, W. S., Braner, M., Calinescu, O., Hatahet, L., & Fendler, K. (2013). Introduction to solid supported membrane based electrophysiology. *Journal of Visualized Experiments*, *75*, e50230. http://dx.doi.org/10.3791/50230.

Bazzone, A., Madej, M. G., Kaback, H. R., & Fendler, K. (2016). pH regulation of electrogenic sugar/H+ symport in MFS sugar permeases. *PloS One*, *11*(5), e0156392. http://dx.doi.org/10.1371/journal.pone.0156392.

Burzik, C., Kaim, G., Dimroth, P., Bamberg, E., & Fendler, K. (2003). Charge displacements during ATP-hydrolysis and synthesis of the Na+-transporting FoF1-ATPase of Ilyobacter tartaricus. *Biophysical Journal*, *85*, 2044–2054.

Calinescu, O., Danner, E., Bohm, M., Hunte, C., & Fendler, K. (2014a). Species differences in bacterial NhaA Na+/H+ exchangers. *FEBS Letters*, *588*(17), 3111–3116. http://dx.doi.org/10.1016/j.febslet.2014.05.066.

Calinescu, O., Linder, M., Wohlert, D., Yildiz, O., Kuhlbrandt, W., & Fendler, K. (2016). Electrogenic cation binding in the electroneutral Na+/H+ antiporter of Pyrococcus abyssi. *The Journal of Biological Chemistry*, *291*(52), 26786–26793. http://dx.doi.org/10.1074/jbc.M116.761080.

Calinescu, O., Paulino, C., Kuhlbrandt, W., & Fendler, K. (2014b). Keeping it simple, transport mechanism and pH regulation in Na+/H+ exchangers. *The Journal of Biological Chemistry*, *289*(19), 13168–13176. http://dx.doi.org/10.1074/jbc.M113.542993.

Choudhary, P., Armstrong, E., Jorgensen, C., Piotrowski, M., Barthmes, M., Torella, R., et al. (2017). Discovery of compounds that positively modulate the high affinity choline transporter. *Frontiers in Molecular Neuroscience*, *10*(40). http://dx.doi.org/10.3389/fnmol.2017.00040.

Gaiko, O., Janausch, I., Geibel, S., Vollert, H., Arndt, P., Gonski, S., et al. (2011). Robust electrophysiological assays using solid supported membranes: The organic cation transporter OCT2. *Australian Journal of Chemistry*, *64*, 31–35.

Ganea, C., Pourcher, T., Leblanc, G., & Fendler, K. (2001). Evidence for intraprotein charge transfer during the transport activity of the melibiose permease from Escherichia coli. *Biochemistry*, *40*(45), 13744–13752.

Garcia-Celma, J. J., Dueck, B., Stein, M., Schlueter, M., Meyer-Lipp, K., Leblanc, G., et al. (2008). Rapid activation of the melibiose permease MelB immobilized on a solid-supported membrane. *Langmuir*, *24*(15), 8119–8126.

Garcia-Celma, J. J., Hatahet, L., Kunz, W., & Fendler, K. (2007). Specific anion and cation binding to lipid membranes investigated on a solid supported membrane. *Langmuir*, *23*, 10074–10080.

Garcia-Celma, J. J., Ploch, J., Smirnova, I., Kaback, H. R., & Fendler, K. (2010). Delineating electrogenic reactions during lactose/H+ symport. *Biochemistry*, *49*(29), 6115–6121. http://dx.doi.org/10.1021/bi100492p.

Garcia-Celma, J. J., Smirnova, I. N., Kaback, H. R., & Fendler, K. (2009). Electrophysiological characterization of LacY. *Proceedings of the National Academy of Sciences of the United States of America*, *106*(18), 7373–7378.

Garcia-Celma, J., Szydelko, A., & Dutzler, R. (2013). Functional characterization of a ClC transporter by solid-supported membrane electrophysiology. *The Journal of General Physiology*, *141*(4), 479–491. http://dx.doi.org/10.1085/jgp.201210927.

Geibel, S., Flores-Herr, N., Licher, T., & Vollert, H. (2006). Establishment of cell-free electrophysiology for ion transporters: Application for pharmacological profiling. *Journal of Biomolecular Screening*, *11*(3), 262–268. http://dx.doi.org/10.1177/1087057105285110.

Gramigni, E., Tadini-Buoninsegni, F., Bartolommei, G., Santini, G., Chelazzi, G., & Moncelli, M. R. (2009). Inhibitory effect of Pb2+ on the transport cycle of the Na+,K+-ATPase. *Chemical Research in Toxicology*, *22*(10), 1699–1704. http://dx.doi.org/10.1021/tx9001786.

Gropp, T., Cornelius, F., & Fendler, K. (1998). K+-dependence of electrogenic transport by the Nak-ATPase. *Biochimica et Biophysica Acta - Biomembranes*, *1368*(2), 184–200.

Kelety, B., Diekert, K., Tobien, J., Watzke, N., Dörner, W., Obrdlik, P., et al. (2006). Transporter assays using solid supported membranes: A novel screening platform for drug discovery. *Assay and Drug Development Technologies*, *4*(5), 575–582.

Khafizov, K., Perez, C., Koshy, C., Quick, M., Fendler, K., Ziegler, C., et al. (2012). Investigation of the sodium-binding sites in the sodium-coupled betaine transporter BetP. *Proceedings of the National Academy of Sciences of the United States of America*, *109*(44), E3035–E3044. http://dx.doi.org/10.1073/pnas.1209039109.

Krause, R., Watzke, N., Kelety, B., Dorner, W., & Fendler, K. (2009). An automatic electrophysiological assay for the neuronal glutamate transporter mEAAC1. *Journal of Neuroscience Methods*, *177*(1), 131–141.

Mager, T., Braner, M., Kubsch, B., Hatahet, L., Alkoby, D., Rimon, A., et al. (2013). Differential effects of mutations on the transport properties of the Na+/H+-antiporter NhaA from Escherichia coli. *The Journal of Biological Chemistry*, *288*(34), 24666–24675. http://dx.doi.org/10.1074/jbc.M113.484071.

Mager, T., Rimon, A., Padan, E., & Fendler, K. (2011). Transport mechanism and pH regulation of the Na^{+}/H^{+} antiporter NhaA from Escherichia coli: An electrophysiological study. *The Journal of Biological Chemistry*, 286(26), 23570–23581. http://dx.doi.org/10.1074/jbc.M111.230235.

Mattle, D., Zhang, L., Sitsel, O., Pedersen, L. T., Moncelli, M. R., Tadini-Buoninsegni, F., et al. (2015). A sulfur-based transport pathway in Cu+-ATPases. *EMBO Reports*, 16(6), 728–740. http://dx.doi.org/10.15252/embr.201439927.

Meyer-Lipp, K., Ganea, C., Pourcher, T., Leblanc, G., & Fendler, K. (2004). Sugar binding induced charge translocation in the melibiose permease from Escherichia coli. *Biochemistry*, *43*, 12606–12613.

Meyer-Lipp, K., Sery, N., Ganea, C., Basquin, C., Fendler, K., & Leblanc, G. (2006). The inner interhelix loop 4-5 of the melibiose permease from Escherichia coli takes part in conformational changes after sugar binding. *Journal of Biological Chemistry*, *281*(36), 25882–25892.

Niessen, K. V., Muschik, S., Langguth, F., Rappengluck, S., Seeger, T., Thiermann, H., et al. (2016). Functional analysis of Torpedo californica nicotinic acetylcholine receptors in multiple activation states by SSM-based electrophysiology. *Toxicology Letters*, *247*, 1–10. http://dx.doi.org/10.1016/j.toxlet.2016.02.002.

Obrdlik, P., Diekert, K., Watzke, N., Keipert, C., Pehl, U., Brosch, C., et al. (2010). Electrophysiological characterization of ATPases in native synaptic vesicles and synaptic plasma membranes. *The Biochemical Journal*, *427*(1), 151–159. http://dx.doi.org/10.1042/BJ20091380.

Pintschovius, J., & Fendler, K. (1999). Charge translocation by the Na+/K+-ATPase investigated on solid supported membranes: Rapid solution exchange with a new technique. *Biophysical Journal*, *76*(2), 814–826.

Pintschovius, J., Fendler, K., & Bamberg, E. (1999). Charge translocation by the Na+/K+-ATPase investigated on solid supported membranes: Cytoplasmic cation binding and release. *Biophysical Journal*, *76*(2), 827–836.

Raunser, S., Appel, M., Ganea, C., Geldmacher-Kaufer, U., Fendler, K., & Kuhlbrandt, W. (2006). Structure and function of prokaryotic glutamate transporters from Escherichia coli and Pyrococcus horikoshii. *Biochemistry*, *45*(42), 12796–12805.

Rigaud, J. L., Pitard, B., & Levy, D. (1995). Reconstitution of membrane proteins into liposomes: Application to energy-transducing membrane proteins. *Biochimica et Biophysica Acta*, *1231*(3), 223–246.

Rosen, B. P. (1986). Ion extrusion systems in Escherichia coli. *Methods in Enzymology*, *125*, 328–336.

Rycovska, A., Hatahet, L., Fendler, K., & Michel, H. (2012). The nitrite transport protein NirC from Salmonella typhimurium is a nitrite/proton antiporter. *Biochimica et Biophysica Acta*, *1818*, 1342–1350.

Schulz, P., Dueck, B., Mourot, A., Hatahet, L., & Fendler, K. (2009). Measuring ion channels on solid supported membranes. *Biophysical Journal*, *97*(1), 388–396.

Schulz, P., Garcia-Celma, J. J., & Fendler, K. (2008). SSM-based electrophysiology. *Methods*, *46*, 97–103.

Schulz, P., Werner, J., Stauber, T., Henriksen, K., & Fendler, K. (2010). The G215R mutation in the Cl^-/H^+-antiporter ClC-7 found in ADO II osteopetrosis does not abolish function but causes a severe trafficking defect. *PloS One*, *5*(9), e12585. http://dx.doi.org/10.1371/journal.pone.0012585.

Seifert, K., Fendler, K., & Bamberg, E. (1993). Charge transport by ion translocating membrane proteins on solid supported membranes. *Biophysical Journal*, *64*(2), 384–391.

Siebels, I., & Drose, S. (2016). Charge translocation by mitochondrial NADH:Ubiquinone oxidoreductase (complex I) from Yarrowia lipolytica measured on solid-supported membranes. *Biochemical and Biophysical Research Communications*, *479*(2), 277–282. http://dx.doi.org/10.1016/j.bbrc.2016.09.059.

Tadini-Buoninsegni, F., Bartolommei, G., Moncelli, M. R., Guidelli, R., & Inesi, G. (2006). Pre-steady state electrogenic events of Ca2+/H+ exchange and transport by the Ca2+-ATPase. *The Journal of Biological Chemistry*, *281*(49), 37720–37727. http://dx.doi.org/10.1074/jbc.M606040200.

Tadini-Buoninsegni, F., Bartolommei, G., Moncelli, M. R., Pilankatta, R., Lewis, D., & Inesi, G. (2010). ATP dependent charge movement in ATP7B Cu(+)-ATPase is demonstrated by pre-steady state electrical measurements. *FEBS Letters*, *584*(22), 4619–4622. http://dx.doi.org/10.1016/j.febslet.2010.10.029.

Tadini-Buoninsegni, F., & Fendler, K. (2015). Recording of pump and transporter activity using solid-supported membranes (SSM-based electrophysiology). In R. J. Clarke & M. A. Ali Khalid (Eds.), *Pumps, channels and transporters: Methods of functional analysis.* Wiley, New Jersey, USA.

Tadini-Buoninsegni, F., Moncelli, M. R., Peruzzi, N., Ninham, B. W., Dei, L., & Nostro, P. L. (2015). Hofmeister effect of anions on calcium translocation by sarcoplasmic reticulum Ca(2+)-ATPase. *Scientific Reports*, *5*, 14282. http://dx.doi.org/10.1038/srep14282.

Tadini-Buoninsegni, F., & Smeazzetto, S. (2017). Mechanisms of charge transfer in human copper ATPases ATP7A and ATP7B. *IUBMB Life*, *69*(4), 218–225. http://dx.doi.org/10.1002/iub.1603.

Wacker, T., Garcia-Celma, J. J., Lewe, P., & Andrade, S. L. (2014). Direct observation of electrogenic NH4(+) transport in ammonium transport (Amt) proteins. *Proceedings of the National Academy of Sciences of the United States of America*, *111*(27), 9995–10000. http://dx.doi.org/10.1073/pnas.1406409111.

Watzke, N., Diekert, K., & Obrdlik, P. (2010). Electrophysiology of respiratory chain complexes and the ADP-ATP exchanger in native mitochondrial membranes. *Biochemistry*, *49*(48), 10308–10318. http://dx.doi.org/10.1021/bi1011755.

Zhou, A., Wozniak, A., Meyer-Lipp, K., Nietschke, M., Jung, H., & Fendler, K. (2004). Charge translocation during cosubstrate binding in the Na+/proline transporter of E. coli. *Journal of Membrane Biology*, *343*(4), 931–942.

Zuber, D., Krause, R., Venturi, M., Padan, E., Bamberg, E., & Fendler, K. (2005). Kinetics of charge translocation in the passive downhill uptake mode of the Na^+/H^+ antiporter NhaA of Escherichia coli. *Biochimica et Biophysica Acta*, *1709*, 240–250.

CHAPTER THREE

Saposin-Lipoprotein Scaffolds for Structure Determination of Membrane Transporters

Joseph A. Lyons*, Andreas Bøggild*, Poul Nissen*[,1], Jens Frauenfeld[†]
*DANDRITE, Nordic-EMBL Partnership for Molecular Medicine, Aarhus University, Aarhus, Denmark
[†]Salipro Biotech AB, Stockholm, Sweden
[1]Corresponding author: e-mail address: pn@mbg.au.dk

Contents

Abstract

Membrane proteins depend on their natural lipid environment for function, which makes them more difficult to study in isolation. A number of approaches that mimic the lipid bilayer of biological membranes have been described (nanodiscs, SMALPs), enabling novel ways to assay activity and elucidate structures of this important class of proteins. More recently, the use of saposin A, a protein that is involved in lipid transport, to form Salipro (saposin–lipid–protein) complexes was demonstrated for a range of membrane protein targets (Frauenfeld et al., 2016). The method is fast and requires few resources. The saposin–lipid–scaffold adapts to various sizes of transmembrane regions during self-assembly, forming a minimal lipid nanoparticle. This results in the formation of a well-defined membrane protein–lipid complex, which is desirable for structural characterization.

Here, we describe a protocol to reconstitute the sarco-endoplasmic reticulum calcium ATPase (SERCA) into Salipro nanoparticles. The complex formation is analyzed using negative stain electron microscopy (EM), allowing to quickly determine an initial structure of the membrane protein and to evaluate sample conditions for structural studies using single-particle cryo-EM in a detergent-free environment.

Methods in Enzymology, Volume 594
ISSN 0076-6879
http://dx.doi.org/10.1016/bs.mie.2017.06.035

1. INTRODUCTION

Membrane proteins are among the most challenging proteins for structural characterization, as they are not readily purified into solution, but instead require detergents to solubilize them from their native lipid bilayer. For crystallization, the typical approach is either to attempt crystallization in detergent micelles, add lipids for crystallization in a bilayer-like environment (Faham & Bowie, 2002; Gourdon et al., 2011; Poulos, Morgan, Zimmer, & Faham, 2015), or crystallize in lipidic cubic phase (Caffrey, 2015).

Several methods have been proposed to stabilize membrane proteins in a more natural lipid environment, while keeping them in solution. One such method is to incorporate the target membrane protein into nanodiscs (Bayburt, Grinkova, & Sligar, 2002), which are small sections of a lipid bilayer, encircled by MSPs (membrane scaffold proteins). The dimensions of the MSP-based nanodisc and its ability to incorporate membrane proteins of various sizes are dictated by exploiting MSPs of differing lengths, with reconstitution success often requiring extensive optimization for a given protein target. While successes have been reported for single-particle cryo-electron microscopy (SP cryo-EM) studies of nanodisc reconstituted membrane proteins, these have been limited to larger membrane protein targets (Baker, Fan, & Serysheva, 2015; Frauenfeld et al., 2011; Gao, Cao, Julius, & Cheng, 2016).

A more recent alternative has been reported, which uses styrene maleic acid copolymers (SMAs) to generate SMALPs (SMA lipid particles) that allow for the isolation of membrane proteins directly from their native membrane, without the need for detergents (Jamshad et al., 2011). Negative stain analysis of AcrB reconstituted into SMALPs has provided an initial proof of concept for the use of SMALPs in structural studies (Postis et al., 2015).

Another new technological breakthrough, and the focus of this chapter, exploits the self-assembly of lipids and saposin into a saposin-lipoprotein nanoparticle (Salipro). The saposin protein family comprises four (saposin A–D) small proteins (~10 kDa) that derive from precursor proteins involved as cofactors in the catabolism of sphingolipids. Crystal structures of saposin proteins have been reported with and without bound detergent/lipid (Ahn, Leyko, Alattia, Chen, & Privé, 2006; Popovic, Holyoake, Pomès, & Privé, 2012). Frauenfeld et al. rationalized that these proteins could be exploited as scaffold proteins to reconstitute membrane proteins in a lipid environment and reported on their application to a number of membrane proteins.

Saposin A can be expressed and purified from *Escherichia coli*. Its general versatility lies in its ability to adapt to the size of the membrane protein, adjusting to transmembrane regions of varying size (Frauenfeld et al., 2016). This reduces the need to screen for a variety of scaffold protein constructs with varying protein-to-lipid ratios and thereby reduces the complexity of the reconstitution. The experimental setup is rather straightforward, where in a rapid process the saposin scaffold, lipids, and membrane proteins self-assemble into robust lipid nanoparticles. To date, mainly the human saposin A has been exploited to generate Salipro particles; however, other saposins from the saposin protein family might provide additional versatility for specific uses, such as in combination with certain lipids or classes of membrane proteins.

EM has seen rapid advances over the last few years, in what has been called the "resolution revolution" (Kühlbrandt, 2014). For instance, new detectors allow for high-resolution data (~2 Å) to be collected, while the use of phase plates provides a boost to contrast, allowing smaller particles to be identified in the micrographs (Danev & Baumeister, 2016). These advances coupled with the improvements to sample supports, microscopes, new software, and computer hardware makes working with large cryo-EM datasets easier and allows for the visualization and atomic model building of membrane proteins.

Here, we demonstrate that Salipro reconstitution can be used for structural characterization of P-type ATPases, using SERCA1a natively isolated from the fast-twitch muscle of rabbit as a test case. The approach described here can easily be applied to other transporters, channels, and receptors.

2. MEMBRANE PROTEIN RECONSTITUTION

For the reconstitution into Salipro particles, membrane proteins may originate either from natural sources or from overexpression in appropriate expression systems, e.g., bacterial, yeast, or mammalian systems (Lyons, Shahsavar, Paulsen, Pedersen, & Nissen, 2016). Membrane proteins require solubilization in an appropriate detergent, followed by one or more purification steps. The choice of detergent is often critical as some detergents might severely reduce the stability of the protein. As this is generally established on a case-to-case basis, we will not describe these initial steps here (for review, see Loll, 2014), but instead start with the final purification step before Salipro reconstitution.

Once the sample of pure solubilized membrane protein is available, reconstitution into Salipro can proceed quite straightforwardly, with purification by a final size exclusion chromatography step to separate the reconstituted membrane protein-Salipro particle from free saposin, lipid-only (empty) Salipro particles, and potential aggregates. However, while the reconstitution protocol is fast, it will likely require some screening of the various parameters to find the optimal strategy for reconstitution. The choice of lipid(s) to use could be one such parameter, along with incubation times and temperature, and ratios between the target protein, saposin, and lipid(s).

2.1 Saposin Purification

Saposin proteins are capable of binding lipids. This class of proteins is composed of amphipathic helices with hydrophobic amino acids that build the protein core, and six invariable cysteine residues, which form disulfide bridges that lead to an extraordinary stability of the domains.

2.1.1 Equipment

- Baffled conical flasks
- ÄKTA chromatography system
- HiLoad Superdex 200 16/600 GL
- Centrifuges for the harvesting of *E. coli* cells (at least 6000 × *g*) and a standard refrigerated centrifuge capable of speeds of up to 26,000 × *g*
- Probe sonicator suitable for lysing *E. coli*
- Water bath
- 10-mm (i.d.) open gravity flow column

2.1.2 Buffers and Reagents

- *E. coli* Rosetta gami-2 (DE3) (Novagen)
- Terrific Broth (TB) medium
- Tetracycline, chloramphenicol, kanamycin
- IMAC resin: Ni Sepharose 6 Fast Flow
- Lysis buffer: 20 m*M* HEPES pH 7.5, 150 m*M* NaCl, 20 m*M* imidazole
- Wash buffer: 20 m*M* HEPES pH 7.5, 150 m*M* NaCl, 40 m*M* imidazole
- Elution buffer: 20 m*M* HEPES pH 7.5, 150 m*M* NaCl, 400 m*M* imidazole
- Size exclusion buffer: 20 m*M* HEPES pH 7.5, 150 m*M* NaCl

- TEV protease containing uncleavable His-tag
- Amicon Ultra concentrator (3000 Da MWCO)
- Dialysis bag (1000 Da MWCO)

2.1.3 Procedure

1. Saposin A is expressed using a vector with the coding region for saposin A inserted into a pNIC-Bsa4 plasmid. This plasmid adds an N-terminal hexahistidine tag with an integrated TEV protease cleavage site.
2. Cultures of 1 L TB medium supplemented with tetracycline, chloramphenicol, and kanamycin in 2.5-L baffled conical flasks are inoculated from an overnight culture and grown at 37°C at 200 rpm.
3. Protein expression is induced at OD600 nm of 1.0 with 0.7 m*M* IPTG at 37°C. 3 h after induction, cells are harvested at 12,000 × *g* for 20 min. At this point, cells can be frozen for long-term storage in a −80°C freezer.
4. For purification, cells are resuspended in lysis buffer (1 g of cell wet weight in 5 mL lysis buffer) and disrupted by sonication.
5. Cell debris is removed by centrifugation at 26000 × *g* for 30 min at 4°C.
6. In a water bath, the supernatant is incubated at 85°C for 10 min.
7. The supernatant is subjected to a second centrifugation step at 26000 × *g* for 30 min at 4°C.
8. IMAC resin (Ni Sepharose 6 Fast Flow) is equilibrated with lysis buffer (10 column volumes (CV)) in 10-mm (i.d.) open gravity flow columns.
9. Transfer of IMAC resin to a Falcon tube and incubated with lysate (1 mL of resin for 20 mL lysate) for 30 min at 4°C.
10. After binding of saposin A to the IMAC resin, the chromatography medium is packed in open gravity flow columns.
11. Unbound proteins are removed by washing with 15 CV of lysis buffer, followed by 15 CV of wash buffer.
12. Saposin A is eluted by adding five CV of elution buffer.
13. Saposin A is subsequently dialyzed against size exclusion buffer together with TEV protease overnight in a dialysis bag at 4°C with gentle stirring.
14. To remove TEV protease and uncleaved proteins, the sample is passed over equilibrated IMAC beads again and the flow through collected and concentrated using Amicon Ultra concentrators to a volume of 5 mL.
15. Using an ÄKTA chromatography system, the sample is loaded onto a HiLoad Superdex 200 16/600 GL column, equilibrated with at least two CV of size exclusion buffer, and fractions are collected.

16. Peak fractions are pooled and concentrated to 1.2 mg/mL protein using an Amicon Ultra concentrator.
17. Aliquot and flash-freeze purified saposin A in liquid nitrogen for storage at −80°C.

2.1.4 Notes

1. Typically, a 1-L bacterial culture yields 5–10 mg of purified saposin.

2.2 SERCA Purification

For the purpose of this protocol, it is assumed that a purified membrane protein is available; therefore, specifics pertaining to the expression and purification strategies will not be discussed in detail here. The protocol will begin from the polishing of the detergent-solubilized protein via size exclusion chromatography, along with considerations for use of detergents. SERCA is natively isolated from rabbit fast-twitch muscle (Møller & Olesen, 2016).

The optimal starting point is a monodisperse peak in the size exclusion chromatogram, indicating a stable and homogenous sample (Fig. 1). Assuming the sample is not intended for activity assays, locking the protein in a specific state using an inhibitor or other ligands is advisable.

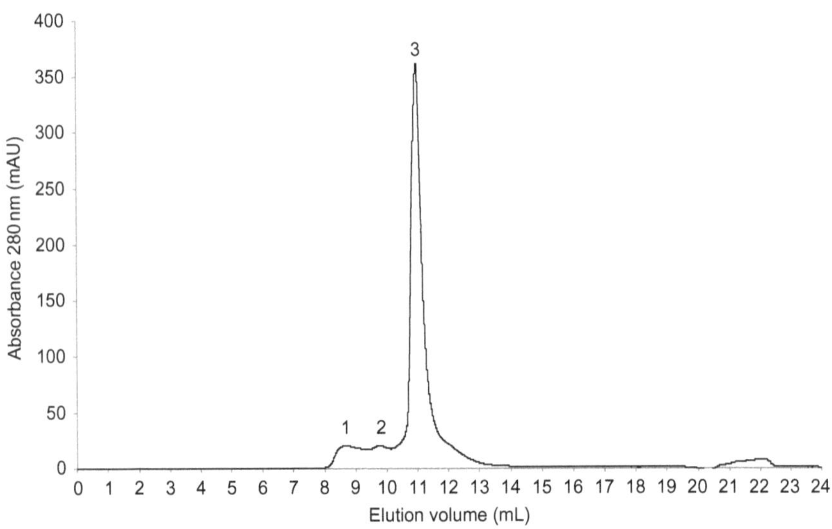

Fig. 1 Elution profile from size exclusion chromatography of SERCA solubilized in DDM. *Peak 1* is at the void volume and was discarded, as was *peak 2* which might correspond to SERCA dimers. *Peak 3* contains purified SERCA monomer and was collected for SERCA-Salipro reconstitution.

2.2.1 Equipment

- ÄKTA Purifier HPLC system in a cold cabinet
- Superdex 200 Increase 10/300 GL column
- DeNovix DS-11 spectrophotometer
- SDS-PAGE equipment and 15% gels
- Optima MAX-XP Ultracentrifuge with a TLA-110 rotor
- Sartorius Vivaspin spin filter with a 100 kDa MW cutoff

2.2.2 Buffers and Reagents

- SERCA solubilized in DDM (*n*-dodecyl β-D-maltoside)
- Size exclusion buffer: 67 m*M* MOPS (pH 6.8), 100 m*M* KCl, 10 m*M* $CaCl_2$, 5 m*M* $MgCl_2$, 0.3 mg/mL DDM
- SDS-PAGE loading buffer

2.2.3 Procedure

1. If needed, concentrate the sample to a volume suitable for size exclusion (50–500 μL), while keeping protein concentration low enough to avoid precipitation. If needed, split the sample into aliquots for multiple injections.
2. Pipette the sample into tubes compatible with ultracentrifugation and spin the sample at 100,000 × *g* at 4°C for 35 min to sediment aggregates.
3. Inject the sample onto a column equilibrated with at least two CV of size exclusion buffer, and collect fractions (250–500 μL).
4. Run samples of the peak fractions on a 15% SDS-PAGE to check purity.
5. Collect pure peak fractions and concentrate to about 3 mg/mL.
6. Use directly for Salipro reconstitution or flash-freeze in liquid nitrogen and store at −80°C.

2.2.4 Notes

1. Consider the choice and the cutoff of the spin filter. If a very low MW cutoff is chosen, not only is the protein concentrated, but also the detergent, due to the size of micelles.

2.3 Reconstitution Into Salipro

The gel filtration profile of a reconstitution usually displays several peaks corresponding to the membrane protein incorporated into Salipro nanoparticles, peaks that originate from smaller lipid-only (empty) Salipro particles and monomeric, unbound saposin.

For further analysis, peak fractions are collected to isolate Salipro-SERCA complexes. In the case of SERCA, a careful collection of fractions might be required, due to the presence of two populations (monomers, dimers) of Salipro-SERCA. However, these are easily discriminated by particle size (Fig. 2).

2.3.1 Equipment

- ÄKTA Purifier HPLC system
- Superdex 200 Increase 10/300 GL column
- DeNovix DS-11 spectrophotometer
- Heat block
- Timer

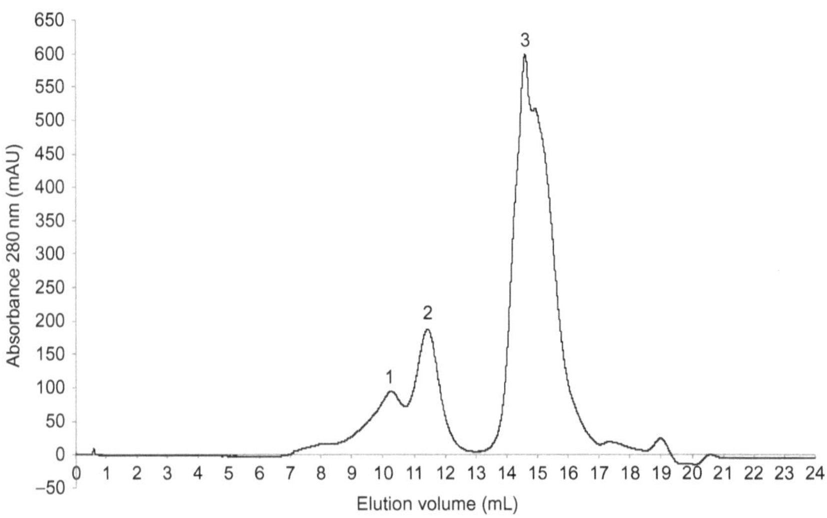

Fig. 2 Elution profile from size exclusion chromatography (SEC) of SERCA-Salipro reconstitution. *Peak 2* contains the SERCA monomers used for negative stain cryo-EM. *Peak 1* likely contains SERCA dimers reconstituted in Salipro, but with careful fractionation a sample of almost pure monomeric form could be obtained, as evident from the micrographs and class averages (Fig. 3). *Peak 3* is composed of lipid-only (empty) Salipro particles. Typically, a peak corresponding to free saposin A is evident at ~17.5 mL; however, concentration of the sample using a 100 kDa MWCO spin filter prior to injection to SEC can decrease the concentration of saposin A and possibly the empty Salipro nanoparticles with respect to the protein of interest.

2.3.2 Buffers and Reagents

- Purified SERCA at 2.9 mg/mL
- Purified saposin at 1.2 mg/mL
- Size exclusion buffer: 67 m*M* MOPS (pH 6.8), 100 m*M* KCl, 10 m*M* $CaCl_2$, 5 m*M* $MgCl_2$
- Brain lipid extract (BLE) (Sigma: B1502) stock at 20 mg/mL in 80 m*M* Tris pH 7.5, 1.5% (w/v) DDM
- Dilution buffer: 67 m*M* MOPS (pH 6.8), 100 m*M* KCl, 10 m*M* $CaCl_2$, 5 m*M* $MgCl_2$

2.3.3 Procedure

1. Dilute BLE stock solution to a final concentration of 5 mg/mL using the dilution buffer, final detergent concentration of 0.375% (w/v).
2. Incubate 20 μL of protein with 40 μL of the 5 mg/mL BLE stock solution at room temperature (20°C) for 10 min.
3. Add 360 μL of saposin A (1.2 mg/mL) and incubate at room temperature for a further 2 min.
4. 500 μL of dilution buffer is added.
5. The above reaction is set up five times, pooled, and concentrated to ~300 μL for injection onto the HPLC column.
6. Inject the sample onto a column (Superdex 200 10/300) equilibrated with at least two CV of size exclusion buffer, and collect fractions (100–200 μL).
7. Appropriate fractions are pooled and concentrated to the desired concentration or can be individually assessed using negative stain EM.

2.3.4 Notes

1. When starting with an entirely new membrane protein target, it is recommended to perform small-scale screens on a Superdex S200 5/150 column first. Typically, three separate reconstitution mixes are set up and screened separately. Here, 1, 3, or 5 μL of the membrane protein (concentration range 0.5–10 mg/mL) are incubated with fixed volumes of brain lipids (5 μL at 5 mg/mL) and saposin (20 μL at 1.2 mg/mL). After the addition of dilution buffer, the samples are subjected to a gel filtration step, which allows to determine the optimal ratios for self-assembly of the Salipro particles.
2. Getting the BLE stock solubilized might require alternating heating (55°C), vortexing and bath sonication.
3. Other lipid mixtures can be used.

4. Screening for an optimum reconstitution protocol can be achieved using 2–10 μL of protein solution depending on the sensitivity of detection of the chromatography system.
5. The throughput of the reconstitution protocol screening can be increased using automation and an autoloader.
6. An excess of saposin A is generally required for optimum reconstitution.
7. For the large-scale preparation described earlier a number of reactions each with a final volume of 1 mL upon dilution (step 4) are set up before pooling and concentrating with a spin filter of appropriate MW cutoff.
8. Adding 5%–10% glycerol to the size exclusion buffer may further stabilize the membrane protein, especially when freezing/thawing; however, the presence of glycerol might hamper the contrast in future cryo-EM studies for high resolution.
9. For larger membrane protein complexes, different column materials (e.g., Superose 6) may prove beneficial.

3. EM OF SALIPRO PARTICLES

Here, we describe how to image the SERCA-Salipro particles using negative stain EM and how to reconstruct a 3D volume from the data. Compared to cryo-EM, negative stain EM has a very limited maximum resolution, revealing only the overall shape of imaged particles. However, sample preparation is quick and the contrast is high, making it well suited for screening samples before moving to high-resolution cryo-EM.

3.1.1 Equipment

- PELCO EasiGlow Glow Discharge Cleaning System
- Tecnai G2 Spirit electron microscope equipped with an LaB_6 filament and operated at an acceleration voltage of 120 kV and a TVIPS F415 CCD camera with 4k × 4k pixels
- Leginon data acquisition software (Suloway et al., 2005)
- Appion data processing suite (Lander et al., 2009)
- EMAN software (Ludtke, Baldwin, & Chiu, 1999)
- RELION version 2.0 (Scheres, 2012)
- Chimera software (Pettersen et al., 2004)
- Computing resources—here we used a GPU cluster (Dell PowerEdge C4130 w/ 2× Xeon E5-2620v4, 128 GB RAM, 4× NVIDIA TESLA K80)

3.1.2 Buffers and Reagents

- Sample dilution buffer: 67 m*M* MOPS (pH 6.8), 100 m*M* KCl, 5 m*M* $MgCl_2$, 4 m*M* EGTA, 0.5 m*M* $BeSO_4$, 10 m*M* NaF
- SERCA1a-Salipro sample at 0.24 mg/mL
- Commercial copper G400-Cu grids overlaid with carbon, glow discharged at 25 mA for 45 s prior to use.
- 2% (w/v) Uranyl formate (Ohi, Li, Cheng, & Walz, 2004)

3.1.3 Procedure

1. SERCA1a-Salipro complex is diluted sixfold using the sample dilution buffer to ~40 μg/mL in the presence of the conformation stabilizing (E2P) inhibitor BeF_x (from $BeSO_4$ and NaF) and incubated on ice for 1 h.
2. 3 μL of protein solution was added to a freshly glow-discharged grid.
3. After ~10 s, the excess protein solution was removed by blotting the edge of the grid against a piece of Whatman grade 4 filter paper.
4. 3 μL of 2% (w/v) uranyl formate stain was immediately added and excess stain was removed by blotting the same edge of the grid as before against the filter paper.
5. Staining was repeated twice more as per step 3.
6. The grid was left to air-dry for a few minutes before use.
7. Images were collected on a Tecnai G2 Spirit microscope using the automated data collection software, Leginon.
8. Particles were picked automatically using dogpicker.py as implemented in Appion, and converted to boxfiles in EMAN.
9. Particles were extracted in RELION 2.0 and submitted to subsequent rounds of 2D and 3D classification (Fig. 3).
10. 3D volumes were manually inspected in chimera and a crystal-derived model of SERCA1a was manually fitted into the volumes (Fig. 4).

3.1.4 Notes

1. If aggregates are observed in the micrographs, subjecting the sample to ultracentrifugation prior to preparing grids might be used to remove them.
2. Protein concentration of the sample applied to the grid will have to be screened to obtain micrographs containing as many particles as possible, while still having them isolated from one another.
3. Data processing can be greatly enhanced and sped up using a GPU-enabled workstation or cluster (Kimanius, Forsberg, Scheres, & Lindahl, 2016).

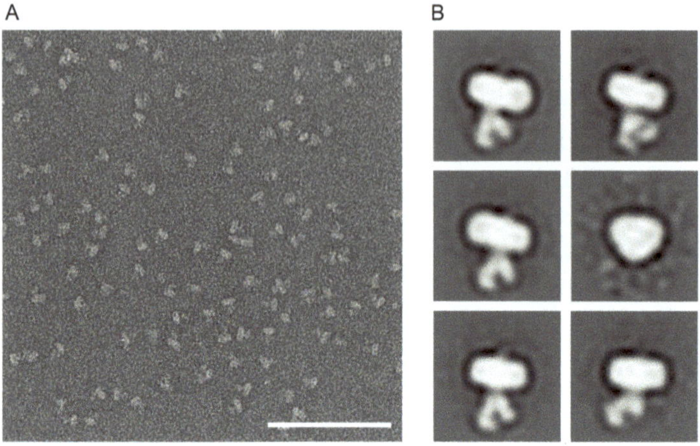

Fig. 3 (A) Micrograph of the negative stain SERCA-Salipro sample. The micrograph is cropped to cover a quarter of the collected micrograph. The scale bar represents 100 nm. The expected shape of Salipro discs with the cytosolic domains of SERCA extending out on one side is clearly visible in most of the particles. Little to no aggregation was observed in the micrographs. (B) Selected 2D class averages as output from a 2D classification in RELION. Class averages are in a 72-pixel box with 3.15 Å/pixel.

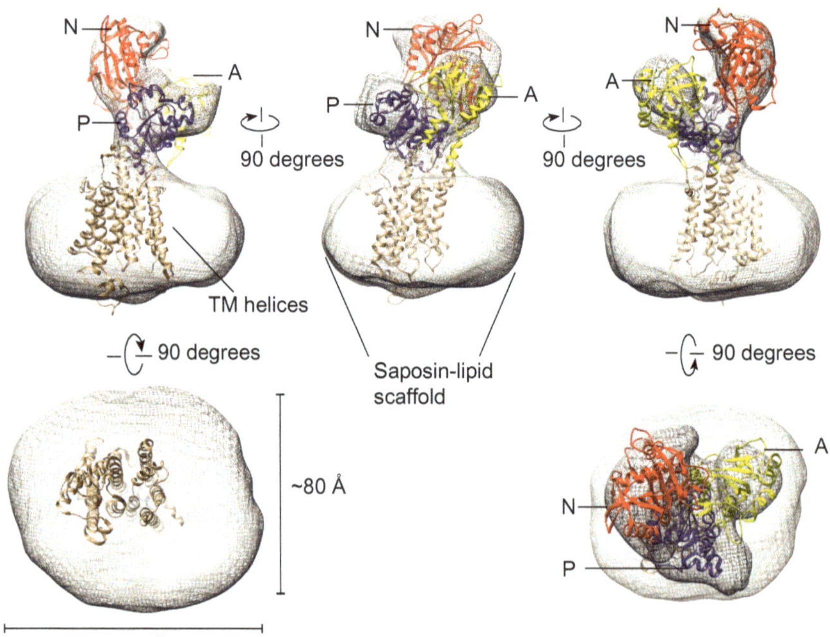

Fig. 4 The 3D envelope as calculated in RELION with a crystallographic model of SERCA (PDB: 3B9B) manually docked into the volume in Chimera. Three side views, 90 degrees apart, and views from the *top* and *bottom* are shown. The transmembrane domain of SERCA is colored in *tan*, the actuator (A) domain in *yellow*, the nucleotide-binding (N) domain in *red*, and the phosphorylation (P) domain in *blue*.

4. SUMMARY AND CONCLUSION

Reconstitution of membrane proteins into Salipro nanoparticles follows a fairly straightforward protocol, and optimizations of, e.g., lipid type, incubation time and temperature, and protein:saposin:lipid ratios can be screened on a small scale. For large-scale production, several methods can be used. One option is to perform several runs of the same reconstitution with subsequent purification on a medium-sized size exclusion column (24 mL), and then pooling the relevant peak fractions from each SEC run and concentrating, as described earlier (Frauenfeld et al., 2016).

Another approach is presented here, where several reconstitution mixtures are pooled and concentrated prior to purification on a medium-sized size exclusion column.

Alternatively, it is possible to perform the reconstitution with higher amounts and volumes (up to 5 mL) and inject the mixture on a size exclusion column with larger volume (120 mL). However, this large-column approach should only be performed if reconstitution conditions are well established on the small and medium scale.

Preparing grids with Salipro samples and collecting data can be performed on the same day as reconstitution, especially with the help of data collection software such as Appion. Nevertheless, for SERCA-Salipro we could also flash-freeze samples and thaw later for grid preparation without significant aggregation observed in the micrographs.

Given the small size of the particle imaged here, and with the use of a GPU cluster to accelerate processing, we were able to go from micrographs to 3D volume in a few days. The generated 3D volume fits the crystallographic model very well, and it is worth noting that SERCA at 110 kDa is still on the small side of what can be characterized to high resolution using cryo-EM.

We have found these procedures to work well for a range of P-type ATPases and other transporters, and are currently exploring the possibilities for also extracting Salipro particles directly from native membranes. There are also interesting applications to small-angle scattering studies, and expressed in *E. coli*, saposin can be subjected to deuteration for neutron scattering studies. It remains to be seen if Salipro complexes of membrane proteins can be crystallized, but given the rather adaptive and dynamic nature of the Salipro interaction it would not seem unfeasible.

ACKNOWLEDGMENTS

This work was in part funded through the Horizon 2020 program of the European Union (iNEXT Grant No. 653706). J.A.L. is supported by a postdoctoral fellowship from the Lundbeck Foundation. Molecular graphics and analyses were performed with the UCSF Chimera package. Chimera is developed by the Resource for Biocomputing, Visualization, and Informatics at the University of California, San Francisco (supported by NIGMS P41-GM103311).

REFERENCES

Ahn, V. E., Leyko, P., Alattia, J.-R., Chen, L., & Privé, G. G. (2006). Crystal structures of saposins A and C. *Protein Science: A Publication of the Protein Society*, *15*(8), 1849–1857. http://dx.doi.org/10.1110/ps.062256606.

Baker, M. R., Fan, G., & Serysheva, I. I. (2015). Single-particle cryo-EM of the ryanodine receptor channel in an aqueous environment. *European Journal of Translational Myology*, *25*(1), 35–48. http://dx.doi.org/10.4081/ejtm.2015.4803.

Bayburt, T. H., Grinkova, Y. V., & Sligar, S. G. (2002). Self-assembly of discoidal phospholipid bilayer nanoparticles with membrane scaffold proteins. *Nano Letters*, *2*(8), 853–856. http://dx.doi.org/10.1021/nl025623k.

Caffrey, M. (2015). A comprehensive review of the lipid cubic phase or in meso method for crystallizing membrane and soluble proteins and complexes. *Acta Crystallographica. Section F, Structural Biology Communications*, *71*(Pt. 1), 3–18. http://dx.doi.org/10.1107/S2053230X14026843.

Danev, R., & Baumeister, W. (2016). Cryo-EM single particle analysis with the Volta phase plate. *eLife*, *5*(March 2016), 1–14. http://dx.doi.org/10.7554/eLife.13046.

Faham, S., & Bowie, J. U. (2002). Bicelle crystallization: A new method for crystallizing membrane proteins yields a monomeric bacteriorhodopsin structure. *Journal of Molecular Biology*, *316*(1), 1–6. http://dx.doi.org/10.1006/jmbi.2001.5295.

Frauenfeld, J., Gumbart, J., van der Sluis, E. O., Funes, S., Gartmann, M., Beatrix, B., et al. (2011). Cryo-EM structure of the ribosome-SecYE complex in the membrane environment. *Nature Structural & Molecular Biology*, *18*(5), 614–621. http://dx.doi.org/10.1038/nsmb.2026.

Frauenfeld, J., Löving, R., Armache, J.-P., Sonnen, A. F.-P., Guettou, F., Moberg, P., et al. (2016). A saposin-lipoprotein nanoparticle system for membrane proteins. *Nature Methods*, *13*(4), 345–351. http://dx.doi.org/10.1038/nmeth.3801.

Gao, Y., Cao, E., Julius, D., & Cheng, Y. (2016). TRPV1 structures in nanodiscs reveal mechanisms of ligand and lipid action. *Nature*, *534*(7607), 347–351. http://dx.doi.org/10.1038/nature17964.

Gourdon, P., Andersen, J. L., Hein, K. L., Bublitz, M., Pedersen, B. P., Liu, X.-Y., et al. (2011). HiLiDe—Systematic approach to membrane protein crystallization in lipid and detergent. *Crystal Growth & Design*, *11*(6), 2098–2106. http://dx.doi.org/10.1021/cg101360d.

Jamshad, M., Lin, Y.-P., Knowles, T. J., Parslow, R. A., Harris, C., Wheatley, M., et al. (2011). Surfactant-free purification of membrane proteins with intact native membrane environment. *Biochemical Society Transactions*, *39*(3), 813–818. http://dx.doi.org/10.1042/BST0390813.

Kimanius, D., Forsberg, B. O., Scheres, S. H., & Lindahl, E. (2016). Accelerated cryo-EM structure determination with parallelisation using GPUs in RELION-2. *eLife*, *5*, 1–21. http://dx.doi.org/10.7554/eLife.18722.

Kühlbrandt, W. (2014). The resolution revolution. *Science (New York, N.Y.)*, *343*(6178), 1443–1444. http://dx.doi.org/10.1126/science.1251652.

Lander, G. C., Stagg, S. M., Voss, N. R., Cheng, A., Fellmann, D., Pulokas, J., et al. (2009). Appion: an integrated, database-driven pipeline to facilitate EM image processing. *Journal of Structural Biology*, *166*(1), 95–102. http://dx.doi.org/10.1016/j.jsb.2009.01.002.

Loll, P. J. (2014). Membrane proteins, detergents and crystals: What is the state of the art? *Acta Crystallographica. Section F, Structural Biology Communications*, *70*(Pt. 12), 1576–1583. http://dx.doi.org/10.1107/S2053230X14025035.

Ludtke, S. J., Baldwin, P. R., & Chiu, W. (1999). EMAN: Semiautomated software for high-resolution single-particle reconstructions. *Journal of Structural Biology*, *128*(1), 82–97. http://dx.doi.org/10.1006/jsbi.1999.4174.

Lyons, J. A., Shahsavar, A., Paulsen, P. A., Pedersen, B. P., & Nissen, P. (2016). Expression strategies for structural studies of eukaryotic membrane proteins. *Current Opinion in Structural Biology*, *38*, 137–144. http://dx.doi.org/10.1016/j.sbi.2016.06.011.

Møller, J. V., & Olesen, C. (2016). Preparation of $Ca2^+$-ATPase1a enzyme from rabbit sarcoplasmic reticulum. In M. Bublitz (Ed.), *P-type ATPases: Methods and protocols* (pp. 11–17). New York, NY: Springer New York. http://dx.doi.org/10.1007/978-1-4939-3179-8_3.

Ohi, M., Li, Y., Cheng, Y., & Walz, T. (2004). Negative staining and image classification—Powerful tools in modern electron microscopy. *Biological Procedures Online*, *6*(1), 23–34. http://dx.doi.org/10.1251/bpo70.

Pettersen, E. F., Goddard, T. D., Huang, C. C., Couch, G. S., Greenblatt, D. M., Meng, E. C., et al. (2004). UCSF Chimera—A visualization system for exploratory research and analysis. *Journal of Computational Chemistry*, *25*(13), 1605–1612. http://dx.doi.org/10.1002/jcc.20084.

Popovic, K., Holyoake, J., Pomès, R., & Privé, G. G. (2012). Structure of saposin A lipoprotein discs. *Proceedings of the National Academy of Sciences of the United States of America*, *109*(8), 2908–2912. http://dx.doi.org/10.1073/pnas.1115743109.

Postis, V., Rawson, S., Mitchell, J. K., Lee, S. C., Parslow, R. A., Dafforn, T. R., et al. (2015). The use of SMALPs as a novel membrane protein scaffold for structure study by negative stain electron microscopy. *Biochimica et Biophysica Acta*, *1848*(2), 496–501. http://dx.doi.org/10.1016/j.bbamem.2014.10.018.

Poulos, S., Morgan, J. L. W., Zimmer, J., & Faham, S. (2015). Bicelles coming of age: An empirical approach to bicelle crystallization. *Methods in Enzymology*, *557*, 393–416. http://dx.doi.org/10.1016/bs.mie.2014.12.024.

Scheres, S. H. W. (2012). RELION: Implementation of a Bayesian approach to cryo-EM structure determination. *Journal of Structural Biology*, *180*(3), 519–530. http://dx.doi.org/10.1016/j.jsb.2012.09.006.

Suloway, C., Pulokas, J., Fellmann, D., Cheng, A., Guerra, F., Quispe, J., et al. (2005). Automated molecular microscopy: The new Leginon system. *Journal of Structural Biology*, *151*(1), 41–60. http://dx.doi.org/10.1016/j.jsb.2005.03.010.

CHAPTER FOUR

Single-Molecule Fluorescence Studies of Membrane Transporters Using Total Internal Reflection Microscopy

Joris M.H. Goudsmits, Antoine M. van Oijen[1,2], Dirk J. Slotboom[2]
University of Groningen, Groningen, The Netherlands
[2]Corresponding authors: e-mail address: vanoijen@uow.edu.au; d.j.slotboom@rug.nl

Contents

Abstract

Cells are delineated by a lipid bilayer that physically separates the inside from the outer environment. Most polar, charged, or large molecules require proteins to reduce the energetic barrier for passage across the membrane and to achieve transport rates that are relevant for life. Here, we describe techniques to visualize the functioning of membrane transport proteins with fluorescent probes at the single-molecule level. First, we

[1] Present address: School of Chemistry, University of Wollongong, Wollongong, New South Wales, Australia.

Methods in Enzymology, Volume 594
ISSN 0076-6879
http://dx.doi.org/10.1016/bs.mie.2017.05.009

explain how to produce membrane-reconstituted transporters with fluorescent labels. Next, we detail the construction of a microfluidic flow cell to image immobilized proteoliposomes on a total internal reflection fluorescence microscope. We conclude by describing the methods that are needed to analyze fluorescence movies and obtain useful single-molecule data.

1. INTRODUCTION

Cells are the fundamental units of life, both structurally and functionally. The interior of a cell is enclosed by a membrane comprised of a lipid bilayer with a thickness of approximated 5 nm in which proteins are embedded. The membrane bilayer is a dynamic and fluid structure. Although the (protein-free) lipid bilayer is impermeable to macromolecules, some small and nonpolar molecules, such as O_2 and CO_2 can partition into the lipid bilayer, and thus diffuse across (Missner & Pohl, 2009). Uncharged polar molecules, such as H_2O and glycerol will also diffuse through, tough at (much) lower rates. Charged molecules, such as inorganic ions, are usually unable to pass the hydrophobic core of the bilayer because the energetic cost of dehydration is too high.

In order to precisely control intracellular solute concentrations and to allow transport of polar and charged molecules at useful rates, cells employ dedicated, membrane-integrated transport proteins. The high portion of genes encoding for transport proteins reflects their importance: 20%–30% of genes encodes for membrane proteins (Krogh, Larsson, von Heijne, & Sonnhammer, 2001), half of which are transporters (Ren & Paulsen, 2005). Transport proteins are divided in two main classes: passive and active transporters (Saier, Reddy, Tamang, & Vastermark, 2014). Passive transporters are catalysts that allow substrate molecules to flow in the direction of their chemical gradient across the membrane. Active transporters use different sources of energy to transport a specific molecule or class of molecules against a gradient. They can for example use ATP, the electrochemical gradient of a cosubstrate (for instance protons or Na^+), or light.

Extensive structural and biochemical investigations already have provided great understanding of membrane transporters (Drew & Boudker, 2016; Groeneveld & Slotboom, 2010). These techniques provide ensemble-averaged readouts and thereby conceal many aspects of the dynamics of individual members of the unsynchronized population.

Single-molecule techniques overcome this limit by directly probing individual proteins, allowing observation of short-lived intermediate states and rare events, and thereby providing new insights into important structural and functional properties. The first single-molecule observations were made in the 1970s by the detection of currents in single ion channels (Neher & Sakmann, 1976). Advances such as the detection of single fluorescent molecules under biologically relevant conditions (Brooks Shera, Seitzinger, Davis, Keller, & Soper, 1990) and single fluorescent proteins (Dickson, Cubitt, Tsien, & Moerner, 1997), as well as the continuous improvement of high-sensitivity and low-noise photon detectors have made single-molecule fluorescence techniques a powerful tool to study membrane proteins in vitro and in vivo (Akyuz, Altman, Blanchard, & Boudker, 2013; Erkens, Hanelt, Goudsmits, Slotboom, & van Oijen, 2013; Sako, Minoghchi, & Yanagida, 2000; Zhao et al., 2010). Other methods such as atomic force microscopy have been established as well to probe the dynamics of single membrane proteins and transporters (Ruan et al., 2017; Whited & Park, 2014), however fluorescence techniques allow for the parallel observation of many single transport proteins under dynamic environmental conditions. In this chapter, we focus on in vitro fluorescence techniques that we will describe in more detail later.

Fluorescence is the radiative decay, in the form of light, from a molecular singlet state back into the ground state from which the molecule has been excited (Lakowicz, 2006). The wavelength of the emitted light is typically higher than the wavelength of the light used for excitation; a phenomenon called Stokes shift. Fluorophores—or dyes—are specifically designed molecules with optimized fluorescence properties such as stability, brightness, and pH sensitivity (Gust et al., 2014). When chemically linked to a protein (or other molecule), their readouts provide information that can provide insight in the protein's structure and function in several ways, some of which will be described later (Weiss, 1999) (Fig. 1). First, they can be utilized in fluorescence microscopy to determine the spatial position with a potential accuracy of a few tens of nanometers. Although the point source of light produces a spot (point spread function—PSF) with a diameter that is fundamentally limited by the wavelength of the light, the center can be localized much more accurately (Schmidt, Schuetz, Baumgartner, Gruber, & Schindler, 1995) (Fig. 1A). When two host molecules are labeled with fluorophores with different spectral properties (i.e., different colors), their possible colocation can be studied. Second, intensity fluctuations of a single dye can

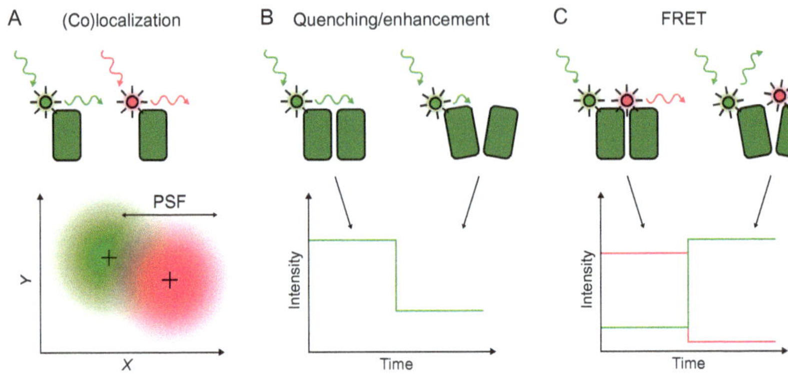

Fig. 1 Fluorescence detection techniques. Several fluorescence techniques; principle (*top*) and observable (*bottom*). (A) By fitting the PSF (typically a few 100nm wide), the fluorophore can be localized (crosshairs) with an accuracy of a few 10nm. Using multiple colors can hint for colocalization. (B) Changes in local environment, which can depend on protein conformation, of the fluorophore can cause quenching or enhancement. (C) FRET can be used to measure small distance changes between roughly 20 and 100Å.

report on the activity or dynamics of proteins (Lu, Xun, & Xie, 1998) (Fig. 1B). A change in the local environment of the dye caused by protein function may quench or enhance fluorescence. Lastly, fluorescence (or Förster) resonance energy transfer (FRET) can provide information on protein dynamics and mechanistic insight in protein function. FRET is a mechanism based on the nonradiative transfer of excitation energy from a donor to an acceptor fluorophore (Ha et al., 1996). This transfer results in a decreased fluorescence intensity of the donor, and an increased intensity of the acceptor. The efficiency E of the transfer depends on the distance R between donor and acceptor:

$$E = \frac{1}{1 + (R/R_0)^6}.$$

Here R_0 is the Förster distance at which the efficiency is 50%. The Förster distance depends on the properties of the pair of dyes and is typically between 50 and 80Å. The anticorrelated intensity changes of donor and acceptor dyes are the hallmark of FRET transients (Fig. 1C). When both dyes are positioned appropriately, it allows the detection of the interaction between molecules or structural changes within a biomolecule on the length scale of a few nanometers.

In this chapter, we focus on the single-molecule imaging of liposome-reconstituted membrane transport proteins using total internal reflection fluorescence (TIRF) microscopy. In the following sections, we will explain the relevant techniques and give a step-by-step description on how to setup, perform, and interpret single-molecule experiments.

2. PREPARING MEMBRANE PROTEINS AND LIPID MEMBRANES

Visualizing inter- and intraprotein dynamics requires the specific fluorescence labeling of subunits in the protein complex. We make use of the thiol-maleimide coupling reaction, which is highly specific at pH values below 7.5, to label cysteines in the folded state of the protein with the appropriate dye molecule, but other labeling chemistries can be used as long as they allow specific and efficient labeling. A cysteine mutant of the protein is designed to specifically position one or more labels in order to observe a change in fluorescence properties of the labels. Any native cysteine that is accessible to undesired maleimide labeling should be replaced by a suitable amino acid, for example serine. Remaining cysteines that cannot be removed without loss of function must be checked for undesirable labeling. Before labeling of engineered cysteines, DTT (dithiothreitol) or TCEP (*tris*-(2-carboxyethyl)phosphine) can be added as a reducing agent to break unwanted disulfide bonds that may have formed. However, excess of these agents must be removed by dialysis prior to adding the reactive dye molecule. Detergent-solubilized membrane proteins are labeled when attached to an affinity-tag column, which is used in the purification procedure. Subsequently, size-exclusion chromatography is performed, in order that remaining unreacted dye molecules are separated, and afterward the purified protein is reconstituted in a lipid bilayer (see later). Sometimes, a 100% labeling efficiency is required to visualize all components. However, occasionally only lower labeling efficiencies can be achieved. Also, homooligomeric proteins sometimes demand lower efficiencies: for example, a dimer might need only one fluorescent label, or a trimer exactly two (different) labels for FRET measurements. Optimizing dye concentrations and dye incubation times yields the desired average, but postacquisition single-molecule analysis (see later) will discern the stoichiometry.

Owing to its size and water solubility, a fluorescent label can alter the properties of the host protein. Therefore, it is crucial to assess the function of the labeled protein complex in detergent micelles or a lipid bilayer (see later) with for example ATP and transport activity essays, which are outside of the scope of this chapter (Geertsma, Nik Mahmood, Schuurman-Wolters, & Poolman, 2008). Ideally, labeling should not affect the transport activity, but often leads to somewhat decreased activity. It is important to exclude that the remaining activity after labeling is not the result of the fraction of unlabeled protein. The fractional activity after labeling must therefore be higher than the fraction of unlabeled proteins.

Membrane proteins are naturally embedded in a lipid bilayer, providing a complex and dynamic environment. The proteins have to be removed from their native surroundings by detergent solubilization (Helenius & Simons, 1975), before they can be purified. Later, they can be analyzed further in the detergent solution, or, after reconstitution, in a (synthetic) lipid environment. Mimicking the complex and dynamic native environment is non-trivial. Biological membranes consist of multiple components that determine properties such as thickness, strength, and tension (Seddon, Curnow, & Booth, 2004). At the same time, membrane proteins interact with the surrounding lipids (Lee, 2003; Marsh, 2008), affecting protein folding and functioning (Lee, 2004; Phillips, Ursell, Wiggins, & Sens, 2009). Carefully choosing the lipids used for protein reconstitution minimizes any negative effects. Prokaryotic transporters are typically reconstituted in a synthetic mixture of dioleoyl-phosphatidylethanolamine (DOPE), dioleoyl-phosphatidylglycerol (DOPG), and dioleoyl-phosphatidylcholine (DOPC), as they often require a significant fraction of nonbilayer (i.e., cone shaped, as opposed to cylinder shaped) lipids (DOPE) and anionic lipids (DOPG) (Geertsma et al., 2008). For eukaryotic systems, typically a sterol is included as well.

A synthetic lipid bilayer reconstituted in vitro resembles the native environment much more closely than a detergent micelle—the latter may alter the protein's function as reflected by various studies (Akyuz et al., 2013; Dorwart, Wray, Brautigam, Jiang, & Blount, 2010; Erkens et al., 2013; Goudsmits, Slotboom, & van Oijen, 2017; Lewinson, Lee, Locher, & Rees, 2010; Mi et al., 2008). Membrane proteins can be reconstituted into nanodiscs (Ritchie et al., 2009) or liposomes (Rigaud & Levy, 2003) of various sizes (Fig. 2). Nanodiscs are small disc-shaped patches of membrane surrounded by scaffolding proteins acting as a belt; the diameter of around 10 nm can be controlled by the length of the scaffolding proteins that define the circumference of the disc (Fig. 2A). Embedded integral membrane proteins are experimentally accessible from both sides. Liposomes, in contrast,

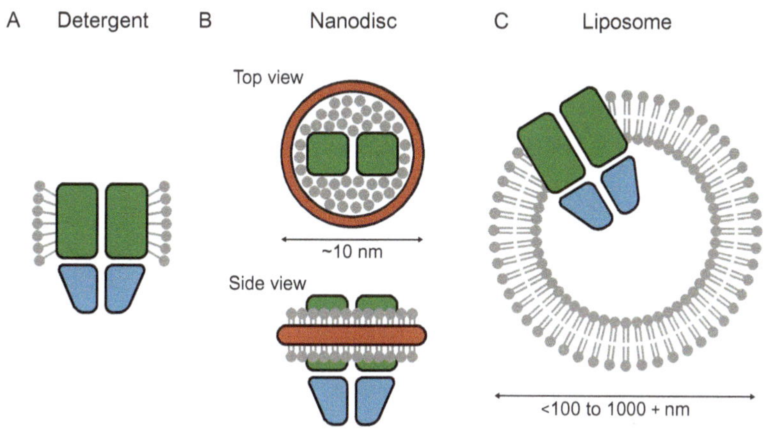

Fig. 2 Environment of membrane proteins. (A) In an aqueous environment, membrane proteins can be stabilized in detergent. Here the hydrophobic tails of the detergent molecules cover the protein and mimic the bilayer. Nanodiscs (B) and liposomes (C) resemble the native environment most closely. (B) In nanodiscs, a scaffolding protein (*red*) holds a patch of lipids together while the membrane protein is accessible from either side. (C) Liposomes on the other hand allow for gradients and allow only the outside to be accessible during the measurement, which is beneficial for studying transport.

have a luminal compartment that is not accessible to membrane-impermeable compounds, but it is possible to preload the inside with desired constituents during preparation of the liposomes (Fig. 2B). The diameters of liposomes vary from below 100 nm to tens of micrometers; unilaminar liposomes are typically in the lower size range. The compartmentalization offered by liposomes allows the establishment of membrane gradients, which makes them of great interest for studying transport. In this chapter, we focus on transport proteins reconstituted in 100-nm liposomes.

Protein reconstitution into liposomes involves mixing of the constituents—proteins and lipids—in the appropriate ratio, followed by detergent removal (for instance by adsorption to polystyrene beads). For single-molecule studies, the number of membrane complexes per liposome is a trade-off between high yield (not too many empty liposomes) and a low probability of multiple complexes per liposome. The optimal average value is around 0.1, resulting in—based on the Poisson distribution—roughly 10% of the liposomes with one complex, 0.5% of the liposomes (meaning 5% of the protein-containing liposomes) with more than one complex, and the remaining 90.5% being empty.

2.1 Protein Purification and Labeling

The following protocol is carried out at 4°C. We set out with membrane vesicles, thawed from a −80°C stock. For immobilization of the protein, we use an *octa*-histidine tag (His_8 tag) on the protein in combination with nickel affinity chromatography here, but any other suitable method can be substituted. Buffers used are 50 m*M* KPi, 200 m*M* KCl, pH 7.5 (buffer A), or the same buffer supplemented with 0.03% dodecyl maltoside (DDM, Anatrace) (buffer B).

1. Solubilize membrane vesicles in buffer A + 1% DDM; mix well and incubate for 1 h.
2. Centrifuge for 20 min at 80,000 rpm to separate insoluble material from detergent-solubilized proteins: decant supernatant in clean tube.
3. Add 500 μL (bed volume) of Ni Sepharose beads (GE Healthcare) washed in buffer B.
4. Incubate for 1 h while gently moving.
5. Add mixture to an empty disposable column and discard flow through.
6. Wash with 10 mL buffer B + 50 m*M* imidazole to remove nonspecifically bound proteins; discard flow through.
7. Label proteins with 0.5 mL dye solution (50 μg dye in 0.5 mL buffer B), incubate for 10–100 min (depending on desired labeling efficiency), mix well from time to time. In case a protein with two labeling sites has to be labeled with two different labels, typically a one-to-one molar ratio of dyes is used; the proper label stoichiometry is selected in postacquisition processing.
8. Wash with 10 mL buffer B to remove unreacted dye molecules; discard flow through.
9. Elute the protein with 2 mL buffer B + 500 m*M* imidazole and collect 500 μL fractions (four fractions in total). The second fraction contains the labeled protein.
10. Perform size-exclusion chromatography (Superdex-200 column, GE Healthcare) to further purify protein from aggregates and remaining free dye molecules, and directly use for reconstitution into liposomes (see later).

In case a soluble protein needs to by copurified (for example a soluble binding protein), the protein can be purified separately with the protocol as described earlier (starting from step 3 with cell extract, and with buffer A instead of buffer B) and mixed with the membrane protein before or after size-exclusion chromatography.

2.2 Lipid Mixture Preparation

Lipid mixtures are prepared separately and can be stored for up to several months in liquid nitrogen to prevent oxidation. They are prepared from the following lipids (dissolved in chloroform): 1,2-dioleoyl-sn-glycero-3-phosphocholine (DOPC); 1,2-dioleoyl-sn-glycero-3-phosphoethanolamine (DOPE); 1,2-dioleoyl-sn-glycero-3-phospho-(19-rac-glycerol) (DOPG); and 1,2-dioleoyl-sn-glycero-3-phosphoethanolamine-*N*-(cap biotinyl) (biotin-DOPE).

1. Mix (w/w) 40% DOPC, 30% DOPE, 30% DOPG, and 1% biotin-DOPE (used for surface immobilization, see later), in total 80 mg. The total amount can be scaled as desired.
2. Evaporate chloroform using rotary evaporation.
3. Add 4 mL buffer A (20 mg/mL) and incubate on ice for 30 min.
4. Dissolve lipids by shaking and aliquot in 400 μL (8 mg).
5. *Optional*: Lipids can be stained with lipophilic tracers (for example DiO, DiI, and DiD) in order to visualize liposomes in microscopy experiments. To this end, mix lipids with the appropriate marker in a 10,000:1 molar ratio, yielding approximately 9 dye molecules per 100-nm liposome (a 100-nm liposome contains roughly 9×10^4 lipid molecules).
6. Subject the mixture to four freeze–thaw cycles (frozen in liquid nitrogen and thawed at room temperate) to homogenize the lipid (and the tracer if applicable) distribution.

2.3 Membrane Reconstitution

A thawed lipid mixture aliquot (8 mg lipids in 400 μL) is used with fresh protein (at least 200 μL of 0.1 μ*M* membrane protein is required) to start a reconstitution. Soluble protein, if required on the inside of liposomes, is added in the appropriate concentration. The following steps are carried out at 4°C.

1. Extrude lipid mixture with 400-nm filter (Mini-Extruder with polycarbonate filter, Avanti Polar Lipids) to create unilamellar liposomes (note that 400 nm is not yet the final size of the liposomes). Extrude an uneven number of times; 11 times is sufficient.
2. Mix empty liposomes (10 mg/mL, final), membrane protein (0.02 μ*M*, final), soluble protein (if applicable; 3.2 μ*M* final equates to on average 1 molecule of soluble protein in a 100-nm liposome), and Triton X-100 (Sigma-Aldrich, 0.3%, final). For 100-nm liposomes, above concentrations lead to an average of 0.1 transport proteins per liposome. In

case liposomes of different size were used, concentrations were adjusted accordingly to maintain the protein/liposome ratio.

3. Add BioBeads SM2 (BioRad, 40 mg/mL mixture) to remove detergent and gently agitate. Fresh BioBeads are added after the following periods (old beads remain in the solution): 15 min, 15 min, 30 min, overnight, and 60 min. Next, remove BioBeads and store the solution in 100 μL aliquots in liquid nitrogen. One aliquot can be thawed on the experimental day and suffices for 1 day of single-molecule experiments (see later).
4. *Optional*: In order to incorporate additional molecules, such as substrates or nucleotides, in the lumen of the vesicles, mix proteoliposomes with these components prior to the experiment. It should be noted that a concentration of 3.2 μ*M* equates to on average 1 molecule in a 100-nm liposome (see step 2).
5. Subject the solution to four freeze–thaw cycles to homogenize the sample and extrude proteoliposomes at 100 nm (or the relevant size).
6. *Optional*: Excess extraluminal components will be washed away during microscopy experiments (see later), however, an extra washing step (i.e., centrifugation of the proteoliposomes followed by resuspension) can aid this removal process. In case soluble His-tagged protein was used throughout the procedure, excess extraluminal protein can be removed by incubating the solution with Ni-NTA magnetic agarose beads (Qiagen) for 1 h in a reaction tube and extracting them with a magnet.

3. SINGLE-MOLECULE FLUORESCENCE MICROSCOPY

Although chemical and photophysical properties of fluorophores are constantly being improved (van der Velde et al., 2016), they do have a limited rate and number of photons emitted before they permanently bleach into a dark state (Ha & Tinnefeld, 2012). These limitations result in a requirement for efficient and low-background detection of emission light. The first need is realized by high-quality optics and high-sensitivity single-photon detectors. Minimizing the amount of background photons, mainly originating from scattering and impurities, can be done in two ways. In confocal imaging, the detection volume is minimized to volumes on the order of femtoliters (Pawley, 2006), thereby typically allowing only one molecule to be measured. The molecule can be either freely diffusing or immobilized on a surface. A point detector captures all photons originating from the detection volume. An alternative method to reduce background photons, which we also employ here, is to decrease the excitation volume

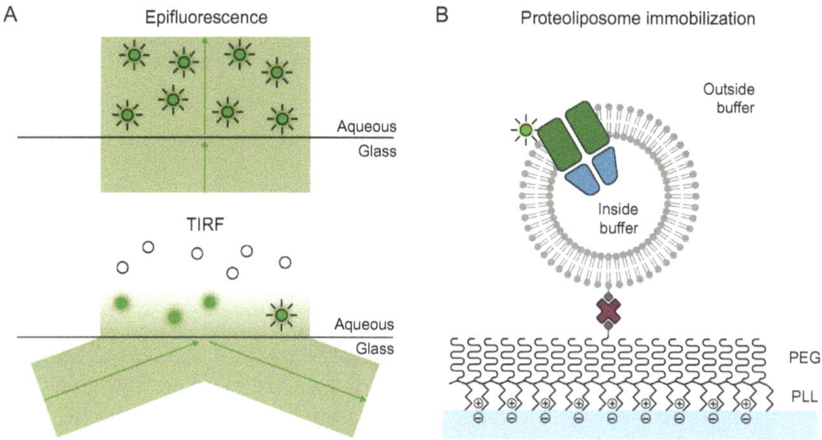

Fig. 3 Microscope setup. (A) Microscopy principles. In epifluorescence (*top*), the entire aqueous solution is excited by the light beam, whereas in TIRF (*bottom*) only a shallow layer (~100–200 nm) on top of the glass surface is excited. (B) Surface immobilization of proteoliposomes. Extensively cleaned glass (negatively charged) is coated with PEG-PLL (positively charged). A small fraction of the PEGs is modified with biotin to allow for immobilization of proteoliposomes containing biotinylated lipids via streptavidin.

by using TIRF microscopy (Axelrod, 2001) (Fig. 3A). As opposed to epifluorescence, where the excitation light beam crosses through the entire sample, TIRF relies on an incident excitation light beam that is totally reflected inside a glass coverslip, thereby creating an evanescent wave of approximately 100–200 nm thick in the medium above the glass. As a result, only fluorophores that are close to the surface will be excited. Here, we use streptavidin to immobilize proteoliposomes containing biotinylated lipids to a glass cover slide coated with poly-L-lysine (PLL) grafted with biotinylated polyethylene glycol (PEG) (see later) (Fig. 3B). Emitted photons are projected on a two-dimensional camera, which enables the recording of multiple fluorophores in parallel. In order to separate fluorescence FRET signals and record two colors simultaneously, we make use of a two-channel simultaneous imaging system that is placed in front of the camera. On top of the glass cover slide used for imaging, a flow cell is constructed that allows for rapid variation of environmental conditions such as pH or substrate concentration, as described in more detail later.

3.1 Glass Functionalization and Flow Cell Construction

A flow cell is constructed based on a functionalized glass cover slide (Fig. 4A). Functionalization of the glass surface is adapted from Tabaei,

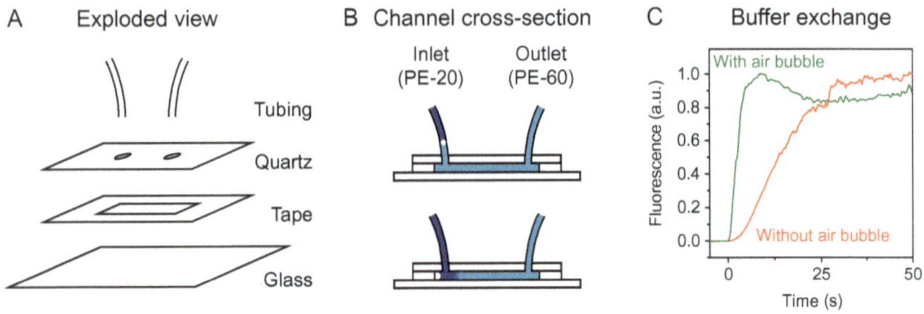

Fig. 4 Flow cell design for rapid buffer exchange. (A) Flow cells were constructed by sandwiching double-sided tape, with a channel (2 cm × 5 mm) excised, between an extensively cleaned glass slip and a reusable quartz plate. (B) Mixing of the buffers is prevented before entering the flow cell by allowing a tiny air bubble, which physically separates the two buffers, in the inlet tube. Upon entering the channel, the air bubble gets trapped, allowing the two buffers to form one continuous phase. (C) The sharp interface only starts to diffuse inside the channel. As a result, at the measurement point, a sharp buffer transition is observed (80% increase within 2 s), as opposed to the case when diffusion already takes place in the inlet tube when no air bubble is present. In this calibration experiment, rhodamine B is used to stain the liquid.

Rabe, Zetterberg, Zhdanov, and Hook (2013). Extensively cleaned glass with a negative surface charge binds to positively charged PLL grafted with polyethylene glycol (PLL-PEG, a fraction of which is biotinylated) to form a biocompatible monolayer that exhibits minimal interaction with proteins and lipids. Cleaned glass surfaces can be stored under vacuum for a few weeks and can later be used to construct a flow cell. The flow cell, which allows for a fast exchange in buffer with a steep gradient (see later) (Fig. 4B and C), can be prepared the day prior to the experiment. Functionalization of the glass surface and attachment of the liposomes happens immediately before performing the single-molecule experiment.

1. Clean microscope cover glass (no. 1.5 precision coverslips, Karl Roth) by sonication in isopropanol for 20 min.
2. Rinse with ultrapure water.
3. Clean microscope cover glass by sonication in acetone for 20 min.
4. Blow-dry with nitrogen gas.
5. Perform oxygen plasma cleaning (Plasma Etch, PE-50) of dried glasses for 10 min to remove remaining organic impurities. Inside the plasma cleaner, the clean side of the glass faces up.

As indicated earlier, cleaned glass slides can be stored in vacuum for some time, before using it to construct a flow cell. All rinsing and incubation steps described later are performed by connecting a syringe to the outlet tube and creating a light vacuum—the inlet tube resides in a tube with the desired buffer.

6. A channel, 2 cm long and 5 mm wide, is cut out of biocompatible double-sided tape (Grace BioLabs, SecureSeal; 170 μm thick) which is sandwiched between the extensively cleaned thin glass slide and a 1-mm thick quartz plate (Technical Glass) (Fig. 4A). The plate is reusable after soaking the flow cell in acetone to dissolve tape and epoxy and subsequent thorough cleaning of the quartz with acetone. In the quartz plate, two holes are predrilled which tightly fit inlet and outlet tubing of the flow cell. Note that the channel extends beyond the inlet, which allows for fast buffer exchange (described later).
7. Seal the flow cell around the upper quartz plate with generic-brand quick-dry epoxy glue.
8. Insert polyethylene inlet tubing (PE-20, inner and outer diameter 0.4 mm and 1.1 mm, respectively) and outlet tubing (PE-60, inner and outer diameter 0.8 mm and 1.2 mm, respectively) in the holes inside the top quartz plate. Support the tubing and seal the bases of the tubes with epoxy glue and let dry overnight.
9. Incubate the flow cell with 1 *M* KOH for 30 min to remove hydroxyl groups from the glass surface, making it negatively charged.
10. Rinse the channel with copious amounts of ultrapure water and wash with 5 mL PBS buffer.
11. Add a mixture (250 μL should be sufficient) of 1 mg/mL PLL-PEG (PLL(20)-g[3.5]-PEG(2), SuSoS) and 2 μg/mL PLL-PEG-biotin (PLL(20)-g[3.5]-PEG(2)/PEG(3.4)-biotin (20%), SuSoS) and incubate for 30 min.
12. Wash the channel with 5 mL PBS buffer. Purge the channel of any air by pulling a strong vacuum in the syringe connected to the outlet tube; flick the tubing to remove any air bubbles stuck to the inner surface. From this point onward, ensure no unwanted air bubbles enter into the tube/channel.

3.2 Experimental Setup and Image Acquisition

At this point, the flow cell can be mounted on the TIRF microscope. We use an Olympus IX-71 inverted microscope using a 100× high-numerical aperture TIRF objective to visualize immobilized proteoliposomes. The outlet tube of the flow cell is now connected to a programmable syringe pump (NE-1000, New Era Pump Systems, Inc.) that precisely controls the flow. The inlet tube resides in a reaction tube with the desired buffer. For a rapid response of the pump, no air should be allowed in the outlet tubing and syringe. A 488, 532, and 637 nm laser were installed to excite the dyes. We use an EM-CCD camera (Hamamatsu Photonics) at the highest EM-gain and a frame rate of 5–10 Hz in combination with a two-channel simultaneous imaging system (DV2, Photometrics) (also termed dual view) to record fluorescence simultaneously in multiple colors. The filters in the dual view are optimized for the excitation and emission wavelengths of the dyes. In case of a FRET acquisition, information is needed about the stoichiometry of donor and acceptor dyes. Therefore, FRET acquisitions are preceded and followed by a 5 s period of acceptor only excitation (Kapanidis et al., 2005).

In order to decrease photobleaching and reduce blinking of the dyes (Ha & Tinnefeld, 2012), we use freshly degassed buffer A (by placing it for 30 min in a vacuum desiccator) with the GODCAT oxygen scavenging system (Blanchard, Kim, Gonzalez, Puglisi, & Chu, 2004) (7.5 U/mL glucose oxidase (glucose oxidase from *Aspergillus niger* type VII, Sigma-Aldrich), 450 U/mL catalase (catalase from bovine liver, Sigma-Aldrich), and 0.8% (w/v) glucose monohydrate) and the antiblinking agent Trolox (1 m*M*, prepared as described previously by Cordes, Vogelsang, & Tinnefeld, 2009) (imaging buffer).

1. Incubate the channel with 1.0 mg/mL bovine serum albumin in PBS (buffer C) for 5 min.
2. Incubate with 0.1 mg/mL streptavidin (streptavidin from *Streptomyces avidinii*, Sigma-Aldrich) in buffer C for 5 min.
3. Wash out unbound streptavidin with 5 mL buffer C.
4. Introduce fresh proteoliposomes diluted to 100 μg/mL (1.5 n*M* 100-nm liposomes) in buffer A and incubate for 10 min. The concentration and incubation time can be optimized for good surface coverage—around three spots per 10 μm^2 results a good yield and is not too dense such that single spots cannot be clearly discriminated anymore.
5. Wash out unbound proteoliposomes with 1 mL of buffer A.
6. Prepare for imaging by introducing 1 mL of imaging buffer.
7. Start imaging. During image acquisition, imaging buffer complemented with other components such as substrate or ATP can rapidly be introduced by allowing a small air bubble (~0.2 μL), which will not traverse the flow cell, between the old and new buffer (Fig. 4B and C). This air bubble can be introduced in the inlet tube and moved to a fixed position before starting the acquisition. To do so, remove the inlet tube from the

reaction tube and raise the inlet tube such that a concave meniscus emerges at the end of the tube. Now, put the inlet tube (held steady) into a new reaction tube by moving the reaction tube up. Run the pump for a short time to move the air bubble to the desired position. During acquisition, start the pump to rapidly introduce the new buffer. The exact moment the new buffer arrives at the measurement point can be found with a proper calibration experiment, for example with a rhodamine B solution (Fig. 4C).

4. DATA ANALYSIS

A recording typically contains fluorescence traces of a few hundred spots (Fig. 5A). Before these data can be analyzed further, they have to be processed to correct for artifacts introduced by components in the setup such as lasers, microscope, and camera. The following subsections explain the removal of these artifacts step by step. All algorithms are implemented in in-house written software for ImageJ (Schneider, Rasband, & Eliceiri, 2012). All steps of the algorithms are depicted later, however, an in-depth review of the implementation is beyond the scope of this chapter.

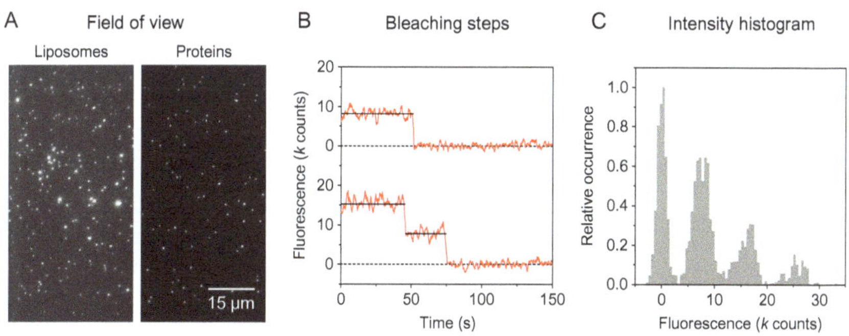

Fig. 5 Typical experimental data. (A) Typical field of view (raw, uncorrected data) with dual view system, allowing for simultaneous acquisition of two colors. Left channel: stained liposomes (DiI). Right channel: soluble proteins in the lumen of these liposomes (BtuF, labeled with Alexa Fluor 647). (B) Bleaching steps of BtuF originating from two spots in (A). Steps are identified manually here and intermediate intensity levels are fitted (*solid black line*). (C) Creating a histogram based on data from a bleaching experiment (11 traces) reveals that single fluorophores have an intensity between 5000 and 12,000 counts, which corresponds to a step size as observed in (B).

4.1 Subtraction of Electronic Offset

The EM-CCD camera introduces a constant hardware electronic offset to all pixels such that noise will never generate negative values (the computer files are stored in positive integer values only). The value of the offset can readily be obtained by closing the shutter in front of the camera and minimizing the exposure time. The average value of a single frame can then be used as offset to subtract from the movie, which is now stored in a format that allows signed values.

4.2 Laser Profile Correction

The Gaussian profile of the laser beam is clearly visible in the emission intensities of the spots in the field of view. A map of intensity values, normalized to one and with the same dimensions as the recording, is used to divide each frame of the movie. This map can be created by a long exposure image, while the field of view is continuously moved in both lateral directions. The image can be smoothened by a 32-pixel radius Gaussian filter. The resulting map, of course, has to be corrected for electronic offset as well.

4.3 Dual View Alignment

The two-channel simultaneous imaging system splits the image on the camera in two halves, each for light at a different wavelength. Owing to chromatic aberrations and imperfections in the optics of the system, coordinates in one half are not merely a lateral translation of coordinates in the other half. A second-order two-dimensional polynomial is used to map points from one halve to the other:

$$\Delta x = a_0 + a_1 x + a_2 y + a_3 x^2 + a_4 xy + a_5 y^2,$$
$$\Delta y = b_0 + b_1 x + b_2 y + b_3 x^2 + b_4 xy + b_5 y^2.$$

Here, Δx and Δy are the lateral displacements of the corresponding coordinates (x, y) and (x', y') in each halve:

$$x' = x + \Delta x, \ y' = y + \Delta y.$$

The parameters a_i and b_i of the polynomial are determined by using a dataset that has signals originating from the same point simultaneously in both channels. Peak positions are identified independently in both channels (although most signals should originate from the same spot, not all points necessarily have to form pairs) using MaximumFinder in ImageJ. Subsequently, two binary-valued images (one for each channel) are created with single-pixel peaks on a black background and registered using a combination of mixture of Gaussians (Jian & Vemuri, 2005) and two-dimensional fitting using the Levenberg–Marquardt algorithm. In a next step, overlapping fluorescent points can easily be identified in both channels, which is needed in, e.g., FRET measurements or colocalization experiments with stained liposomes and fluorescently labeled proteins.

4.4 Drift Correction

Long-duration acquisitions of several minutes can be affected by stage drift. Subsequent images in the recording can therefore be registered with the first image by applying FFT-based techniques for subpixel translational image registration (Guizar-Sicairos, Thurman, & Fienup, 2008; Reddy & Chatterji, 1996).

4.5 Data Analysis

Peaks are detected by MaximumFinder in ImageJ. Fluorescence traces, corrected for all artifacts as described earlier, are constructed by plotting the integrated intensity in a 7-pixel diameter circle around a peak. Integrated background intensity, extracted from a nearby area devoid of peaks and normalized to a 7-pixel diameter circle, is subtracted after a 25-frame temporal median filter is applied. Shot noise and electronic noise are removed from the resulting traces by applying a 5-frame temporal mean filter. Traces are finally corrected for cross-talk between the donor and acceptor channel: bleeding of donor signal into the acceptor channel and vice versa. These values can be derived from the spectral properties of the filters and fluorophores, and it can easily be verified experimentally by having only donor or only acceptor molecules present.

The stoichiometry of fluorescent labels, both in single and dual-color images, is important for further analysis. For example, in FRET measurements, it is important to have one donor and one acceptor. Furthermore, the number of labeled proteins in the membrane or lumen of liposomes can also contain valuable information. The number of dye molecules can be obtained by determination of discrete bleaching steps—the hallmark of single-molecule traces (Fig. 5B). This procedure requires the detection of change points in the fluorescence trace, which we will describe later. If these data are not available, or bleaching is unwanted, the intensity of a single fluorophore needs to be determined. After creating a histogram of all fluorescence intensity values, it is possible to assign a number of fluorophores to the amount of fluorescence (Fig. 5C). This information can be used later in experiments with identical fluorescent dyes and laser power to assess the number of dyes or proteins.

Change points in fluorescence transients are indicative for conformational changes of the labeled protein or bleaching as described earlier. They can be detected in several ways. The most common and straightforward way is by thresholding. If prior knowledge exists about the different fluorescence states, for example by examining the histogram of all intensity values of dynamic traces, each time point can be assigned to a distinct fluorescence intensity level. Without prior information, a jump can be detected by identifying for example a change, between two adjacent time points, which magnitude exceeds three times (this value is adjustable) the standard deviation of the signal before the change. Often, this simple and fast algorithm does not perform well, as slow or small changes will not be detected. A more sophisticated technique that can be applied is a method that uses a generalized likelihood ratio test to find change points (Watkins & Yang, 2005). For FRET time series, if the protein complex transitions between stable conformations, modeling can be done by a hidden Markov model (HMM). Several software packages have been developed that adapt HMM to analyze FRET traces (Bronson, Fei, Hofman, Gonzalez, & Wiggins, 2009; McKinney, Joo, & Ha, 2006). Further analysis of the data, which depends on the system being studied, is highly specific and is therefore beyond the scope of this chapter.

ACKNOWLEDGMENTS

D.J.S. would like to acknowledge funding from the Netherlands Organization for Scientific Research (NWO) (Vici grant 865.11.001) and the European Research Council (ERC) (ERC Starting Grant 282083). A.v.O. acknowledges support by the European Research Council (ERC) (ERC Starting Grant 281098), the Netherlands Organization for Scientific Research (NWO) (Vici grant 680.47.607), and the Australian Research Council (Laureate Fellowship FL140100027).

REFERENCES

Akyuz, N., Altman, R. B., Blanchard, S. C., & Boudker, O. (2013). Transport dynamics in a glutamate transporter homologue. *Nature*, *502*(7469), 114–118. http://dx.doi.org/10.1038/nature12265.

Axelrod, D. (2001). Total internal reflection fluorescence microscopy in cell biology. *Traffic*, *2*(11), 764–774.

Blanchard, S. C., Kim, H. D., Gonzalez, R. L., Jr., Puglisi, J. D., & Chu, S. (2004). tRNA dynamics on the ribosome during translation. *Proceedings of the National Academy of Sciences of the United States of America*, *101*(35), 12893–12898. http://dx.doi.org/10.1073/pnas.0403884101.

Bronson, J. E., Fei, J., Hofman, J. M., Gonzalez, R. L., Jr., & Wiggins, C. H. (2009). Learning rates and states from biophysical time series: A Bayesian approach to model selection and single-molecule FRET data. *Biophysical Journal*, *97*(12), 3196–3205. http://dx.doi.org/10.1016/j.bpj.2009.09.031.

Brooks Shera, E., Seitzinger, N. K., Davis, L. M., Keller, R. A., & Soper, S. A. (1990). Detection of single fluorescent molecules. *Chemical Physics Letters*, *174*(6), 553–557. http://dx.doi.org/10.1016/0009-2614(90)85485-U.

Cordes, T., Vogelsang, J., & Tinnefeld, P. (2009). On the mechanism of Trolox as antiblinking and antibleaching reagent. *Journal of the American Chemical Society*, *131*(14), 5018–5019. http://dx.doi.org/10.1021/ja809117z.

Dickson, R. M., Cubitt, A. B., Tsien, R. Y., & Moerner, W. E. (1997). On/off blinking and switching behaviour of single molecules of green fluorescent protein. *Nature*, *388*(6640), 355–358. http://dx.doi.org/10.1038/41048.

Dorwart, M. R., Wray, R., Brautigam, C. A., Jiang, Y., & Blount, P. (2010). *S. aureus* MscL is a pentamer in vivo but of variable stoichiometries in vitro: Implications for detergent-solubilized membrane proteins. *PLoS Biology*, *8*(12), e1000555. http://dx.doi.org/10.1371/journal.pbio.1000555.

Drew, D., & Boudker, O. (2016). Shared molecular mechanisms of membrane transporters. *Annual Review of Biochemistry*, *85*, 543–572. http://dx.doi.org/10.1146/annurev-biochem-060815-014520.

Erkens, G. B., Hanelt, I., Goudsmits, J. M., Slotboom, D. J., & van Oijen, A. M. (2013). Unsynchronised subunit motion in single trimeric sodium-coupled aspartate transporters. *Nature*, *502*(7469), 119–123. http://dx.doi.org/10.1038/nature12538.

Geertsma, E. R., Nik Mahmood, N. A., Schuurman-Wolters, G. K., & Poolman, B. (2008). Membrane reconstitution of ABC transporters and assays of translocator function. *Nature Protocols*, *3*(2), 256–266. http://dx.doi.org/10.1038/nprot.2007.519.

Goudsmits, J. M. H., Slotboom, D. J., & van Oijen, A. M. (2017) Single-molecule visualization of the vitamin B12 transport mechanism of the ABC importer BtuCD-F.

Groeneveld, M., & Slotboom, D. J. (2010). Na(+): Aspartate coupling stoichiometry in the glutamate transporter homologue Glt(Ph). *Biochemistry*, *49*(17), 3511–3513. http://dx.doi.org/10.1021/bi100430s.

Guizar-Sicairos, M., Thurman, S. T., & Fienup, J. R. (2008). Efficient subpixel image registration algorithms. *Optics Letters*, *33*(2), 156–158.

Gust, A., Zander, A., Gietl, A., Holzmeister, P., Schulz, S., Lalkens, B., et al. (2014). A starting point for fluorescence-based single-molecule measurements in biomolecular research. *Molecules*, *19*(10), 15824–15865. http://dx.doi.org/10.3390/molecules191015824.

Ha, T., Enderle, T., Ogletree, D. F., Chemla, D. S., Selvin, P. R., & Weiss, S. (1996). Probing the interaction between two single molecules: Fluorescence resonance energy transfer between a single donor and a single acceptor. *Proceedings of the National Academy of Sciences of the United States of America*, *93*(13), 6264–6268.

Ha, T., & Tinnefeld, P. (2012). Photophysics of fluorescent probes for single-molecule biophysics and super-resolution imaging. *Annual Review of Physical Chemistry*, *63*, 595–617. http://dx.doi.org/10.1146/annurev-physchem-032210-103340.

Helenius, A., & Simons, K. (1975). Solubilization of membranes by detergents. *Biochimica et Biophysica Acta*, *415*(1), 29–79.

Jian, B., & Vemuri, B. C. (2005). A robust algorithm for point set registration using mixture of Gaussians. *Proceedings/IEEE International Conference on Computer Vision*, *2*, 1246–1251. http://dx.doi.org/10.1109/ICCV.2005.17.

Kapanidis, A. N., Laurence, T. A., Lee, N. K., Margeat, E., Kong, X., & Weiss, S. (2005). Alternating-laser excitation of single molecules. *Accounts of Chemical Research*, *38*(7), 523–533. http://dx.doi.org/10.1021/ar0401348.

Krogh, A., Larsson, B., von Heijne, G., & Sonnhammer, E. L. (2001). Predicting transmembrane protein topology with a hidden Markov model: Application to complete genomes. *Journal of Molecular Biology*, *305*(3), 567–580. http://dx.doi.org/10.1006/jmbi.2000.4315.

Lakowicz, J. R. (2006). *Principles of fluorescence spectroscopy* (3rd ed.). New York: Springer.

Lee, A. G. (2003). Lipid–protein interactions in biological membranes: A structural perspective. *Biochimica et Biophysica Acta*, *1612*(1), 1–40.

Lee, A. G. (2004). How lipids affect the activities of integral membrane proteins. *Biochimica et Biophysica Acta*, *1666*(1–2), 62–87. http://dx.doi.org/10.1016/j.bbamem.2004.05.012.

Lewinson, O., Lee, A. T., Locher, K. P., & Rees, D. C. (2010). A distinct mechanism for the ABC transporter BtuCD-BtuF revealed by the dynamics of complex formation. *Nature Structural & Molecular Biology*, *17*(3), 332–338. http://dx.doi.org/10.1038/nsmb.1770.

Lu, H. P., Xun, L., & Xie, X. S. (1998). Single-molecule enzymatic dynamics. *Science*, *282*(5395), 1877–1882.

Marsh, D. (2008). Protein modulation of lipids, and vice-versa, in membranes. *Biochimica et Biophysica Acta*, *1778*(7–8), 1545–1575. http://dx.doi.org/10.1016/j.bbamem.2008.01.015.

McKinney, S. A., Joo, C., & Ha, T. (2006). Analysis of single-molecule FRET trajectories using hidden Markov modeling. *Biophysical Journal*, *91*(5), 1941–1951. http://dx.doi.org/10.1529/biophysj.106.082487.

Mi, L. Z., Grey, M. J., Nishida, N., Walz, T., Lu, C., & Springer, T. A. (2008). Functional and structural stability of the epidermal growth factor receptor in detergent micelles and phospholipid nanodiscs. *Biochemistry*, *47*(39), 10314–10323. http://dx.doi.org/10.1021/bi801006s.

Missner, A., & Pohl, P. (2009). 110 years of the Meyer–Overton rule: Predicting membrane permeability of gases and other small compounds. *ChemPhysChem*, *10*(9–10), 1405–1414. http://dx.doi.org/10.1002/cphc.200900270.

Neher, E., & Sakmann, B. (1976). Single-channel currents recorded from membrane of denervated frog muscle fibres. *Nature*, *260*(5554), 799–802.

Pawley, J. B. (2006). *Handbook of biological confocal microscopy* (3rd ed.). New York: Springer.

Phillips, R., Ursell, T., Wiggins, P., & Sens, P. (2009). Emerging roles for lipids in shaping membrane–protein function. *Nature*, *459*(7245), 379–385. http://dx.doi.org/10.1038/nature08147.

Reddy, B. S., & Chatterji, B. N. (1996). An FFT-based technique for translation, rotation, and scale-invariant image registration. *IEEE Transactions on Image Processing*, *5*(8), 1266–1271. http://dx.doi.org/10.1109/83.506761.

Ren, Q., & Paulsen, I. T. (2005). Comparative analyses of fundamental differences in membrane transport capabilities in prokaryotes and eukaryotes. *PLoS Computational Biology*, *1*(3), e27. http://dx.doi.org/10.1371/journal.pcbi.0010027.

Rigaud, J. L., & Levy, D. (2003). Reconstitution of membrane proteins into liposomes. *Methods in Enzymology*, *372*, 65–86. http://dx.doi.org/10.1016/S0076-6879(03)72004-7.

Ritchie, T. K., Grinkova, Y. V., Bayburt, T. H., Denisov, I. G., Zolnerciks, J. K., Atkins, W. M., et al. (2009). Chapter 11—Reconstitution of membrane proteins in phospholipid bilayer nanodiscs. *Methods in Enzymology*, *464*, 211–231. http://dx.doi.org/10.1016/S0076-6879(09)64011-8.

Ruan, Y., Miyagi, A., Wang, X., Chami, M., Boudker, O., & Scheuring, S. (2017). Direct visualization of glutamate transporter elevator mechanism by high-speed AFM. *Proceedings of the National Academy of Sciences of the United States of America*, *114*(7), 1584–1588. http://dx.doi.org/10.1073/pnas.1616413114.

Saier, M. H., Jr., Reddy, V. S., Tamang, D. G., & Vastermark, A. (2014). The transporter classification database. *Nucleic Acids Research, 42*(Database issue), D251–D258. http://dx.doi.org/10.1093/nar/gkt1097.

Sako, Y., Minoghchi, S., & Yanagida, T. (2000). Single-molecule imaging of EGFR signalling on the surface of living cells. *Nature Cell Biology, 2*(3), 168–172. http://dx.doi.org/10.1038/35004044.

Schmidt, T., Schuetz, G. J., Baumgartner, W., Gruber, H. J., & Schindler, H. (1995). Characterization of photophysics and mobility of single molecules in a fluid lipid membrane. *The Journal of Physical Chemistry, 99*(49), 17662–17668. http://dx.doi.org/10.1021/j100049a030.

Schneider, C. A., Rasband, W. S., & Eliceiri, K. W. (2012). NIH image to ImageJ: 25 years of image analysis. *Nature Methods, 9*(7), 671–675.

Seddon, A. M., Curnow, P., & Booth, P. J. (2004). Membrane proteins, lipids and detergents: Not just a soap opera. *Biochimica et Biophysica Acta, 1666*(1–2), 105–117. http://dx.doi.org/10.1016/j.bbamem.2004.04.011.

Tabaei, S. R., Rabe, M., Zetterberg, H., Zhdanov, V. P., & Hook, F. (2013). Single lipid vesicle assay for characterizing single-enzyme kinetics of phospholipid hydrolysis in a complex biological fluid. *Journal of the American Chemical Society, 135*(38), 14151–14158. http://dx.doi.org/10.1021/ja4046313.

van der Velde, J. H., Oelerich, J., Huang, J., Smit, J. H., Aminian Jazi, A., Galiani, S., et al. (2016). A simple and versatile design concept for fluorophore derivatives with intramolecular photostabilization. *Nature Communications*, 7, 10144. http://dx.doi.org/10.1038/ncomms10144.

Watkins, L. P., & Yang, H. (2005). Detection of intensity change points in time-resolved single-molecule measurements. *The Journal of Physical Chemistry. B, 109*(1), 617–628. http://dx.doi.org/10.1021/jp0467548.

Weiss, S. (1999). Fluorescence spectroscopy of single biomolecules. *Science, 283*(5408), 1676–1683.

Whited, A. M., & Park, P. S. (2014). Atomic force microscopy: A multifaceted tool to study membrane proteins and their interactions with ligands. *Biochimica et Biophysica Acta, 1838*(1 Pt. A), 56–68. http://dx.doi.org/10.1016/j.bbamem.2013.04.011.

Zhao, Y., Terry, D., Shi, L., Weinstein, H., Blanchard, S. C., & Javitch, J. A. (2010). Single-molecule dynamics of gating in a neurotransmitter transporter homologue. *Nature, 465*(7295), 188–193. http://dx.doi.org/10.1038/nature09057.

CHAPTER FIVE

The MEMbrane Protein Single ShoT Amplification Recipe: MemStar

Abdul Aziz-Qureshi, Pascal F. Meier, Chiara Lee, David Drew[1]
Stockholm University, SE-106 91 Stockholm, Sweden
[1]Corresponding author: e-mail address: ddrew@dbb.su.se

Contents

Abstract

Here, we present a simple overexpression condition for high-throughput screening of membrane proteins in *Escherichia coli*. For the vast majority of bacterial membrane protein targets tested the MEMbrane protein Single shoT Amplification Recipe—MemStar—leads to high production yields of target protein. The use of MemStar has facilitated structural studies of several transport proteins.

Methods in Enzymology, Volume 594
ISSN 0076-6879
http://dx.doi.org/10.1016/bs.mie.2017.05.010

1. INTRODUCTION

Escherichia coli is by far the most widely used host for the production of bacterial membrane proteins (Bill et al., 2011; Schlegel, Hjelm, Baumgarten, Vikstrom, & de Gier, 2014). Most often, different strains, culture conditions, and target gene expression regimes are screened for in an attempt to identify the optimum production strategy that yields sufficient material for functional and structural studies. Alternatively, many clones are screened in a single platform for those that express to levels that are sufficiently high enough for further biochemical and structural investigation. However, with this later strategy, it is very important that the single platform used is maximal for as many targets as possible. The primary reason for this is that there is no clear correlation between membrane protein production levels and detergent-solubilized stability (Kang, Lee, & Drew, 2013; Sonoda et al., 2011; Tate, 2012). Thus, if the platform used is suboptimal, poorly expressing membrane proteins might be discarded, despite the fact that these targets might, in fact, be the most stable in detergent solution and therefore the most suitable for structural investigation (Kang et al., 2013; Sonoda et al., 2011; Tate, 2012).

In our search for better alternatives than the currently used *E. coli*-based membrane protein production approaches, we previously evaluated the production of 10 model membrane protein targets in the widely used *E. coli* membrane protein production strain C43(DE3) (Miroux & Walker, 1996). As model targets for our reevaluation exercise, we used only membrane proteins with known crystal structures Mhp1, NhaA, EmrD, GlpG, AsbTNM, YiiP, GlpT, AmtB, LacY, and TehA (Lee et al., 2014a). This allowed us to verify the quality of the membrane proteins produced by crystallography. To rapidly and accurately monitor production levels of the 10 model targets, they were C-terminally fused to a His_8-tagged green fluorescent protein (GFP) (Drew, von Heijne, Nordlund, & de Gier, 2001; Lee et al., 2014a). We compared the production of these 10 membrane proteins in C43(DE3) cultured in Lysogeny Broth (LB, a.k.a. Luria Bertani broth), Terrific Broth (TB), and the autoinduction media ZYM-5052 and PASM-5052 (Studier, 2005). We found that the media PASM-5052 gave the highest expression levels. PASM-5052 is an autoinduction media tailored for selenomethionine expression, although in our testing we replaced selenomethionine with methionine (Lee et al., 2014a). Surprisingly, we found that even higher expression levels in PASM-5052 media were obtainable by including an IPTG induction step at mid-log phase (Fig. 1).

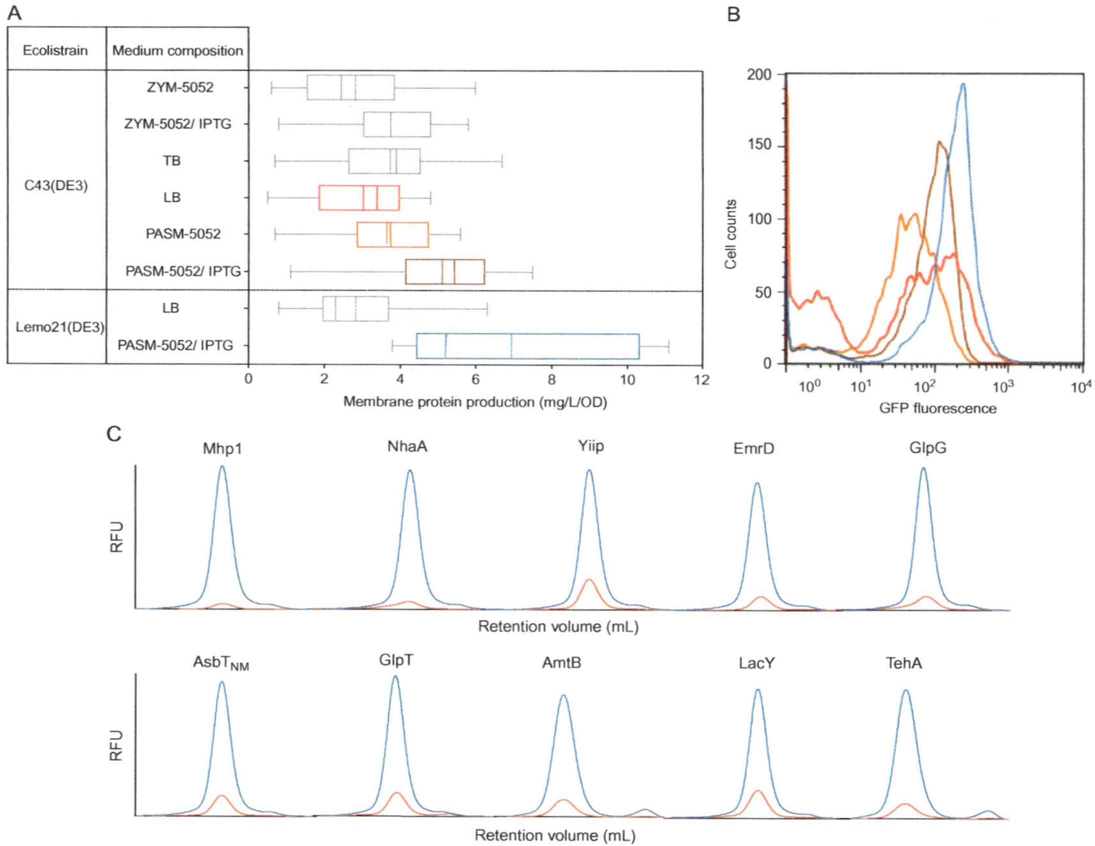

Fig. 1 See legend on next page.

Lemo21(DE3) is an BL21(DE3) strain that contains the plasmid pLemo (Wagner et al., 2008). The pLemo plasmid harbors the gene encoding T7 lysozyme, a natural inhibitor of T7 RNA polymerase, which is under the control of the tuneable L-rhamnose promoter (Studier & Moffatt, 1986; Fig. 2). By testing different amounts of L-rhamnose in the culture medium, one can rapidly identify the target gene expression, leading to the optimal production of a particular membrane protein target (Wagner et al., 2008). Indeed, by combining the PASM-5052 media with IPTG induction and the Lemo21(DE3) strain a further 35% mean improvement was obtainable (Lee et al., 2014a). Flow cytometry analysis showed that the incremental improvements from C43(DE3) cells cultured in LB (red line) through to Lemo21(DE3) cells cultured in PASM-5052 with IPTG induction (blue line). In essence, the final culture/strain combination increased both the number *E. coli* overproducing cells and the amount each cell produced (Fig. 1). To verify quality, membranes were isolated from either C43(DE3)/LB or Lemo21(DE3)/PASM-5052 with IPTG induction, and the 10 model proteins solubilized in the same amount of dodecyl-β-D-maltopyranoside (DDM) detergent. The quality of the DDM-solubilized fusions was then assessed by fluorescence-detection size-exclusion chromatography (FSEC) (Kawate & Gouaux, 2006; Lee et al., 2014a). Indeed, we could show that with

Fig. 1 Setting the stage for the development of a MEMbrane protein Single shoT Amplification Recipe: MemStar. (A) Box-and-whisker plots illustrating the distribution of membrane protein production (mg/L) normalized to cell density (OD_{600}) for the 10 model membrane protein targets in each of the different culture conditions as labeled. The *bottom* and *top* of *each box* represents the 25th and 75th percentile with the whisker lines, indicating maximum and minimum production levels. The *solid vertical line* represents the median and the *dashed vertical line* the mean in each condition. The *colored boxes* represent the culture conditions further tested. With Lemo21(DE3) strain, protein production levels are presented using the optimum L-rhamnose concentration. Unless otherwise stated, the optimum rhamnose concentration is applied throughout all further experiments using the Lemo21(DE3) strain. (B) Flow cytometry analysis of $AsbT_{NM}$ production in: C43(DE3) cells grown in LB (*red*), PASM-5052 (*yellow*), PASM-5052 with IPTG induction (*brown*), and Lemo21(DE3) cells grown in PASM-5052 with IPTG induction (*blue*). The colored culture conditions correspond with the box-and-whisker plots in (A). Note: $AsbT_{NM}$ was selected due to its close membrane protein production levels (mg/L/OD) and mean production levels of the 10 model targets. (C) Quality of produced control proteins as judged by the raw FSEC profiles of DDM-solubilized membranes isolated from either C43(DE3)/LB (*red*) or Lemo21(DE3)/PASM-5052 with IPTG induction (*blue*) conditions (which is later referred to as MemStar). *From Lee, C., et al. (2014). MemStar: A one-shot Escherichia coli-based approach for high-level bacterial membrane protein production.* FEBS Letters, 588*(20), 3761–3769.*

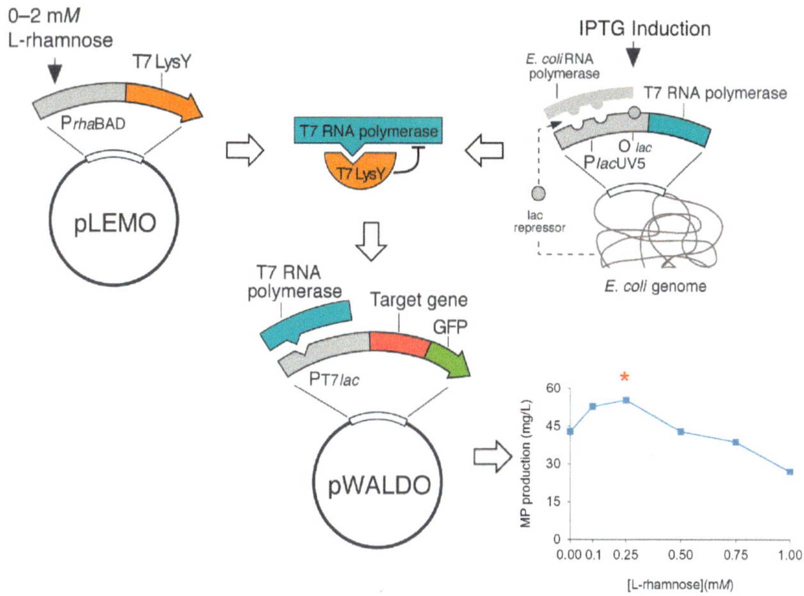

Fig. 2 Tuning T7 RNA polymerase activity to optimize protein production levels using Lemo21(DE3). Illustration showing the tuning of T7 RNA polymerase activity with T7 lysozyme, whose gene is under the control of the *rha*BAD promoter following induction by the addition of an L-rhamnose titration series. Production levels of the GFP-fusion protein target are quantified from fluorescence, with the optimum L-rhamnose concentration (*red asterisk*) determined for large-scale protein production. *From Lee, C., et al. (2014). MemStar: A one-shot Escherichia coli-based approach for high-level bacterial membrane protein production.* FEBS Letters, 588*(20), 3761–3769.*

a new strain and culture combination, on average, ~10-fold more monodisperse protein was obtainable (Lee et al., 2014a).

As a last step, we found that by using 0.25 m*M* L-rhamnose in the medium, apart from one outlier, the production levels were at least ~65% of the (maximal) level that would have otherwise been achievable with the optimal amount of L-rhamnose. We called the combination of Lemo21(DE3) cells grown in PASM-5052 media with an IPTG induction step and 0.25 m*M* L-rhamnose in the culture condition an MEMbrane protein Single shoT Amplification Recipe, or for short, MemStar (Lee et al., 2014a). We further tested MemStar on a previously constructed bacterial membrane protein library consisting of 24 membrane proteins ranging from low (0.6 mg/L) to high (27 mg/L) production levels in C43(DE3)/LB. The normalized mean production levels increased similar to test proteins, with 17 of the 24 membrane proteins reaching levels of more than 12.5 mg/L (Lee et al., 2014a).

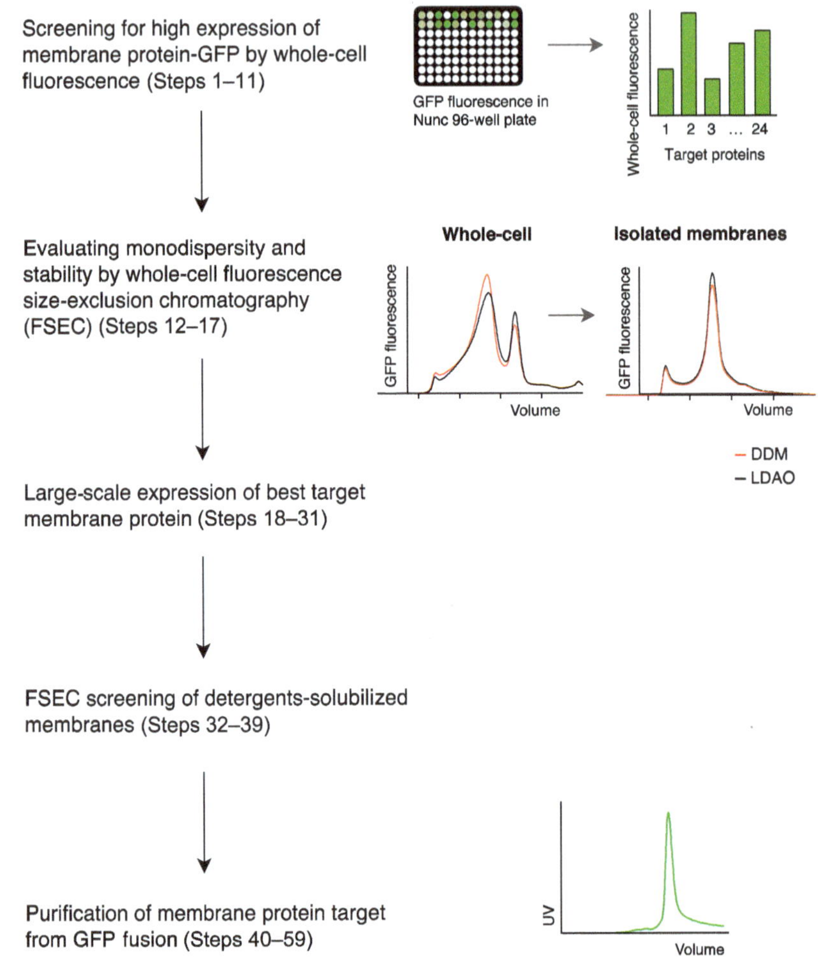

Fig. 3 Flowchart illustrating the target screening, large-scale expression, and membrane isolation in *E. coli* according to the MemStar protocol.

Here, we outline the MemStar protocol and downstream purification, which has been tailored for efficient screening of many bacterial targets in parallel (Fig. 3).

2. MATERIALS

2.1 Equipment

- 24 deep well blocks (Invitrogen, CS15124)
- 2.5 L baffled Tunair® shaker flasks (Sigma, Z710822)

- Nunc 96-well black optical bottom plate (Nunc)
- Cell disruptor (Constant Systems)
- Prominence HPLC (Shimadzu)
- Poly-Prep/glass Econo-Column chromatography columns (Bio-Rad)
- Superose 6 10/300 GL gel filtration column (GE Healthcare)
- Superdex 200 10/300 GL gel filtration column (GE Healthcare)
- SpectraMax Gemini EM microplate spectrofluorometer (Molecular devices)
- Vivaspin 2, 100,000 MWCO PES (VS0242, Sartorius)
- Vivaspin 20, 50,000 MWCO PES (VS2032, Sartorius)

2.2 Buffers and Reagents

- Lemo21(DE3)-competent cells (NEB)
- PASM-5052 autoinduction media: 50 m*M* Na_2HPO_4, 50 m*M* KH_2PO_4, 25 m*M* $(NH_4)_2SO_4$, 2 m*M* $MgSO_4$, 0.2 × metals, 0.5% (v/v) glycerol, 0.05% (w/v) glucose, 0.2% (w/v) α-lactose, 200 μg/mL each of 17 amino acids (tyrosine and cysteine not included), 500 μg/mL methionine, 125 μg/mL selenomethionine (Generon) (optional), 100 n*M* vitamin B_{12}

(*All components are from Sigma if not otherwise indicated*)

- Isopropyl β-D-1-thiogalactopyranoside (IPTG) (Sigma)
- *n*-Dodecyl-β-D-maltopyranoside (DDM) (Anatrace)
- *n*-Decyl-β-D-maltopyranoside (DM) (Anatrace)
- *n*-Nonyl-β-D-maltopyranoside (NM) (Anatrace)
- *n*-Dodecyl-*N,N*-dimethylamine-N-oxide (LDAO) (Anatrace)
- Phosphate-buffered saline (PBS) (Components from Sigma)
- Kanamycin (Sigma)
- Chloramphenicol (Sigma)
- L-Rhamnose monohydrate (Sigma)
- Purified GFP (variants S64L and S65T)

2.3 Reagent Setup

PASM-5052 media

We recommend preparing the following stock solutions as described by Studier (2005):

20 × P—1 × P = (50 m*M* Na_2HPO_4, 50 m*M* KH_2PO_4), 25 m*M* $(NH_4)_2SO_4$

50 × 5052—1 × 5052 = 0.5% glycerol, 0.05% glucose, 0.2% α-lactose

1000 × metals for a total volume of 100 mL

36 mL sterile H_2O	2 mL 0.1 *M* $CuCl_2 \cdot 2H_2O$
50 mL 0.1 *M* $FeCl_3$ in 0.12 *M* HCl	1 mL 0.2 *M* $NiCl_2 \cdot 6H_2O$
2 mL 1 *M* $CaCl_2$	2 mL 0.1 *M* $Na_2MoO_4 \cdot 2H_2O$
1 mL 1 *M* $MnCl_2 \cdot 4H_2O$	2 mL 0.1 *M* $Na_2SeO_3 \cdot 5H_2O$
1 mL 1 *M* $ZnSO_4 \cdot 7H_2O$	2 mL 0.1 *M* H_3BO_3
1 mL 0.2 *M* $CoCl_2 \cdot 6H_2O$	
2 mL 0.1 *M* $CuCl_2 \cdot 2H_2O$	

Except $FeCl_3$, all stock solutions of individual metals were autoclaved. Solutions were stored at room temperature. A slight precipitate forms in 1000 × metals over time.

0.1 *M* $FeCl_3$ in ~0.12 *M* HCl—99 mL sterile H_2O, 1 mL conc. HCl (~12 *M*), 0.1 *M* $FeCl_3 \cdot 6H_2O$

50 × 18 amino acid mixture (without Cys and Tyr)—25 g of Met and 10 g of other each amino acid (except Cys and Tyr) dissolved in 1 L. Heating is required while stirring to dissolve the amino acids. The solution is filtered and stored at 4°C.

1 *M* $MgSO_4$

0.24 m*M* vitamin B12

2.3.1 Preparation of 1 L of PASM-5052 Media

900 mL sterilized H_2O

2 mL 1 *M* $MgSO_4$

0.2 mL 1000 × metals

20 mL 50 × 5052

50 mL 20 × P

20 mL 50 × 18 amino acid mix

0.417 mL of 0.24 m*M* vitamin B_{12}

2.3.2 Antibiotic

50 μg/mL kanamycin or 100 μg/mL ampicillin is used for transformations and in media.

34 μg/mL of chloramphenicol is used for Lemo21(DE3)-competent cells containing the Lemo plasmid.

3. PROCEDURE

3.1 Screening Membrane Protein Targets for High Expression Levels and Stability

The protocol below is designed for screening at least 24 membrane proteins at a time, but can easily be upscaled to screen many more than this. To take advantage of MemStar, we recommend cloning into a T7-based vector and using a C-terminal GFP fusion as a folding indicator. As such, check the membrane protein is likely to have a cytoplasmic C-terminus (C_{in}-topology), as GFP is inefficiently translocated across the *E. coli* inner membrane in C_{out} membrane proteins (Drew et al., 2002). Online tools, such as TMHMM, can be used to predict topology, if it is not been experimentally established. Lastly, by including a purification tag on the C-terminus of GFP and a protease cleavage site between the membrane protein and GFP fusion, one can easily purify the expressed fusions. See Drew, Lerch, Kunji, Slotboom, and de Gier (2006) and Drew et al. (2008) as examples for construct design and downstream purification.

1. Transform the ~24 plasmids harboring the cloned membrane protein-GFP fusions into Lemo21(DE3) cells with the appropriate antibiotic(s). Incubate LB-agar plates overnight at 37°C.
2. Pick a single colony from each of the freshly transformed plates and inoculate 5 mL of LB containing the appropriate antibiotics in 15 mL tubes. Incubate cultures overnight while shaking at 220 r.p.m. at 37°C.
3. Dispense of 5 mL of PASM-5052 media (as described in Section 2.3) with the appropriate antibiotic(s) and 0.25 m*M* L-rhamnose into a 24-well block plate.
4. Add 100 μL (50 × dilution) of the overnight inoculum into each of the 24 wells.
   - *Note*: 0.25 m*M* of L-rhamnose is the best compromise for screening many targets, but for optimal expression a full L-rhamnose titration from 0 to 1.0 m*M* should be tested for routine larger scale cultures.
5. Seal the plate with an airpore tape sheet (Qiagen) and shake at 220 r.p.m. at 37°C.
6. Upon reaching an OD_{600} of 0.5 (after 2–3 h), reduce the temperature to 25°C and add (final v/v) 0.4 m*M* IPTG.
7. Harvest 1 mL cells after 16–18 h post-IPTG induction by centrifugation at 14,000 × *g* for 10 min.

8 Remove the supernatant and add 1 mL 1 × PBS and centrifuge at 14,000 × *g* for 10 min.
9 Remove the supernatant and resuspend the pellet in 80 μL 1 × PBS buffer.
10 Transfer 100 μL of cell suspension to a 96-well black optical bottom plate and measure GFP fluorescence (RFU) using a microplate spectrofluorometer.
11 RFU values from GFP fluorescence can then be converted to mg/L for membrane protein expression levels as described earlier (Drew et al., 2006)
- *Note*: We have calibrated our microplate spectrofluorometer using varying amounts of purified GFP. With our microplate spectrofluorometer, 12,000 RFU counts correspond to 0.03 mg/mL of purified GFP.

3.2 Whole-Cell FSEC

Once the whole-cell GFP fluorescence is measured, the next step is to evaluate the stability of the expressed fusions by whole-cell FSEC. Since there is no correlation between expression levels and stability of the solubilized fusions, this step, rather than expression, should be used as the primary selection criterion.

12 Recentrifuge the 1 × PBS resuspended cells from Step 11 at 14,000 × *g* for 10 min.
13 Discard the supernatant and resuspend the cell pellet in 1 × PBS, 150 m*M* NaCl, and 1% DDM or 1% LDAO. Make the volume to 1 mL.
14 Incubate the mixture while gently mixing at 4°C for 1 h.
15 Centrifuge the mixture at 14,000 × *g* for 10 min to remove any unsolubilized material.
16 Inject 100 μL of the detergent-solubilized supernatant on a Superose 6 column (10/300, GE-healthcare) preequilibrated with 20 m*M* Tris–HCl pH 7.5, 150 m*M* NaCl, 0.03% (w/v) DDM at a flow rate of 0.5 mL/min.
17 Using an excitation wavelength of 488 nm and an emission wavelength of 512 nm, select the best targets by the quality of the FSEC traces.

3.3 Large-Scale Expression

The targets with the best-quality FSEC traces from Steps 12 to 16 are selected for large-scale expression and membrane isolation.

18 From a freshly transformed plate of the best membrane protein target(s), inoculate 150 mL LB containing appropriate antibiotic. Incubate overnight while shaking at 220 r.p.m. at 37°C.

19 To six sterilized 2.5 L Tunair flasks, add 100 mL of PASM-5052 media, 900 mL autoclaved water, and the appropriate antibiotics.

20 To each flask add 0.25 m*M* of L-rhamnose followed by 20 mL of the overnight culture (50-fold dilution) and shake at 260 r.p.m. at 37°C.

21 Upon reaching an OD_{600} of 0.5 (after 2–3 h), reduce the temperature to 25°C and add 0.4 mM IPTG for 16–18 h growth.

22 Centrifuge 1 mL cell culture at 14,000 × *g* for 10 min and estimate the production levels by GFP fluorescence as described previously in Steps 9–11.

23 Harvest the cells by centrifugation at 5000 × *g* for 15 min. Resuspend the cell pellet in 1 × PBS.

- *Pause point*: Cell suspension can be rapidly frozen in liquid N_2 and stored at −80°C.

24 Completely resuspend cells in 1 × PBS and make up the volume up to 400 mL for good cell breakage efficiency.

25 Add 1 m*M* $MgCl_2$ and a small spatula amount of DNase and mix the cell suspension by stirring at 4°C for 15–20 min.

26 For cell breakage pass the cell suspension twice through a Constant Systems cell disruptor (or equivalent) at 25 kpsi.

27 Take 100 μL of broken cells and measure GFP fluorescence as outlined in Step 11.

28 Centrifuge the broken cell suspension at 22,000 × *g* at 4°C for 15 min to remove the cell debris.

29 Take 100 μL of the supernatant to measure GFP fluorescence as outlined in Step 11. Estimate cell breakage efficiency.

30 Transfer supernatant to six ultracentrifugation tubes and collect membranes by ultracentrifugation of the supernatant at 195,000 × *g* at 4°C for 2 h.

31 Discard the supernatant and resuspend the membrane pellet in ice-cold 1 × PBS using a disposable 10 mL syringe with a 21-gauge needle. Aim to make the final volume of resuspended membranes 80–100 mL. Take out a 1-mL aliquot and freeze separately (see next section). Measure the GFP fluorescence as outlined in Step 11. Estimate the quantity of membrane protein available for purification.

- *Pause point*: Isolated membranes can be rapidly frozen in liquid N_2 and store at −80°C.

3.4 Membrane Solubilized FSEC

Prior to proceeding to purification, it is strongly recommended to repeat the FSEC from the detergent-solubilized membranes. Typically these FSEC traces are more accurate than those from crude whole-cell samples as one can more accurately work with the correct detergent:lipid ratio.

32 Determine the concentration of total protein in the isolated membranes using the BCA (bicinchoninic acid) protein assay kit following manufacturer's instructions.

33 Dilute the total protein concentration to 3.5 mg/mL (final) in 1 mL mixture containing 1 × PBS, 150 m*M* NaCl, and 1% DDM or 1% LDAO in a 1.5-mL tube.

34 Solubilize the protein with gentle mixing at 4°C for 1 h.

35 Take 100 μL of solubilized membrane protein and measure GFP fluorescence as outlined in Step 11.

36 Ultracentrifuge the 1.5-mL tubes at 120,000 × *g* at 4°C for 45 min to remove the unsolubilized material.

37 Take 100 μL of the supernatant to measure GFP fluorescence as outlined in Step 11. Estimate solubilization efficiency.

38 Inject 100 μL of the detergent-solubilized supernatant on a size-exclusion column for FSEC analysis, as outlined in Step 16.

39 Overlay FSEC traces in DDM and LDAO. Very stable membrane proteins have FSEC traces that look almost the same in both these detergents and, if they crystallize, are more likely to yield well-diffracting crystals.

3.5 Purification of Membrane Protein From Isolated Membranes

Membrane protein purification from isolated membranes takes 2 days. We recommend preparing the buffers fresh each time. It is also important to keep all samples on ice and use precooled buffers at 4°C.

3.6 Buffers and Reagents

Day 1:

Wash buffer (1 L)—1 × PBS, 150 m*M* NaCl, 0.1% DDM

Dialysis buffer (3 L)—20 m*M* Tris pH 7.5, 150 m*M* NaCl, 0.03% DDM

Day 2:

Reverse-IMAC and Gel filtration buffer (500 mL)—20 m*M* Tris pH 7.5, 150 m*M* NaCl, 0.03% DDM

40 In a 500-mL beaker, dilute the resuspended isolated membranes in 1 × PBS, 150 m*M* NaCl, and 1% DDM to a final volume of 350 mL.

41 Incubate the mixture while gently stirring at 4°C for 1 h.

42 Transfer supernatant to ultracentrifugation tubes and remove non-solubilized material by ultracentrifugation at 195,000 × *g* at 4°C for 45 min. Measure GFP fluorescence before and after ultracentrifugation to calculate the solubilization efficiency and the amount of membrane protein-GFP fusion. Transfer the supernatant to a fresh 500-mL beaker and add 10 m*M* imidazole (final concentration) to the detergent-solubilized membranes.

43 Add 1 mL of preequilibrated Ni-NTA resin in 1 × PBS per 1 mg of membrane protein-GFP fusion. Incubate the mixture with gentle stirring at 4°C for 2–3 h.

44 Pour the slurry into a glass Econo-Column (Bio-Rad) and collect the flow-through. Measure GFP fluorescence from 100 μL of the flow-through to calculate Ni-NTA binding efficiency.

45 Wash the column with 200 mL of wash buffer containing the following final four final concentrations of imidazole: 10, 20, 35, and 45 m*M*. The imidazole can be added to the wash buffer from a 1 *M* stock pH 7.5.

46 Elute the membrane protein-GFP fusion using wash buffer supplemented with 250 m*M* imidazole.

47 Measure GFP fluorescence with 100 μL of elute and calculate the amount of fusion.

48 Add 1 mg of purified His_6-tagged TEV (or appropriate His_6-tagged protease) per 1 mg of fusion. Transfer to dialysis tubing and dialyze in 3 L of dialysis buffer containing 20 m*M* Tris–HCl pH 7.5, 150 m*M* NaCl, 0.03% DDM overnight at 4°C.

49 Remove the dialyzed sample and filter through a 0.22-μ*M* filter.

50 To confirm GFP-His_6 cleavage efficiency subject sample to FSEC analysis in buffer containing 20 m*M* Tris–HCl pH 7.5, 150 m*M* NaCl, 0.03% DDM. This step is optional; however, it is good practice when either purifying the protein for the first time.

51 Reverse-IMAC is the second purification step to remove the His-tagged TEV protease, cleaved GFP-His_8 tag and coeluting contaminating proteins. Equilibrate a 5-mL His-Trap column in 20 m*M* Tris–HCl pH 7.5, 150 m*M* NaCl, 0.03% DDM using a peristaltic pump.

52 Pass the dialyzed sample through the equilibrated His-Trap at a flow rate of 2 mL/min collecting the flow-through into a 50-mL falcon tube.

- *Note*: The bound material can later be eluted off using the same buffer containing 250 m*M* imidazole.

53 Concentrate the cleaved membrane protein flow-through using a 50 kDa centrifugal concentrator (Vivaspin 20). Centrifuge at 3000 × *g* for 10 min and remix the solution above the filter by gentle pipetting up and down in order to keep the protein in solution and avoid protein aggregation at the bottom. Repeat this step until the final volume of concentrated membrane protein is 400–500 μL.

54 Inject the concentrated protein onto a Superdex 200 10/300 column preequilibrated with 20 m*M* Tris pH 7.5, 150 m*M* NaCl, 0.03% DDM at 4°C. Run it at a flow rate of 0.5 mL/min. Collect 0.2 mL fractions. Analyze the UV trace to determine the fractions containing cleaved membrane protein.

55 Collect and pool the membrane protein peak using 100-kDa centrifugal concentrator (Vivaspin 2) twice at 3000 × *g* for 10 min with gentle pipetting of the concentrated protein to avoid protein aggregation at the bottom.

56 Analyze the concentrated protein by SDS-PAGE.

57 Determine the protein concentration using Nanodrop or BCA protein assay kit.

58 The optimum concentration for setting up crystallization trials is 10 mg/mL.

4. SUMMARY AND CONCLUSION

MemStar offers a one-step approach that may significantly increase the amount of purified membrane protein that can be otherwise isolated from existing *E. coli*-based membrane protein production platforms with equivalent detergent usage. MemStar does not only help reduce both screening time and costs for purification because of the higher yields, but it also minimizes excessive detergent concentration, which is a common bottleneck known to hinder crystallization efforts. With protein purified from all of the 10 model targets, we have since solved the dimeric structure of the sodium/proton antiporter NhaA from *E. coli* (Lee et al., 2014b), a new ligand-bound conformational state of the hydantoin transporter Mhp1 from *Microbacterium liquefaciens* (Simmons et al., 2014), and reproduced the structure of the anion channel TehA (Chen et al., 2010) from *Haemophilus influenzae* with data to 1.5 Å resolution. Furthermore, using MemStar we were able to improve the production of the sodium/proton antiporter NapA from *Thermus thermophilus*, where levels from 2.5 to 17 mg/L (Lee et al., 2014a). The increased production yields significantly facilitated purification and crystallization, especially in small-sized detergents, and have been able to obtain several structures of NapA (Coincon et al., 2016; Lee et al., 2013).

To our knowledge, MemStar provides the highest production levels in a simple one-shot expression approach. MemStar is a broadly applicable approach facilitating structural and functional studies of bacterial membrane proteins, which have proven to be ideal model systems to understand general mechanistic principles of mammalian counterparts (Wang et al., 2013). Lastly, the ability of growing high protein levels should facilitate X-feL experiments that require much more protein than standard crystallography experiments (Hunter & Fromme, 2011).

ACKNOWLEDGMENTS

This work was funded by grants from the Swedish Research Council (D.D.) and The Knut and Alice Wallenberg Foundation (D.D). D.D. acknowledges the support from EMBO through the Young Investigator Program (YIP).

REFERENCES

Bill, R. M., et al. (2011). Overcoming barriers to membrane protein structure determination. *Nature Biotechnology*, *29*(4), 335–340.

Chen, Y. H., et al. (2010). Homologue structure of the SLAC1 anion channel for closing stomata in leaves. *Nature*, *467*(7319), 1074–1080.

Coincon, M., et al. (2016). Crystal structures reveal the molecular basis of ion translocation in sodium/proton antiporters. *Nature Structural & Molecular Biology*, *23*(3), 248–255.

Drew, D., Lerch, M., Kunji, E., Slotboom, D. J., & de Gier, J. W. (2006). Optimization of membrane protein overexpression and purification using GFP fusions. *Nature Methods*, *3*(4), 303–313.

Drew, D. E., von Heijne, G., Nordlund, P., & de Gier, J. W. (2001). Green fluorescent protein as an indicator to monitor membrane protein overexpression in *Escherichia coli*. *FEBS Letters*, *507*(2), 220–224.

Drew, D., et al. (2002). Rapid topology mapping of *Escherichia coli* inner-membrane proteins by prediction and PhoA/GFP fusion analysis. *Proceedings of the National Academy of Sciences of the United States of America*, *99*(5), 2690–2695.

Drew, D., et al. (2008). GFP-based optimization scheme for the overexpression and purification of eukaryotic membrane proteins in *Saccharomyces cerevisiae*. *Nature Protocols*, *3*(5), 784–798.

Hunter, M. S., & Fromme, P. (2011). Toward structure determination using membrane-protein nanocrystals and microcrystals. *Methods*, *55*(4), 387–404.

Kang, H. J., Lee, C., & Drew, D. (2013). Breaking the barriers in membrane protein crystallography. *The International Journal of Biochemistry & Cell Biology*, *45*(3), 636–644.

Kawate, T., & Gouaux, E. (2006). Fluorescence-detection size-exclusion chromatography for precrystallization screening of integral membrane proteins. *Structure*, *14*(4), 673–681.

Lee, C., et al. (2013). A two-domain elevator mechanism for sodium/proton antiport. *Nature*, *501*, 573–577.

Lee, C., et al. (2014a). MemStar: A one-shot *Escherichia coli*-based approach for high-level bacterial membrane protein production. *FEBS Letters*, *588*(20), 3761–3769.

Lee, C., et al. (2014b). Crystal structure of the sodium-proton antiporter NhaA dimer and new mechanistic insights. *The Journal of General Physiology*, *144*(6), 529–544.

Miroux, B., & Walker, J. E. (1996). Over-production of proteins in *Escherichia coli*: Mutant hosts that allow synthesis of some membrane proteins and globular proteins at high levels. *Journal of Molecular Biology*, *260*(3), 289–298.

Schlegel, S., Hjelm, A., Baumgarten, T., Vikstrom, D., & de Gier, J. W. (2014). Bacterial-based membrane protein production. *Biochimica et Biophysica Acta*, *1843*, 1739–1749.

Simmons, K. J., et al. (2014). Molecular mechanism of ligand recognition by membrane transport protein, Mhp1. *The EMBO Journal*, *33*(16), 1831–1844.

Sonoda, Y., et al. (2011). Benchmarking membrane protein detergent stability for improving throughput of high-resolution X-ray structures. *Structure*, *19*(1), 17–25.

Studier, F. W. (2005). Protein production by auto-induction in high density shaking cultures. *Protein Expression and Purification*, *41*(1), 207–234.

Studier, F. W., & Moffatt, B. A. (1986). Use of bacteriophage T7 RNA polymerase to direct selective high-level expression of cloned genes. *Journal of Molecular Biology*, *189*(1), 113–130.

Tate, C. G. (2012). A crystal clear solution for determining G-protein-coupled receptor structures. *Trends in Biochemical Sciences*, *37*(9), 343–352.

Wagner, S., et al. (2008). Tuning *Escherichia coli* for membrane protein overexpression. *Proceedings of the National Academy of Sciences of the United States of America*, *105*(38), 14371–14376.

Wang, H., et al. (2013). Structural basis for action by diverse antidepressants on biogenic amine transporters. *Nature*, *503*(7474), 141–145.

CHAPTER SIX

Efficient Screening and Optimization of Membrane Protein Production in *Escherichia coli*

Jacopo Marino*, Katharina Holzhüter†,1, Benedikt Kuhn†,1, Eric R. Geertsma†,2

*University of Zurich, Zurich, Switzerland
†Goethe University Frankfurt, Frankfurt/Main, Germany
[2]Corresponding author: e-mail address: geertsma@em.uni-frankfurt.de

Contents

Abstract

Escherichia coli is one of the most widely used expression hosts for membrane proteins. However, establishing conditions for its recombinant production of membrane proteins remains difficult. Attempts to produce membrane proteins frequently result in either no expression or expression as misfolded aggregates. We developed an efficient pipeline for improving membrane protein overexpression in *E. coli* that is based on two approaches. The first involves transcriptional fusions, small additional RNA sequences upstream of the target open reading frame, to overcome no or poor overall expression

[1] Equal contribution.

Methods in Enzymology, Volume 594
ISSN 0076-6879
http://dx.doi.org/10.1016/bs.mie.2017.05.011

levels. The other is based on a tunable promoter in combination with a fusion to green fluorescent protein serving as a reporter for the folding state of the target membrane protein. The latter combination allows adjusting the membrane protein expression rate to the downstream folding capacity, in order to decrease the formation of protein aggregates. This pipeline has proven successful for the efficient and parallel optimization of a diverse set of membrane proteins.

ABBREVIATIONS

FP fusion partner
GFP green fluorescent protein
HRV human rhinovirus
ORF open reading frame
SD sequence Shine-Dalgarno sequence

1. INTRODUCTION

A basic and immediate challenge for membrane protein researchers lies in the production of high-quality protein in significant quantities. Membrane proteins are notorious for their recalcitrancy toward overexpression. In *Escherichia coli*, the most popular prokaryotic host for heterologous protein production (Hattab, Warschawski, Moncoq, & Miroux, 2015; Raman, Cherezov, & Caffrey, 2006), expression trials can have a variety of undesired outcomes: from no or poor expression of functionally folded protein to high expression levels of aggregated material in inclusion bodies. Under which conditions the recombinant production of a particular membrane protein will be successful remains, except for a few general guidelines (Grisshammer, 2006; Grisshammer & Tate, 1995), difficult to predict. In the past years, several strategies have been developed to increase the success rates for functional overexpression of membrane proteins in sufficient amounts. Herein, two of those are described in more detail—transcriptional fusions to enhance the production levels of otherwise poorly expressed membrane proteins, and the combination of a tunable promoter and a C-terminal fusion of green fluorescent protein (GFP) to prevent formation of misfolded protein.

The biogenesis of membrane proteins is comparably complex and involves the targeting of nascent polypeptide chains to the membrane, followed by their cotranslational insertion into and assembly in the lipid bilayer (Elvekrog & Walter, 2015; Luirink, Yu, Wagner, & de Gier, 2012). Each of these steps requires unique components. For several membrane proteins, correct folding is achieved at intermediate expression levels, while high expression levels lead to formation of aggregated material, suggesting that the required cellular pathways can be saturated (Geertsma, Groeneveld, Slotboom, & Poolman, 2008; Wagner et al., 2008). This makes an efficient system necessary for tuning the expression rate and monitoring the folding state to establish optimal expression conditions. Other factors potentially involved in membrane protein misfolding, such as nonoptimal translation, elongation speed, and cotranslational folding processes (Buhr et al., 2016; Spencer, Siller, Anderson, & Barral, 2012; Yu et al., 2015), are currently less understood and therefore more difficult to correct for.

For improving low overall production levels of membrane proteins, the translational fusion of a soluble domain to the N-terminus of the protein of interest has been one of the most successful strategies (Esposito & Chatterjee, 2006; Waugh, 2005). Recent complete codon alteration studies of open reading frames (ORFs) have now revealed that the basis of this effect lies in less stabile mRNA secondary structure elements in the vicinity of the start codon, resulting in an increased efficiency of translation initiation (Boel et al., 2016; Gu, Zhou, & Wilke, 2010; Kozak, 2005; Kudla, Murray, Tollervey, & Plotkin, 2009; Norholm et al., 2013). This recognition provides means to improve low production levels without N-terminal fusion proteins.

Here, we detail an approach to overcome poor expression levels of membrane proteins in *E. coli* using transcriptional fusions affecting such mRNA secondary structures. This method exploits the presence of an additional RNA sequence upstream of the mRNA sequence of the target membrane protein, resulting in a bicistronic mRNA (Fig. 1). As advantage to translational fusion strategies, the resulting bicistronic mRNA does neither extend the membrane protein ORF nor require enzymatic removal of the fusion protein. We subsequently discuss an efficient cloning and GFP-based expression screening approach that allows one to rapidly identify the most optimal constructs and cultivation conditions. All strategies in combination have proven successful for increasing the functional overexpression of membrane proteins of different origin and topology in *E. coli* (Marino, Hohl, Seeger, Zerbe, & Geertsma, 2015).

1.1 Transcriptional Fusions to Enhance Membrane Protein Expression Levels

Protein fusions to the N-terminus of a target protein often result in improved expression levels (Esposito & Chatterjee, 2006; Waugh, 2005).

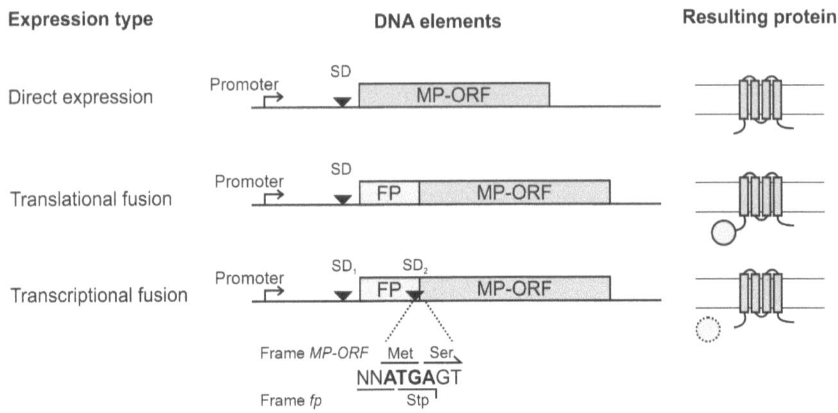

Fig. 1 Schematic representation of common approaches for expressing membrane proteins in *E. coli*. *Direct expression* refers to a situation where the promoter and a Shine-Dalgarno (SD) sequence precede the open reading frame of the membrane protein (MP-ORF). *Translational fusion* refers to classical N-terminal fusion partners (FPs) that are part of the same protein chain as the membrane protein. Following protein production, these fusion proteins can be separated from the membrane protein provided a proteolytic cleavage site is available. *Transcriptional fusions* do not lead to a physical linkage of the fusion protein and the membrane protein, avoiding the need for proteolytic cleavage. A SD sequence placed within the sequence coding for the fusion partner precedes the ORF of the membrane protein. The ORFs of the FP and membrane protein overlap on their stop and start codon, respectively. *Modified with permission from Marino, J., Hohl, M., Seeger, M. A., Zerbe, O., & Geertsma, E. R. (2015). Bicistronic mRNAs to enhance membrane protein overexpression.* Journal of Molecular Biology, 427*(4), 943–954. doi:10.1016/j.jmb.2014.11.002.*

Such fusion proteins represent independent folding domains that are mostly hydrophilic and vary in size from 10 to 60 kDa (Loughran & Walls, 2017; Young, Britton, & Robinson, 2012). Commonly used fusion proteins are MBP (maltose-binding protein from *E. coli*) (Kapust & Waugh, 1999), Trx (*E. coli* thioredoxin) (LaVallie, Lu, Diblasio-Smith, Collins-Racie, & McCoy, 2000), SUMO (small ubiquitin-like modifier from *Saccharomyces cerevisiae*) (Peroutka III, Orcutt, Strickler, & Butt, 2011), Mistic (membrane-integrating sequence for translation of inner membrane proteins from *Bacillus subtilis*) (Roosild et al., 2005), and YbeL, a small hydrophilic protein domain from *E. coli* (Leviatan, Sawada, Moriyama, & Nelson, 2010). Fusions to the N-terminus of target proteins are generally used because of their favorable properties in enhancing expression levels (Young et al., 2012), a feature attributed to improved translation initiation efficiency due to more favorable mRNA structures (Haberstock et al., 2012; Kudla et al., 2009). After serving its beneficial role, removal of the fusion protein by proteolytic cleavage is often desired as fusion proteins can interfere with the activity of the target protein (Majtan & Kraus, 2012; Sabaty et al., 2013; Singh et al., 2013) or its crystallization (Smyth, Mrozkiewicz, McGrath, Listwan, & Kobe, 2003). However, efficient and complete cleavage cannot always be achieved (Vergis & Wiener, 2011).

We explored transcriptional fusions as an alternative to N-terminal fusion proteins (Marino et al., 2015). Transcriptional fusions are similar to bicistronic mRNAs and involve an RNA sequence preceding the ORF of the target protein (Fig. 1). Such transcriptional fusions are very common in prokaryotes and result in a transcript containing multiple ORFs that are separately translated (Blattner et al., 1997). We have observed that transcriptional fusions of either *mstX* or *ybeL*, coding for Mistic (Roosild et al., 2005) and YbeL (Leviatan et al., 2010), respectively, led to improved expression levels of functional membrane proteins not achieved by standard methods (Fig. 2). For well-expressed membrane proteins, transcriptional fusions

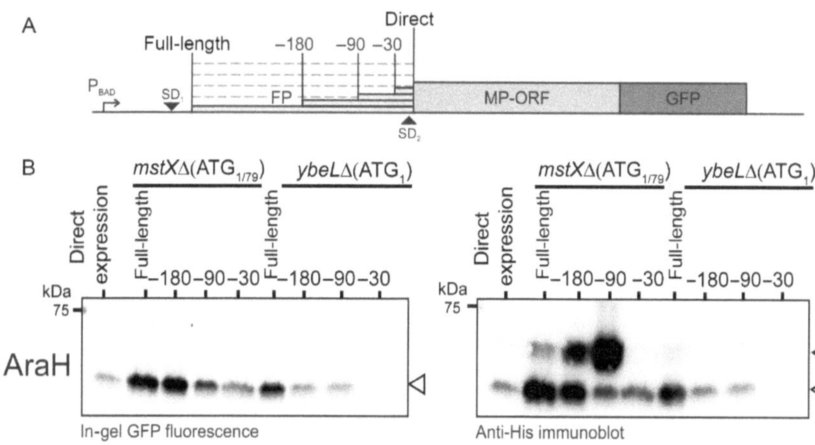

Fig. 2 Analysis of the production level of the membrane protein AraH expressed as transcriptional fusion to *mstX*- or *ybeL* derivatives. (A) Schematic representation of the nucleotide starting positions of *mstX* and *ybeL* 5′-truncations. Sequences coding for transcriptional fusion partners, membrane proteins, and His-tagged GFP are indicated by FP, MP-ORF, and GFP, respectively. (B) AraH with a C-terminal GFP as folding reporter was expressed directly or as transcriptional fusions to *mstX* or *ybeL* derivatives devoid of start codons. On the *left panel*, the protein is detected by in-gel GFP fluorescence; on the *right panel* an anti-His immunoblot of the same gel is shown. The lower fluorescent band (indicative of well-folded protein) and the upper nonfluorescent band (indicative of misfolded material) are indicated by *white* and *black arrows*, respectively. *Modified with permission from Marino, J., Hohl, M., Seeger, M. A., Zerbe, O., & Geertsma, E. R. (2015). Bicistronic mRNAs to enhance membrane protein overexpression.* Journal of Molecular Biology, 427*(4), 943–954. doi:10.1016/j.jmb.2014.11.002.*

resulted in only modest increases in the amount of functionally folded protein, but did lead to a decrease of the fraction of aggregated material.

The beneficial effects of transcriptional fusions do not rely on translation of the preceding ORFs, as removal of the respective start codons in *mstX* and *ybeL* had no effect. A strong ribosomal-binding site in the sequence of the transcriptional fusion near the start codon of the target gene is also not required. However, its presence is recommended as this increases the efficiency of translation initiation and allows initiation to result from both reinitiation by ribosomes that remain bound after translating the preceding ORF and de novo initiation by ribosomes recruited from the cytoplasmic pool (Andre et al., 2000; Das & Yanofsky, 1984; Kozak, 2005). Finally, truncating the ORFs of the fusion proteins from their 5′-side indicated that the enhanced expression levels could not be explained by a more favorable secondary structure involving exclusively the nucleotides nearby the SD sequence preceding the ORF of the target protein (Fig. 2). Rather this suggests an important role of long-range interactions in the mRNA in establishing conditions for improved protein expression.

1.2 C-Terminal GFP Fusions to Report on Membrane Protein Quantity and Quality

During expression screening of membrane proteins, it is important to monitor both expression level and folding state of the target protein. Membrane proteins are easily produced as misfolded aggregates in the cytoplasm. Especially for α-helical membrane proteins, the isolation of functional protein from such inclusion bodies using a refolding procedure is challenging (Harris & Booth, 2012).

Various methods allow evaluation of the folding state of a membrane protein, e.g., the determination of the fraction of membrane-inserted protein or activity assays. However, both quality and quantity of the produced membrane protein are most rapidly and efficiently assessed by using a C-terminal GFP fusion as a reporter (Drew, von Heijne, Nordlund, & de Gier, 2001; Geertsma et al., 2008; Waldo, Standish, Berendzen, & Terwilliger, 1999). The productive folding of GFP, required for the formation of its fluorophore, critically depends on the correct folding of its preceding fusion protein. Misfolding and aggregation of the upstream membrane protein interfere with GFP folding and result in a nonfluorescent membrane protein GFP fusion. A prerequisite for this application of GFP is a cytoplasmic location of the C-terminus of the target protein, but this condition is met for the majority of inner membrane proteins (Daley et al., 2005).

While the fluorescence of GFP fusion proteins is a quantitative indicator for well-folded fusion protein produced, the fraction of nonfluorescent GFP fusion reports on the amount of misfolded material. The two species can be visualized based on their differential migration during SDS-PAGE (Geertsma et al., 2008). The GFP domain of well-folded GFP fusion proteins remains folded and fluorescent during electrophoresis (Drew, Lerch, Kunji, Slotboom, & de Gier, 2006) and migrates at an apparent molecular weight of 10–15 kDa lower than nonfluorescent GFP fusion proteins. Subsequent immunodetection using a generic antibody allows simultaneous determination of both species (Fig. 2). The ratio of the intensities of the upper nonfluorescent band to the lower fluorescent band corresponds to the ratio of aggregated to folded GFP fusion proteins produced in the cell. Combining this GFP-based folding reporter system with a tunable promoter, such as the *araBAD* promoter (Guzman, Belin, Carson, & Beckwith, 1995), allows the adjustment of the protein expression rate to the processing capacity to reduce or prevent protein misfolding.

1.3 Efficient Expression Screening Using Fragment Exchange Cloning

Efficient and broad screening of constructs depends on efficient and reliable methods for creating expression vectors. Popular methods relying on recombination such as Gateway (Hartley, Temple, & Brasch, 2000), or on annealing of complementary single-stranded overhangs such as ligation-independent cloning (Aslanidis & de Jong, 1990), have unique favorable properties, for example the ability to do subcloning or being economical in use. However, these specific positive characteristics were thus far not combined in one technique but instead accompanied by less desirable features such as the extension of the ORF with long cloning sequences (Geertsma, 2013). Fragment exchange (FX) cloning was developed as an alternative to these approaches, unifying all their attractive features in one single cloning method (Geertsma & Dutzler, 2011). FX cloning allows subcloning of sequence-verified ORFs, extends ORFs minimally, requires little handling, and is robust and economic in its use.

FX cloning is based on the type IIS restriction enzyme *Sap*I that cleaves DNA at a defined position outside of its nonpalindromic recognition sequence (Szybalski, Kim, Hasan, & Podhajska, 1991). The overall cloning strategy is outlined in Fig. 3. The target ORF is amplified by PCR using a primer pair containing *Sap*I sites. These restriction sites are oriented in such a way that they are physically separated from the ORF upon cleavage (Fig. 3C). Furthermore, the resulting single-stranded overhangs differ sufficiently in their sequence to allow directional cloning. The cleaved ORF can be cloned into an intermediate sequencing vector (Fig. 3A) or immediately into expression vectors. Both vector classes contain the same pair of *Sap*I sites flanking the strong *ccdB* counterselection marker (Bernard, Gabant, Bahassi, & Couturier, 1994), but in the sequencing vector pINITIAL the sites have the same orientation as in the PCR product and the *Sap*I recognition sequence remains thus associated with the plasmid after cleavage. In the expression vector the direction is reversed and the recognition sequences are lost after digestion (Fig. 3D). Following cleavage, both classes of vectors contain complementary overhangs that hybridize with the insert and are finally joined by ligation. Subsequent transformation of the ligation mix into a non-*ccdB*-resistant *E. coli* strain leads to exclusive selection of daughter plasmids containing the target ORF. The strong *ccdB* counterselection marker eliminates the need for intermediate purifications and allows the entire reaction to be performed in one cup.

2. PREPARATION OF EXPRESSION VECTORS AND CLONING

2.1 Construction of Expression Vectors for Transcriptional Fusions

Our approach combines transcriptional fusions with efficient screening for membrane protein overexpression using the AraC/P_{BAD} expression system (Guzman et al., 1995), GFP as folding indicator (Drew et al., 2001; Geertsma et al., 2008) and the FX cloning method (Geertsma & Dutzler, 2011). ORFs of tentative transcriptional fusions are first cloned at the 5′-side of the FX cloning cassette of the dedicated vector pBXXhoIC3GH (Fig. 4). We obtained good results using ORFs with an internal GA-rich sequence at their 3′-side, such as *mstX* and *ybeL* (Marino et al., 2015). To determine the presence of such a sequence and predict its efficiency, we have used RBSDesigner (Na & Lee, 2010). RBSDesigner takes into account the folding of the mRNA near the SD sequence, the efficiency of binding of the ribosomal 16S rRNA to the SD sequence and the translation efficiency determined by the length of the spacer

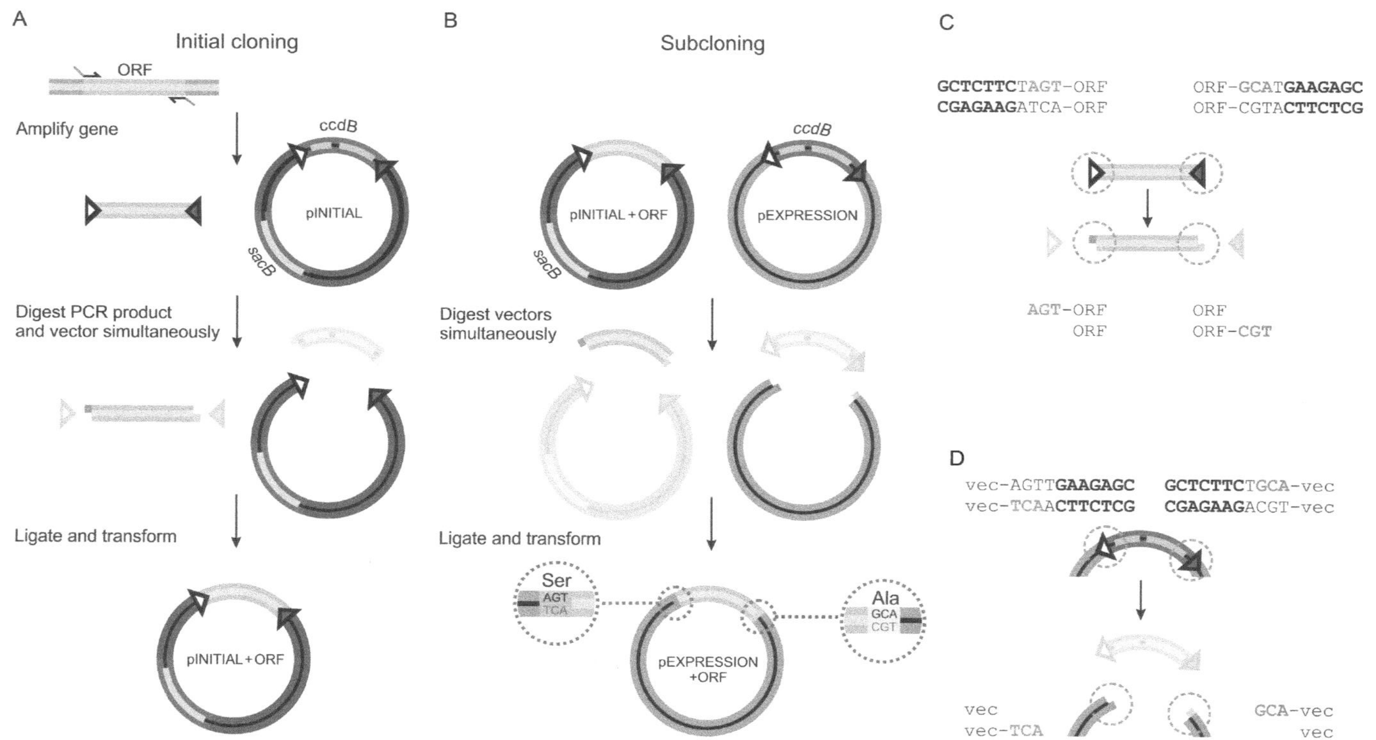

Fig. 3 Schematic overview of the FX cloning method. (A) Cloning of a PCR product into pINITIAL. The direction of the *Sap*I sites is represented by arrows colored corresponding to their respective overhangs generated after cleavage. (B) Subcloning of a sequence-verified ORF into an expression vector. The three nucleotide overhangs as a result from the cloning procedure are shown as *insets*. (C) Orientation of the *Sap*I cleavage sites in the PCR product and pINITIAL and (D) in expression vectors. *Modified with permission from Geertsma, E. R., & Dutzler, R. (2011). A versatile and efficient high-throughput cloning tool for structural biology.* Biochemistry, *50(15), 3272–3278. doi:10.1021/bi200178z.*

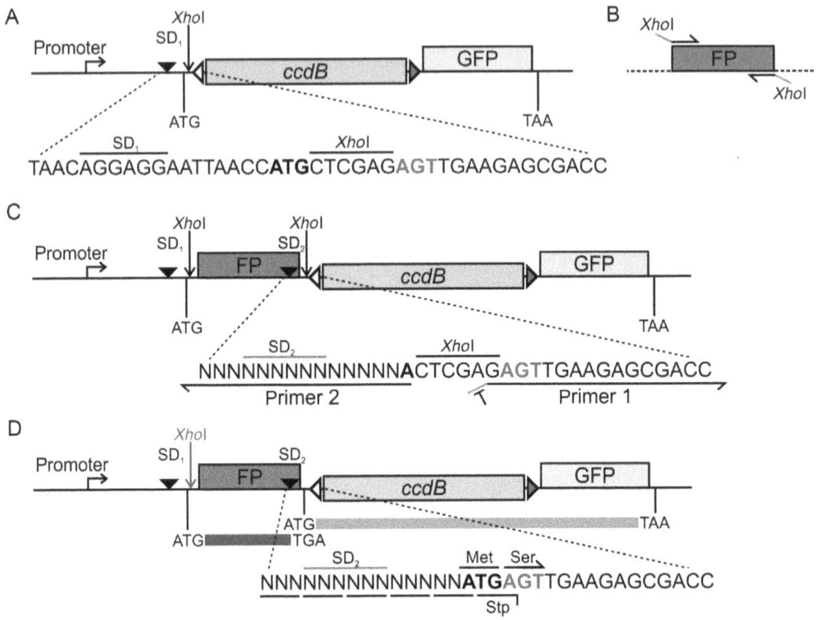

Fig. 4 Schematic overview of the cloning strategy to construct expression vectors for transcriptional fusions. (A) A unique *Xho*I site present in the pBXXhoIC3GH plasmid is used to insert the sequence of the fusion partner (FP) between the start codon and the FX cloning cassette. The FX cloning cassette and the sequence coding for a HRV 3C protease site, GFP, and His-tag are indicated by *ccdB* and GFP, respectively. (B) The FP is amplified with primers containing *Xho*I sites. (C) The PCR product is cloned into the *Xho*I site of the pBXXhoIC3GH vector. After verifying that the FP-insert is in the correct orientation, an overlapping stop/start codon preceding the FX cloning cassette is introduced by PCR. (D) The FX cloning cassette in the final vector can subsequently be replaced with the ORF of a membrane protein. This allows the expression of the target proteins as transcriptional fusion to the fusion partner and with a C-terminal HRV 3C protease site, GFP, and His-tag.

between the SD sequence and the start codon. For both *mstX* and *ybeL*, efficient translation efficiencies were predicted.

Following the insertion of the ORF for the transcriptional fusion, a further whole-plasmid amplification is performed to generate overlapping stop/start codons (ATGA; Fig. 4D). The penultimate codon of the ORF for the transcriptional fusion must end with an adenine nucleobase (A) to achieve the ATGA sequence.

2.1.1 Equipment

- PCR machine
- DNA agarose gel-electrophoresis setup

- Nanodrop
- Incubator

2.1.2 Buffers and Reagents

- Primers containing *Xho*I-sites for amplification of the transcriptional fusion ORF
- Primers for whole-plasmid amplification to generate transcriptional fusions are 5′-phosphorylated. Forward primer: 5′-tgagttgaagagcgacctgcagactggctg tg; Reverse primer: complement of FP-ORF, final 5′-nucleotide must be "T" (Fig. 4C)
- Template DNA for the fusion gene of interest (e.g., genomic DNA or plasmid)
- 10 m*M* dNTPs
- Phusion DNA polymerase and corresponding buffer
- TAE buffer and TAE agarose gel
- Agarose gel extraction kit
- pBXXhoIC3GH vector
- *Xho*I restriction enzyme and corresponding buffer
- Calf intestinal alkaline phosphatase (CIP)
- T4 DNA ligase
- *ccdB*-resistant *E. coli* strain (e.g., DB3.1; Hartley et al., 2000; see Note 1 in Section 2.1.4)
- Miniprep kit
- LB medium and LB-agar plate supplemented with the appropriate antibiotic
- Sequencing primers

2.1.3 Procedure

1. Design primers for amplification of the N-terminal fusion of interest.
2. PCR amplifying the gene of interest. Prepare a 50-μL PCR reaction and add the DNA polymerase immediately prior to starting the reaction. Use a touch-down program (see Note 2 in Section 2.1.4), e.g., (1) 30 s at 98°C; (2) 10 s at 98°C; (3) 15 s at 61°C (annealing temperature decreased by 0.5°C per cycle); (4) 15–30 s/kb at 72°C; repeat (2)–(4) 14 times; (5) 10 s at 98°C; (6) 15 s at 53°C; (7) 15–30 s/kb at 72°C; repeat (5)–(7) 14 times; (8) 120 s at 72°C; (9) unlimited at 10°C.
3. Analyze the product on a preparative TAE agarose gel. Purify the relevant band of the correct size using a DNA gel extraction kit and quantify the DNA spectrophotometrically.

4. Digest pBXXhoIC3GH with *Xho*I while simultaneously dephosphorylating with CIP. Digest the PCR product with *Xho*I in a separate tube.
5. Purify the linearized vector from a preparative TAE agarose gel. Purify the insert over a column using a PCR product purification kit. Quantify the DNA spectrophotometrically.
6. Mix the vector and insert in a molar ratio of 1:3. Supplement with T4 DNA ligation buffer and T4 DNA ligase. Incubate for 1 h at room temperature and subsequently heat inactivate the T4 DNA ligase for 20 min at 65°C. Transform the ligation mix to chemically competent *E. coli* DB3.1 cells and plate the cells (see Note 1 in Section 2.1.4).
7. Isolate the plasmid and determine the orientation of the insert by restriction analysis and sequencing. Continue with a plasmid holding the transcriptional fusion ORF in frame with the start codon on the vector.
8. PCR to amplify the whole plasmid and obtain overlapping start/stop codons using phosphorylated primers; use a touch-down program, e.g., (1) 60 s at 98°C; (2) 15 s at 98°C; (3) 15 s at 55°C (annealing temperature decreased by 0.5°C per cycle); (4) 30 s/kb at 72°C; repeat (2)–(4) 20 times; (5) 15 s at 98°C; (6) 15 s at 51°C; (7) 30 s/kb at 72°C; repeat (5)–(7) 15 times; (8) 180 s at 72°C; (9) unlimited at 10°C.
9. Remove the template by *Dpn*I treatment and self-ligate the vector.
10. Transform chemically competent *E. coli* DB3.1 cells and plate the cells.
11. Isolate the plasmid and verify it by restriction analysis and sequencing.

2.1.4 Notes

1. Few *E. coli* strains are resistant to CcdB, which is poison for the *E. coli* topoisomerase II. *E. coli* DB3.1 carries a mutation in GyrA protein that renders it resistant (Bernard & Couturier, 1992). CcdB is a very efficient counterselection marker. Accidental use of CcdB-sensitive strains will result in the absence of any transformants.
2. Touch-down PCR enhances the production of the desired product compared to products resulting from unspecific priming (Don, Cox, Wainwright, Baker, & Mattick, 1991). Additionally, the varying annealing temperature makes the PCR program compatible with parallelization of different reactions without specific fine-tuning.

The resulting validated vector for expression of membrane proteins using transcriptional fusions can be used in the following step for (high-throughput) cloning of membrane protein ORFs with the FX cloning method.

2.2 FX Cloning and Subcloning

FX cloning allows the quick exchange of a cloning cassette holding the *ccdB* counterselection marker with the ORF of a target protein. Primer design for FX cloning and sequence analysis of target ORFs can be done automated online (Geertsma, 2014). Following PCR amplification and purification, a "one pot" procedure transfers the target ORFs to an expression vector or to an intermediate vector (pINITIAL) that allows subcloning of sequence-verified ORFs. The latter is advantageous if many different expression constructs are to be prepared. Subsequent subcloning from pINITIAL to expression vectors follows a similar procedure as the initial cloning step of the PCR product (Fig. 3).

2.2.1 Equipment

- PCR machine
- DNA agarose gel-electrophoresis setup
- Nanodrop
- Incubator

2.2.2 Buffers and Reagents

- FX cloning compatible forward and reverse primers targeting the gene of interest
- Template DNA for the gene of interest (e.g., genomic DNA or plasmid)
- 10 m*M* dNTPs
- Phusion DNA polymerase and corresponding buffer
- TAE buffer and TAE agarose gel
- Agarose gel extraction kit
- pINITIAL plasmid (Addgene plasmid 46858; see Note 1 in Section 2.2.4)
- FX cloning expression vectors for transcriptional fusions. Additional vectors are available in Addgene kit 1000000039
- *Sap*I restriction enzyme and corresponding buffer
- T4 DNA ligase
- 10 m*M* ATP: 10 m*M* Na_2-ATP, 10 m*M* $MgSO_4$. Dissolve in 50 m*M* KPi, pH 7.0 and adjust to pH 6.5–7.0 with NaOH (see Note 2 in Section 2.2.4). Store in small aliquots at −20°C.
- *CcdB*-sensitive *E. coli* strain (e.g., MC1061; Casadaban & Cohen, 1980; see Note 3 in Section 2.2.4)

- Miniprep kit
- LB medium and LB-agar plate supplemented with the appropriate antibiotic
- Sequencing primers

2.2.3 Procedure

1. Design FX cloning compatible primers targeting your gene of interest. This is most conveniently done online at the http://www.fxcloning.org website. Order the optimized primer set devoid of stable hairpin structures. Use the smallest synthesis scale and mere desalting as purification.
2. PCR to amplify the gene of interest. Prepare a 50-μL PCR reaction and add the DNA polymerase immediately prior to starting the reaction. Use a touch-down program (see Note 4 in Section 2.1.4), e.g., (1) 30 s at 98°C; (2) 10 s at 98°C; (3) 15 s at 61°C (annealing temperature decreased by 0.5°C per cycle); (4) 15–30 s/kb at 72°C; repeat (2)–(4) 14 times; (5) 10 s at 98°C; (6) 15 s at 53°C; (7) 15–30 s/kb at 72°C; repeat (5)–(7) 14 times; (8) 120 s at 72°C; (9) unlimited at 10°C.
3. Analyze the product on a preparative TAE agarose gel. Purify the relevant band of the correct size using a DNA gel extraction kit and quantify the DNA spectrophotometrically.
4. Mix 50 ng of pINITIAL with the extracted insert to a molar ratio of vector:insert of 1:5. Add 1 μL 10 × *Sap*I buffer and adjust the volume to 9 μL with ultrapure water. Add 1 μL *Sap*I (2 U) and incubate for 1 h at 37°C.
5. Heat inactivate the *Sap*I for 20 min at 65°C. Let the sample cool down and add 1.25 μL 10 m*M* ATP and 1.25 μL T4 DNA ligase (1.25 U). Incubate for 1 h at room temperature.
6. Heat inactivate the T4 DNA ligase for 20 min at 65°C and transform 5 μL of the ligation mix to 100 μL chemically competent *E. coli* MC1061 cells (CcdB-sensitive strain).
7. Plate 1% and 10% aliquots on LB-agar plates containing the appropriate antibiotic. Alternatively, 5 μL aliquots of undiluted, 10-, 100-, and 1000-fold diluted cells can be spotted to accommodate several transformation reactions on one plate. Incubate the plates overnight at 37°C.
8. Isolate the plasmid using a miniprep kit, determine the concentration spectrophotometrically, and verify the insert by DNA sequencing.

Instead of subcloning a sequence-verified ORF from pINITIAL to an expression vector as detailed below, PCR products can also be cloned immediately into an expression vector following the procedure described earlier. This requires sequence verification of each expression vector. If multiple expression vectors are planned, subcloning prevents excessive sequencing.

1. Mix 50 ng of the FX cloning expression vector with pINITIAL carrying the insert to a molar ratio of vector:pINITIAL of 1:5. Add 1 μL 10 × *Sap*I buffer and adjust the volume to 9 μL with ultrapure water. Add 1 μL *Sap*I (2 U) and incubate for 1 h at 37°C.
2. Heat inactivate the *Sap*I for 20 min at 65°C. Let the sample cool down and add 1.25 μL 10 m*M* ATP and 1.25 μL T4 DNA ligase (1.25 U). Incubate for 1 h at room temperature.
3. Heat inactivate the T4 DNA ligase for 20 min at 65°C and transform 5 μL of the ligation mix to 100 μL chemically competent *E. coli* MC1061 cells.
4. Plate 1% and 10% aliquots on LB-agar plates containing the appropriate antibiotic (see Note 4 in Section 2.2.4). Incubate the plates over night at 37°C.
5. Use a single colony to inoculate 5 mL LB containing the appropriate antibiotic and cultivate overnight at 37°C.
6. Archive the culture as a glycerol stock at −80°C. This stock can serve for inoculation of expression cultures based on the *araBAD* promoter. For other expression systems that require fresh transformations for expression cultures, e.g., those based on the T7 promoter, the stock serves as a plasmid source. No additional verification by DNA sequencing is required.

2.2.4 Notes

1. All original FX cloning plasmids still containing the *ccdB* counter-selection marker need to be maintained in a CcdB-insensitive strain such as *E. coli* DB3.1 (Hartley et al., 2000).
2. The pH of 10 m*M* ATP can be determined best by spotting small drops on pH indicator paper.
3. The majority of *E. coli* strains are CcdB-sensitive.
4. Ideally, pINITIAL and the expression vector of choice confer resistance against different antibiotics. For this reason different pINITIAL-derivatives holding kanamycin, chloramphenicol, or tetracycline markers are available. Should pINITIAL and the expression vector carry the same marker, the emergence of pINITIAL-containing colonies can be suppressed by plating on low salt (5 g/L) LB-agar supplemented with 7% (w/v) sucrose and the appropriate antibiotic. Each pINITIAL variant carries the *sacB* marker that renders *E. coli* sensitive to sucrose (Recorbet et al., 1993).

3. ANALYSIS OF MEMBRANE PROTEIN PRODUCTION USING GFP

3.1 Initial Expression Screening

Commonly screened cultivation parameters for enhancing membrane protein overproduction involve strain type, growth medium, temperature, and inducer concentration (Wang et al., 2003). Complete initial screening of these variables for a multitude of constructs sacrifices valuable screening capacity, making a limitation therefore appropriate and, based on the following arguments, possible.

The screening of different expression strains is virtually a routine procedure when using the strong inducible T7 promoter (Studier, Rosenberg, Dunn, & Dubendorff, 1990). However, detailed characterization of the popular Walker strains, C41 and C43, originally selected for their ability to survive the production of membrane proteins as inclusion bodies (Miroux & Walker, 1996), revealed that in essence the beneficial effects of using these strains resulted from genomic mutations lowering the promoter strength (Wagner et al., 2008). Others and we have observed that this effect can be mimicked using well-tunable promoters of moderate strength, such as the *araBAD* (Guzman et al., 1995) and *rhaBAD* (Haldimann, Daniels, & Wanner, 1998) promoters (Geertsma et al., 2008; Wagner et al., 2008). Thus, instead of adjusting the expression rate by using different strains, we recommend to vary this parameter using the *araBAD* promoter in combination with a strain unable to metabolize arabinose, such as MC1061 (Casadaban & Cohen, 1980; Guzman et al., 1995). The P_{BAD} promoter is comparably tightly regulated and is therefore well suited for the expression of proteins that have a strong effect on cell physiology, such as membrane proteins (Lewinson, Lee, & Rees, 2008). *E. coli* MC1061 has the additional advantage that it functions as both cloning and expression strain, thereby avoiding additional plasmid isolations and transformations.

Furthermore, given the aim of producing significant amounts of the target protein, a rich growth medium, such as terrific broth (TB) which supports growth to high cell densities, is required (Lee et al., 2014). We interpret the screening of poor and rich media and induction at different growth phases as indirect adjustment of protein production to the folding capacity of the cell, which can be set more straightforwardly using a tunable promoter.

Temperature has a large effect on cell physiology and dramatically impacts membrane protein production. In general, lower temperatures increase the amount of well-folded protein (Wang et al., 2003), but the accompanying reduction in growth rate also leads to a slower increase in biomass. We therefore recommend lowering the postinduction temperature to at least 25°C and extending cultivation times to 12–16 h.

The *araBAD* promoter has a comparably large dynamic range with induction/repression ratios spanning three orders of magnitude. Though the *araBAD* promoter is weaker than promoters based on P_{LAC}, such as P_{TAC} and P_{TRC}, and much weaker than the strong T7 promoter (Guzman et al., 1995), it is still of sufficient strength to saturate the membrane protein folding capacity upon full induction (Geertsma et al., 2008). A balance between too low and too high expression levels resulting in poor amounts of folded protein or high amounts of misfolded material, respectively, needs to be found. For a broad set of different membrane proteins, we have observed that the relevant inducer concentration range for the P_{BAD} promoter lies between 70 and 7000 μ*M* L-arabinose (Ehrnstorfer, Geertsma, Pardon, Steyaert, & Dutzler, 2014; Geertsma et al., 2008, 2015). Although functional expression levels can be significantly increased by fine-tuning the inducer concentration, an L-arabinose concentration of 700 μ*M* is generally a good starting point for initial expression screening.

Identification of well-expressed membrane protein constructs is most efficiently done using analytical cultures. In our experience, membrane protein expression levels that cannot be detected in culture volumes below 3 mL do generally not suffice for structural characterization using crystallography. To facilitate the handling of many different constructs, the use of a 96-deep-well block is recommended. Using high shaking speeds and gas-permeable seals allows the routine cultivation up to high cell densities ($OD_{600} > 10$).

Taken together, our recommendations for initial cultivation conditions aimed to select promising candidate constructs include the use of *E. coli* MC1061 and the *araBAD* promoter, cultivation in TB, induction with 700 μ*M* arabinose and a postinduction temperature of 25°C for approximately 16 h.

3.1.1 Equipment

- Sterile 96-deep-well block
- Gas-permeable seal for 96-well plates

- Multipipets
- Plate shaking incubator
- Tabletop centrifuge for 96-well plates
- Microplate spectrofluorometer
- Black and transparent 96-well plates

3.1.2 Buffers and Reagents

- Terrific broth: 12 g tryptone, 24 g yeast extract, 8.6 mL 86% (w/v) glycerol. Add demineralized water up to 950 mL and autoclave. Supplement with 50 mL of sterile 20 × phosphate solution (1.44 *M* K_2HPO_4, 0.34 *M* KH_2PO_4), 1 mL 1 *M* $MgSO_4$ (see Note 1 in Section 3.1.4), and the appropriate antibiotic before use.
- Filter-sterilized 1 *M* L-arabinose
- 50 m*M* KPi buffer, pH 7.5: prepare a 50 m*M* K_2HPO_4 solution and a 50 m*M* KH_2PO_4 solution and mix in an approximate ratio of 4:1 (K_2HPO_4:KH_2PO_4) under constant stirring while measuring the pH. Set the pH by adjusting with the basic or acidic component. Prepare fresh on the day of use.

3.1.3 Procedure

1. Inoculate 500 μL TB medium in a 96-deep-well block from a glycerol stock (see Note 2 in Section 3.1.4). Cover the block with a gas-permeable seal and cultivate overnight at 37°C.
2. Inoculate 700 μL prewarmed TB medium with 7 μL of the overnight culture using a multipipet. Cover the plate with a gas-permeable seal and grow for 90 min at 37°C in a plate shaker (see Note 3 in Section 3.1.4). Set the temperature to 25°C and grow for one additional hour.
3. Induce the cells with 14 μL 35 m*M* L-arabinose and grow overnight at 25°C (see Note 4 in Section 3.1.4).
4. Pellet the cells by centrifugation for 10 min at 2.500 × *g* at 4°C. Discard the supernatant by inverting the deep-well block and remove remaining liquid by tapping the block powerfully on a paper towel. Resuspend the cell pellets in 300 μL 50 m*M* KPi, pH 7.5. Transfer 150 μL cell suspension to a black 96-well plate and 75 μL to a transparent 96-well plate while preventing bubbles.
5. Determine the optical density at a wavelength of 600 nm using the transparent plate. For few samples, determine the optical density as well using a spectrophotometer in a cuvette with 1-cm path length. Calculate a correction factor for multiplying the optical densities measured in the plate.

This corrected optical density will be used in Section 3.2 to calculate the volume of cell suspension required for further analysis.

6. Determine the GFP fluorescence intensity (excitation wavelength 488 nm, 535 nm emission). Adjust the gain of the photomultiplier to assure high sensitivity while preventing saturation.
7. Recover the cell suspensions from both plates and proceed with the analysis of the folding state.
8. Normalize the GFP fluorescence intensity to the OD_{600} to calculate the amount of GFP fusion protein per cell.

3.1.4 Notes

1. Depending on suppliers and batches, complex media can be deficient in magnesium (Studier, 2005).
2. The AraC/P_{BAD} system is sufficiently tight to allow the use of glycerol stocks for inoculating expression cultures. When using leaky expression systems, such as those using the T7 promoter, highest expression levels are generally obtained using freshly transformed cells.
3. The required rotational speed highly depends on the amplitude of the used shaker. In our hands, 300 rpm at an amplitude of 12.5 mm works best for deep-well blocks with 700 μL cultures.
4. For screening an inducer concentration range during initial experiments, a concentration series of 70 μ*M*, 700 μ*M*, and 7 m*M* is recommended.

Target constructs that display intermediate to high GFP fluorescence levels should be submitted to an additional round of screening to determine the most optimal L-arabinose concentration for induction. Next to a determination of GFP fluorescence in whole cells as described here, the folding state can then be addressed as detailed in the following section.

3.2 Membrane Protein Folding State

While the fluorescence of GFP fusion proteins in whole cells is a good initial indicator of the folding state of the protein immediately following its biogenesis, the robustness of GFP makes it a poor indicator for long-term stability. For example, the GFP moiety often remains fluorescent following proteolytic degradation of the fusion protein. Furthermore, a more detailed analysis of the GFP fusion protein using SDS-PAGE and immunodetection will not only report on the amount of intact protein but also quantify the ratio of aggregated to folded protein (Fig. 2). The latter serves to guide subsequent optimization. Low amounts of unfolded protein suggest that higher inducer concentrations or stronger promoters can be used to increase the expression level even further.

3.2.1 Equipment

- FastPrep-24 benchtop homogenizer
- Screw cap microtubes, 2-mL
- Glass beads, 0.1-mm diameter
- SDS-PAGE and Western blotting setups
- ImageQuant LAS 4000

3.2.2 Buffers and Reagents

- FastPrep buffer: 50 m*M* KPi buffer, pH 7.5, 10% (w/v) glycerol, 1 m*M* $MgSO_4$. Supplement with trace amounts of DNAse I and 1 m*M* PMSF (using a 200-m*M* stock in EtOH) prior to use.

3.2.3 Procedure

1. Express target membrane proteins as described in the previous section. Determine the OD_{600} of the culture. Optical density values determined in plates need to be corrected using a correction factor to approach the OD_{600} value as measured in cuvettes with a 1-cm path length.
2. Assuming an OD_{600} of 1 corresponds to ~0.3 mg total protein per mL of cell culture, pellet 1 mg total protein in 2-mL screw cap microtubes.
3. Discard the supernatant and add 400 μL ice-cold FastPrep buffer and ~300-mg glass beads.
4. Disrupt the cells using the FastPrep-24 homogenizer by bead-beating twice for 20 s at force 6. Cool the samples in between on ice for 5 min.
5. Allow the glass beads to settle. Transfer 100 μL of the supernatant to a new cup and add 25 μL 5 × SDS-PAGE loading buffer to the sample. Homogenize by vortexing shortly. Do not heat the samples. Analyze 10 μL of each sample by SDS-PAGE. Use a prestained molecular weight marker to aid calibration of the images resulting from the in-gel GFP fluorescence and the immunoblot.
6. Transfer the gel from the cassette to the black imaging tray and record the in-gel GFP fluorescence using the ImageQuant LAS 4000. Excite the gel at 460 nm (blue LED) and use a 515-nm emission filter. Collect a series of images exposed for prolonged periods, e.g., 7, 15, 30, 60, 120, and 240 s.

7. Without disturbing the tray, change the settings to visualize the prestained marker by exposure at white light (EPI). Collect an image of the gel.
8. Submit the gel to immunoblotting and mark the bands of the prestained marker with a ballpoint pen after transfer. Use an anti-His antibody-HRP conjugate to develop the immunoblot. Use the ImageQuant LAS 4000 in chemiluminescence mode to measure the signal. Collect a series of images exposed for prolonged periods, e.g., 7, 15, 30, 60, 120, 240, and 480 s.
9. Without disturbing the tray, change the settings to visualize the membrane and the prestained marker bands using the top white light mode. Collect an image of the gel.
10. Use imaging software to correlate the images of the in-gel GFP fluorescence and the immunoblot guided by the bands of the prestained marker.

The differential electrophoretic mobility of folded and misfolded membrane protein GFP fusions serves as a guide to compare different expression conditions during optimization. Optimization should primarily aim to increase the absolute amount of folded protein. Avoiding the formation of unfolded protein is helpful but not required as this material is normally removed during conventional purification steps. Expression conditions optimized using membrane protein GFP fusions are not necessarily optimal expression conditions to express the respective membrane protein without GFP and often require a separate optimization round.

4. SUMMARY AND CONCLUSION

Producing membrane proteins in *E. coli* is not a trivial task and often requires extensive optimization. In this chapter, we have detailed strategies to manage two commonly encountered scenarios when expressing membrane proteins: no or poor expression of folded protein and high expression levels of misfolded material.

Our approach involves transcriptional fusions whose effect of increasing membrane protein expression likely originates from an increased efficiency of translation initiation due to more favorable RNA structures near the start codon. In addition, less understood long-range interactions seem to contribute. Transcriptional fusions are particularly suitable for those researchers that already use conventional N-terminal fusion proteins to overexpress soluble or membrane proteins. The choice of the transcriptional fusion partner is not trivial, and we propose that several sequences, among which *mstX* and *ybeL*, should be screened. To support such broad expression screening, we subsequently detailed FX cloning to facilitate the parallel generation of expression vectors. We further employ the tunable AraC/P_{BAD} system and GFP to efficiently monitor well-folded but also misfolded proteins in whole cells or crude lysates. The variation of the inducer concentration during expression screening allows careful adjustment of the protein expression rate to the capacity of the cell to correctly process the protein, often improving the yield of functionally folded membrane protein significantly. Taken together, our pipeline allows the efficient selection of expression conditions that overcome different bottlenecks limiting the functional production of membrane proteins.

ACKNOWLEDGMENTS

We acknowledge funding from the Cluster of Excellence Frankfurt "Macromolecular Complexes" and the Collaborative Research Center "Transport and Communication across Biological Membranes" (SFB 807). J.M. acknowledges financial support from the Novartis Foundation for Biomedical Research. K.H. acknowledges the Integrated Research Training Group TRAM (SFB 807) for financial support.

REFERENCES

Andre, A., Puca, A., Sansone, F., Brandi, A., Antico, G., & Calogero, R. A. (2000). Reinitiation of protein synthesis in Escherichia coli can be induced by mRNA cis-elements unrelated to canonical translation initiation signals. *FEBS Letters*, *468*(1), 73–78.

Aslanidis, C., & de Jong, P. J. (1990). Ligation-independent cloning of PCR products (LIC-PCR). *Nucleic Acids Research*, *18*(20), 6069–6074.

Bernard, P., & Couturier, M. (1992). Cell killing by the F plasmid CcdB protein involves poisoning of DNA-topoisomerase II complexes. *Journal of Molecular Biology*, *226*(3), 735–745.

Bernard, P., Gabant, P., Bahassi, E. M., & Couturier, M. (1994). Positive-selection vectors using the F plasmid ccdB killer gene. *Gene*, *148*(1), 71–74.

Blattner, F. R., Plunkett, G., 3rd, Bloch, C. A., Perna, N. T., Burland, V., Riley, M., et al. (1997). The complete genome sequence of Escherichia coli K-12. *Science*, *277*(5331), 1453–1462.

Boel, G., Letso, R., Neely, H., Price, W. N., Wong, K. H., Su, M., et al. (2016). Codon influence on protein expression in E. coli correlates with mRNA levels. *Nature*, *529*(7586), 358–363. http://dx.doi.org/10.1038/nature16509.

Buhr, F., Jha, S., Thommen, M., Mittelstaet, J., Kutz, F., Schwalbe, H., et al. (2016). Synonymous codons direct cotranslational folding toward different protein conformations. *Molecular Cell*, *61*(3), 341–351. http://dx.doi.org/10.1016/j.molcel.2016.01.008.

Casadaban, M. J., & Cohen, S. N. (1980). Analysis of gene control signals by DNA fusion and cloning in Escherichia coli. *Journal of Molecular Biology*, *138*(2), 179–207.

Daley, D. O., Rapp, M., Granseth, E., Melen, K., Drew, D., & von Heijne, G. (2005). Global topology analysis of the Escherichia coli inner membrane proteome. *Science*, *308*(5726), 1321–1323. http://dx.doi.org/10.1126/science.1109730.

Das, A., & Yanofsky, C. (1984). A ribosome binding site sequence is necessary for efficient expression of the distal gene of a translationally-coupled gene pair. *Nucleic Acids Research, 12*(11), 4757–4768.

Don, R. H., Cox, P. T., Wainwright, B. J., Baker, K., & Mattick, J. S. (1991). Touchdown' PCR to circumvent spurious priming during gene amplification. *Nucleic Acids Research, 19*(14), 4008.

Drew, D., Lerch, M., Kunji, E., Slotboom, D. J., & de Gier, J. W. (2006). Optimization of membrane protein overexpression and purification using GFP fusions. *Nature Methods, 3*(4), 303–313. http://dx.doi.org/10.1038/nmeth0406-303.

Drew, D. E., von Heijne, G., Nordlund, P., & de Gier, J. W. (2001). Green fluorescent protein as an indicator to monitor membrane protein overexpression in Escherichia coli. *FEBS Letters, 507*(2), 220–224.

Ehrnstorfer, I. A., Geertsma, E. R., Pardon, E., Steyaert, J., & Dutzler, R. (2014). Crystal structure of a SLC11 (NRAMP) transporter reveals the basis for transition-metal ion transport. *Nature Structural & Molecular Biology, 21*(11), 990–996. http://dx.doi.org/10.1038/nsmb.2904.

Elvekrog, M. M., & Walter, P. (2015). Dynamics of co-translational protein targeting. *Current Opinion in Chemical Biology, 29*, 79–86. http://dx.doi.org/10.1016/j.cbpa.2015.09.016.

Esposito, D., & Chatterjee, D. K. (2006). Enhancement of soluble protein expression through the use of fusion tags. *Current Opinion in Biotechnology, 17*(4), 353–358. http://dx.doi.org/10.1016/j.copbio.2006.06.003.

Geertsma, E. R. (2013). FX cloning: A versatile high-throughput cloning system for characterization of enzyme variants. *Methods in Molecular Biology, 978*, 133–148. http://dx.doi.org/10.1007/978-1-62703-293-3_10.

Geertsma, E. R. (2014). FX cloning: A simple and robust high-throughput cloning method for protein expression. *Methods in Molecular Biology, 1116*, 153–164. http://dx.doi.org/10.1007/978-1-62703-764-8_11.

Geertsma, E. R., Chang, Y. N., Shaik, F. R., Neldner, Y., Pardon, E., Steyaert, J., et al. (2015). Structure of a prokaryotic fumarate transporter reveals the architecture of the SLC26 family. *Nature Structural & Molecular Biology, 22*(10), 803–808. http://dx.doi.org/10.1038/nsmb.3091.

Geertsma, E. R., & Dutzler, R. (2011). A versatile and efficient high-throughput cloning tool for structural biology. *Biochemistry, 50*(15), 3272–3278. http://dx.doi.org/10.1021/bi200178z.

Geertsma, E. R., Groeneveld, M., Slotboom, D. J., & Poolman, B. (2008). Quality control of overexpressed membrane proteins. *Proceedings of the National Academy of Sciences of the United States of America, 105*(15), 5722–5727. http://dx.doi.org/10.1073/pnas.0802190105.

Grisshammer, R. (2006). Understanding recombinant expression of membrane proteins. *Current Opinion in Biotechnology, 17*(4), 337–340. http://dx.doi.org/10.1016/j.copbio.2006.06.001.

Grisshammer, R., & Tate, C. G. (1995). Overexpression of integral membrane proteins for structural studies. *Quarterly Reviews of Biophysics, 28*(3), 315–422.

Gu, W., Zhou, T., & Wilke, C. O. (2010). A universal trend of reduced mRNA stability near the translation-initiation site in prokaryotes and eukaryotes. *PLoS Computational Biology, 6*(2), e1000664. http://dx.doi.org/10.1371/journal.pcbi.1000664.

Guzman, L. M., Belin, D., Carson, M. J., & Beckwith, J. (1995). Tight regulation, modulation, and high-level expression by vectors containing the arabinose PBAD promoter. *Journal of Bacteriology, 177*(14), 4121–4130.

Haberstock, S., Roos, C., Hoevels, Y., Dotsch, V., Schnapp, G., Pautsch, A., et al. (2012). A systematic approach to increase the efficiency of membrane protein production in cell-free expression systems. *Protein Expression and Purification, 82*(2), 308–316. http://dx.doi.org/10.1016/j.pep.2012.01.018.

Haldimann, A., Daniels, L. L., & Wanner, B. L. (1998). Use of new methods for construction of tightly regulated arabinose and rhamnose promoter fusions in studies of the Escherichia coli phosphate regulon. *Journal of Bacteriology*, *180*(5), 1277–1286.

Harris, N. J., & Booth, P. J. (2012). Folding and stability of membrane transport proteins in vitro. *Biochimica et Biophysica Acta*, *1818*(4), 1055–1066. http://dx.doi.org/10.1016/j.bbamem.2011.11.006.

Hartley, J. L., Temple, G. F., & Brasch, M. A. (2000). DNA cloning using in vitro site-specific recombination. *Genome Research*, *10*(11), 1788–1795.

Hattab, G., Warschawski, D. E., Moncoq, K., & Miroux, B. (2015). Escherichia coli as host for membrane protein structure determination: A global analysis. *Scientific Reports 5*, 12097. http://dx.doi.org/10.1038/srep12097.

Kapust, R. B., & Waugh, D. S. (1999). Escherichia coli maltose-binding protein is uncommonly effective at promoting the solubility of polypeptides to which it is fused. *Protein Science*, *8*(8), 1668–1674. http://dx.doi.org/10.1110/ps.8.8.1668.

Kozak, M. (2005). Regulation of translation via mRNA structure in prokaryotes and eukaryotes. *Gene*, *361*, 13–37. http://dx.doi.org/10.1016/j.gene.2005.06.037.

Kudla, G., Murray, A. W., Tollervey, D., & Plotkin, J. B. (2009). Coding-sequence determinants of gene expression in Escherichia coli. *Science*, *324*(5924), 255–258. http://dx.doi.org/10.1126/science.1170160.

LaVallie, E. R., Lu, Z., Diblasio-Smith, E. A., Collins-Racie, L. A., & McCoy, J. M. (2000). Thioredoxin as a fusion partner for production of soluble recombinant proteins in Escherichia coli. *Methods in Enzymology*, *326*, 322–340.

Lee, C., Kang, H. J., Hjelm, A., Qureshi, A. A., Nji, E., Choudhury, H., et al. (2014). MemStar: A one-shot Escherichia coli-based approach for high-level bacterial membrane protein production. *FEBS Letters*, *588*(20), 3761–3769. http://dx.doi.org/10.1016/j.febslet.2014.08.025.

Leviatan, S., Sawada, K., Moriyama, Y., & Nelson, N. (2010). Combinatorial method for overexpression of membrane proteins in Escherichia coli. *The Journal of Biological Chemistry*, *285*(31), 23548–23556. http://dx.doi.org/10.1074/jbc.M110.125492.

Lewinson, O., Lee, A. T., & Rees, D. C. (2008). The funnel approach to the pre-crystallization production of membrane proteins. *Journal of Molecular Biology*, *377*(1), 62–73. http://dx.doi.org/10.1016/j.jmb.2007.12.059.

Loughran, S. T., & Walls, D. (2017). Tagging recombinant proteins to enhance solubility and aid purification. *Methods in Molecular Biology*, *1485*, 131–156. http://dx.doi.org/10.1007/978-1-4939-6412-3_8.

Luirink, J., Yu, Z., Wagner, S., & de Gier, J. W. (2012). Biogenesis of inner membrane proteins in Escherichia coli. *Biochimica et Biophysica Acta*, *1817*(6), 965–976. http://dx.doi.org/10.1016/j.bbabio.2011.12.006.

Majtan, T., & Kraus, J. P. (2012). Folding and activity of mutant cystathionine beta-synthase depends on the position and nature of the purification tag: Characterization of the R266K CBS mutant. *Protein Expression and Purification*, *82*(2), 317–324. http://dx.doi.org/10.1016/j.pep.2012.01.019.

Marino, J., Hohl, M., Seeger, M. A., Zerbe, O., & Geertsma, E. R. (2015). Bicistronic mRNAs to enhance membrane protein overexpression. *Journal of Molecular Biology*, *427*(4), 943–954. http://dx.doi.org/10.1016/j.jmb.2014.11.002.

Miroux, B., & Walker, J. E. (1996). Over-production of proteins in Escherichia coli: Mutant hosts that allow synthesis of some membrane proteins and globular proteins at high levels. *Journal of Molecular Biology*, *260*(3), 289–298. http://dx.doi.org/10.1006/jmbi.1996.0399.

Na, D., & Lee, D. (2010). RBSDesigner: Software for designing synthetic ribosome binding sites that yields a desired level of protein expression. *Bioinformatics*, *26*(20), 2633–2634. http://dx.doi.org/10.1093/bioinformatics/btq458.

Norholm, M. H., Toddo, S., Virkki, M. T., Light, S., von Heijne, G., & Daley, D. O. (2013). Improved production of membrane proteins in Escherichia coli by selective codon substitutions. *FEBS Letters*, *587*(15), 2352–2358. http://dx.doi.org/10.1016/j.febslet.2013.05.063.

Peroutka III, R. J., Orcutt, S. J., Strickler, J. E., & Butt, T. R. (2011). SUMO fusion technology for enhanced protein expression and purification in prokaryotes and eukaryotes. *Methods in Molecular Biology*, *705*, 15–30. http://dx.doi.org/10.1007/978-1-61737-967-3_2.

Raman, P., Cherezov, V., & Caffrey, M. (2006). The Membrane Protein Data Bank. *Cellular and Molecular Life Sciences*, *63*(1), 36–51. http://dx.doi.org/10.1007/s00018-005-5350-6.

Recorbet, G., Robert, C., Givaudan, A., Kudla, B., Normand, P., & Faurie, G. (1993). Conditional suicide system of Escherichia coli released into soil that uses the Bacillus subtilis sacB gene. *Applied and Environmental Microbiology*, *59*(5), 1361–1366.

Roosild, T. P., Greenwald, J., Vega, M., Castronovo, S., Riek, R., & Choe, S. (2005). NMR structure of Mistic, a membrane-integrating protein for membrane protein expression. *Science*, *307*(5713), 1317–1321. http://dx.doi.org/10.1126/science.1106392.

Sabaty, M., Grosse, S., Adryanczyk, G., Boiry, S., Biaso, F., Arnoux, P., et al. (2013). Detrimental effect of the 6 His C-terminal tag on YedY enzymatic activity and influence of the TAT signal sequence on YedY synthesis. *BMC Biochemistry*, *14*, 28. http://dx.doi.org/10.1186/1471-2091-14-28.

Singh, P., Sharma, L., Kulothungan, S. R., Adkar, B. V., Prajapati, R. S., Ali, P. S., et al. (2013). Effect of signal peptide on stability and folding of Escherichia coli thioredoxin. *PloS One*, *8*(5), e63442. http://dx.doi.org/10.1371/journal.pone.0063442.

Smyth, D. R., Mrozkiewicz, M. K., McGrath, W. J., Listwan, P., & Kobe, B. (2003). Crystal structures of fusion proteins with large-affinity tags. *Protein Science*, *12*(7), 1313–1322. http://dx.doi.org/10.1110/ps.0243403.

Spencer, P. S., Siller, E., Anderson, J. F., & Barral, J. M. (2012). Silent substitutions predictably alter translation elongation rates and protein folding efficiencies. *Journal of Molecular Biology*, *422*(3), 328–335. http://dx.doi.org/10.1016/j.jmb.2012.06.010.

Studier, F. W. (2005). Protein production by auto-induction in high density shaking cultures. *Protein Expression and Purification*, *41*(1), 207–234.

Studier, F. W., Rosenberg, A. H., Dunn, J. J., & Dubendorff, J. W. (1990). Use of T7 RNA polymerase to direct expression of cloned genes. *Methods in Enzymology*, *185*, 60–89.

Szybalski, W., Kim, S. C., Hasan, N., & Podhajska, A. J. (1991). Class-IIS restriction enzymes—A review. *Gene*, *100*, 13–26.

Vergis, J. M., & Wiener, M. C. (2011). The variable detergent sensitivity of proteases that are utilized for recombinant protein affinity tag removal. *Protein Expression and Purification*, *78*(2), 139–142. http://dx.doi.org/10.1016/j.pep.2011.04.011.

Wagner, S., Klepsch, M. M., Schlegel, S., Appel, A., Draheim, R., Tarry, M., et al. (2008). Tuning Escherichia coli for membrane protein overexpression. *Proceedings of the National Academy of Sciences of the United States of America*, *105*(38), 14371–14376. http://dx.doi.org/10.1073/pnas.0804090105.

Waldo, G. S., Standish, B. M., Berendzen, J., & Terwilliger, T. C. (1999). Rapid protein-folding assay using green fluorescent protein. *Nature Biotechnology*, *17*(7), 691–695. http://dx.doi.org/10.1038/10904.

Wang, D. N., Safferling, M., Lemieux, M. J., Griffith, H., Chen, Y., & Li, X. D. (2003). Practical aspects of overexpressing bacterial secondary membrane transporters for structural studies. *Biochimica et Biophysica Acta*, *1610*(1), 23–36.

Waugh, D. S. (2005). Making the most of affinity tags. *Trends in Biotechnology*, *23*(6), 316–320. http://dx.doi.org/10.1016/j.tibtech.2005.03.012.

Young, C. L., Britton, Z. T., & Robinson, A. S. (2012). Recombinant protein expression and purification: A comprehensive review of affinity tags and microbial applications. *Biotechnology Journal*, 7(5), 620–634. http://dx.doi.org/10.1002/biot.201100155.

Yu, C. H., Dang, Y., Zhou, Z., Wu, C., Zhao, F., Sachs, M. S., et al. (2015). Codon usage influences the local rate of translation elongation to regulate co-translational protein folding. *Molecular Cell*, *59*(5), 744–754. http://dx.doi.org/10.1016/j.molcel.2015.07.018.

CHAPTER SEVEN

Pinning Down the Mechanism of Transport: Probing the Structure and Function of Transporters Using Cysteine Cross-Linking and Site-Specific Labeling

Christopher Mulligan*[,1], Joseph A. Mindell[†]
*School of Biosciences, University of Kent, Canterbury, United Kingdom
[†]Membrane Transport Biophysics Section, Porter Neuroscience Research Center, National Institute of Neurological Disorders and Stroke, National Institutes of Health, Bethesda, MD, United States
[1]Corresponding author: e-mail address: c.mulligan@kent.ac.uk

Contents

Methods in Enzymology, Volume 594
ISSN 0076-6879
http://dx.doi.org/10.1016/bs.mie.2017.05.012

Abstract

Transporters are crucial in a number of cellular functions, including nutrient uptake, cell signaling, and toxin removal. As such, transporters are important drug targets and their malfunction is related to several disease states. Treating transporter-related diseases and developing pharmaceuticals targeting transporters require an understanding of their mechanism. Achieving a detailed understanding of transporter mechanism depends on an integrative approach involving structural and computational approaches as well as biochemical and biophysical methodologies. Many of the elements of this toolkit exploit the unique and useful chemistry of the amino acid cysteine. Cysteine offers researchers a specific molecular handle with which to precisely modify the protein, which enables the introduction of biophysical probes to assess ligand binding and the conformational ensemble of the transporter, to topologically map transporters and validate structural models, and to assess essential conformational changes.

Here, we summarize several uses for cysteine-based labeling and cross-linking in the pursuit of understanding transporter mechanism, the common cysteine-reactive reagents used to probe transporter mechanism, and strategies that can be used to confirm cysteine cross-link formation. In addition, we provide methodological considerations for each approach and a detailed procedure for the cross-linking of introduced cysteines, and a simple screening method to assess cross-link formation.

ABBREVIATIONS

BME β-mercaptoethanol
DTT dithiothreitol
EPR electron paramagnetic resonance
IFS inward-facing state
ISO inside-out
ITC isothermal titration calorimetry
mPEG5K methoxypolyethylene glycol maleimide
OFS outward-facing state
RSO right side-out
SCAM substituted cysteine accessibility method
SDS sodium dodecyl sulfate
SDS-PAGE sodium dodecyl sulfate polyacrylamide gel electrophoresis
smFRET single-molecule Förster resonance energy transfer
TCEP tris(2-carboxyethyl)phosphine

1. INTRODUCTION

Transporters perform crucial cellular functions in all forms of life, from nutrient uptake to signal transduction to toxin and by-product efflux. Understanding the molecular mechanism of transporters is key to manipulating their function and tackling transporter-related human diseases, including Alzheimer's disease, ALS, and diabetes, as well as other threats such as bacterial pathogenesis and antimicrobial resistance. A multidisciplinary approach is required to fully illuminate transporter mechanism. Several critical elements of this toolkit rely on cysteine, a specific molecular handle, enabling attachment of a series of biochemical and biophysical probes and permitting formation of intramolecular cross-links; careful use of these approaches, individually and together, can lead to insight into the conformational dynamics of transport proteins.

Transporters are molecular machines that couple in many instances the transport of a substrate to an energy source. The substrates transported range from single ions, such as chloride and calcium (Accardi & Miller, 2004; Giladi, Shor, Lisnyansky, & Khananshvili, 2016), through to large, complex compounds such as vitamin B12 (Locher, 2016). Various energy sources are harnessed depending on the transporter; the two most frequently tapped are the binding and hydrolysis of ATP, and the electrochemical potential of ion gradients across the membrane (usually H^+ or Na^+). Regardless of the substrate or energy source used, decades of research have revealed transporters to be highly dynamic proteins that sample a variety of conformations in order to function. The mechanisms of all transporters can be described by a simplified alternating access model, which offers a framework to describe how a transporter gets substrate from one side of the membrane to the other (Jardetzky, 1966). In this model, the transporter alternately exposes its binding site to one side of the membrane or the other, accessing a series of essential conformations (Fig. 1). Such a model requires that each transporter has a minimum of two essential conformational states, the inward-facing state (IFS) and the outward-facing state (OFS).

Depending on the transporter family, the conformational interconversions required for transport vary in magnitude from the minimal wobble of the chloride transporters (Basilio, Noack, Picollo, & Accardi, 2014; Khantwal et al., 2016) to the relatively gargantuan movement of the succinate transporter, VcINDY (Mulligan et al., 2016). To date, four different general mechanisms have been described by which transporters achieve their

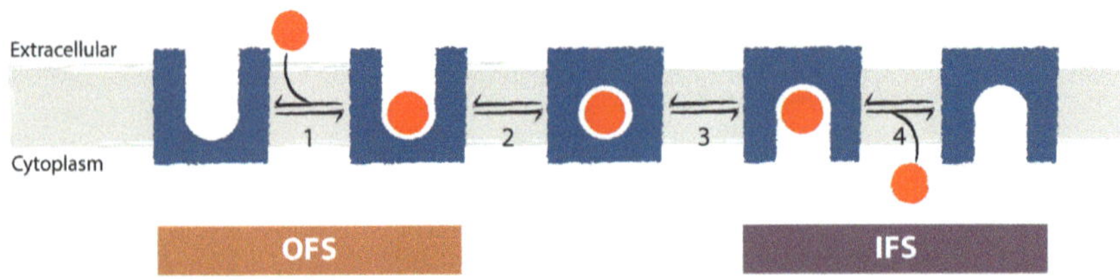

Fig. 1 Conformational states required for a simple alternating access model for a transporter. (1) Substrate (*red circle*) binds to the outward-facing state (OFS) of the transporter (*blue square*), in which the substrate-binding site is exposed to the extracellular side of the membrane. (2) The transporter undergoes a conformational change into an occluded state where the substrate is blocked from both sides. (3) Another conformational change occurs moving the transporter into the inward-facing state (IFS). (4) The substrate is released into the cytoplasm resulting in an empty IFS state. The empty transporter transitions back to the OFS, resetting the system (not shown).

goal of vectorial movement of substrate: the gated-pore mechanism, the rocking bundle mechanism, the elevator mechanism, and the "tumbling" mechanism (Drew & Boudker, 2016; Forrest, Krämer, & Ziegler, 2011; Slotboom, 2014; Swier, Guskov, & Slotboom, 2016).

X-ray crystallography has long been the method *du jour* for obtaining structural information about transporters, and advances in this technique have catalyzed major leaps in our understanding of transporter mechanism. Despite the invaluable insight we have obtained from transporter crystal structures, one cannot rely solely on crystal structures to determine the mechanism of transporters. A crystal structure captures a single conformation of the protein, and it is a difficult task to obtain crystal structures of more than one state for the same transporter. Structures may also be deformed by crystal contacts or the methods that were required to purify and crystallize the protein. Crystal structures, by their very nature, also do not reveal dynamic information, which is key to understanding transporter mechanism. Chemical biology approaches such as site-specific amino acid labeling and chemical cross-linking have the potential to reveal great insight into transport mechanism in the absence of structural information. However, the power of these tools is enhanced when used in conjunction with high-resolution crystal structures.

This chapter will summarize some key applications of cysteine-based site-specific labeling and protein cross-linking that can be used to understand transporter mechanism as well the variety of approaches and methodological considerations that need to be taken into account. We will limit the discussion to cysteine-mediated labeling and cross-linking, as this is currently the most common strategy to label and cross-link membrane proteins. However, there are many commercially available cross-linkers and labeling reagents that recognize other amino acids and protein moieties that may also be useful depending on the experimenter's end goal.

2. THE UNIQUE AND USEFUL CHEMISTRY OF CYSTEINE

Nature has provided the protein tinkerer with several options to choose from when it comes to modifying proteins with reactive groups or introducing cross-links. The most common modifiable functional groups found in amino acids are primary amine groups found at the N-terminus of the polypeptide or in the side chain of lysine residues; carboxyl groups located at the C-terminus of the polypeptide or in the aspartic acid or glutamic acid side chains; and sulfhydryl groups found only in the side chain of cysteine residues.

Targeting cysteine for modification offers a number of advantages over amine and carboxyl groups, especially with regards to integral membrane proteins. Cysteines contain a highly reactive thiol group (−SH, also referred to as the sulfhydryl group) capable of reacting very specifically with a variety of reagents as well as another proximal thiol group to form a disulfide bond. The thiol group is subject to deprotonation to a thiolate ($-S^-$), which makes it substantially more reactive to common cysteine-reactive reagents. As a consequence of this, cysteine reactivity is highly pH dependent and needs to be carefully controlled during experiments. The pK_a of free cysteine is ~8.5, but, when ensconced in a folded protein, this pK_a varies depending on the local environment (Houk, Singh, & Whitesides, 1987). Sulfur is at the heart of cysteine's unique reactivity and can exist in many oxidation states, which also affects this reactivity; the most common in biology are the thiol/thiolate ($-SH/-S^-$) and the disulfide ($-SS^-$). However, higher oxidation states, sulfenic (−SOH), sulfinic (−SOOH), and sulfonic (−SOOOH) acids, are possible after treatment with oxidizing agents; all of these are nonreactive and the latter two are also irreversible (Fig. 2A; Akabas, 2015; Lo Conte & Carroll, 2013). Therefore, to maintain full reactivity of introduced cysteines, care should be taken to prevent unwanted oxidation of the target transporter.

Cysteine residues are small, relatively hydrophobic amino acids that are often buried in the protein core and appear to have no preference for any secondary structural elements (Eilers, Patel, Liu, & Smith, 2002; Marino & Gladyshev, 2010). These features increase the chance that

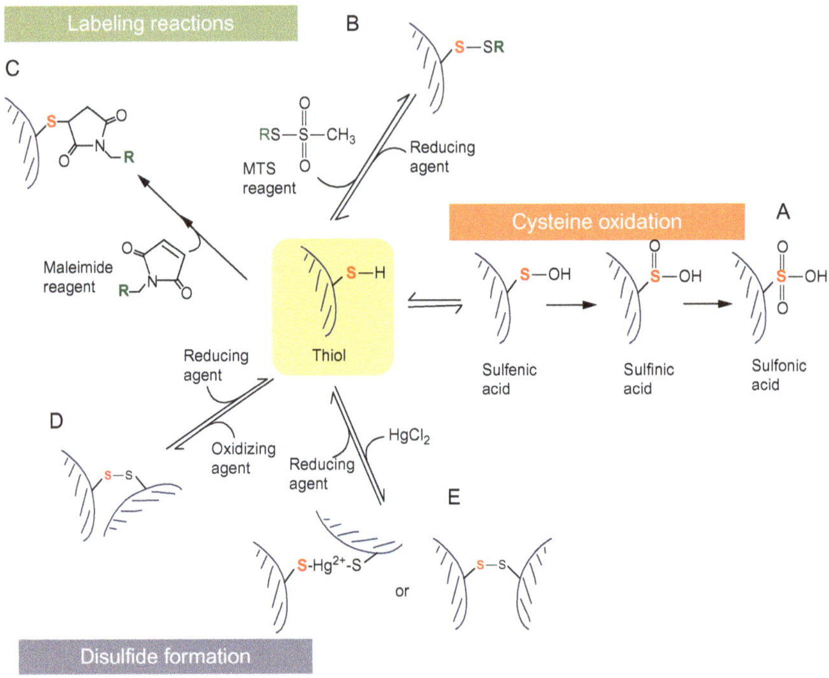

Fig. 2 The thiol group of cysteine can be modified by several reactive agents. Cysteinyl thiol (*yellow box*) can react with several different reactive groups, including: (A) MTS reagents to form a mixed disulfide, (B) maleimide reagents to form a bulky thioester bond, (C) Hg^{2+} ion to form either a disulfide or a Hg^{2+} coordinated cross-link, (D) oxidizing agents such as CuPhen that catalyze disulfide bond formation with adjacent cysteinyl thiols, and (E) ambient oxygen to form higher oxidation states (can also be catalyzed by oxidizing agents). The *red*, emboldened "S" signifies the cysteinyl sulfur group in the cysteine reactant. The *green*, emboldened "R" signifies functional groups to be attached to the thiol.

substitution of native amino acids for cysteines will be tolerated by the transporter. Ideally the target protein will be devoid of native cysteine residues so that the template into which cysteines are introduced is unreactive. Conveniently, cysteines are relatively scarce in membrane proteins, especially bacterial proteins, which, due to their relative tractability, are common subjects in modification experiments. This scarcity of native cysteines makes production of a cysteine-free derivative of the target protein more straightforward, paving the way for very precise posttranslational modifications of only introduced cysteines. Native cysteines are commonly substituted for alanine or serine residues, as these amino acids are similarly sized. However, this side-chain exchange dramatically changes the chemistry of the residue, which may affect transporter stability or function. The functional effects of defanging native cysteines need to be determined empirically as there are many examples of cysteine to serine substitutions that lead to diminution or even complete ablation of transport activity (Mulligan, Fitzgerald, Wang, & Mindell, 2014; Radchenko, Nie, & Lu, 2016; Tsai, Fang, & Miller, 2012; van Iwaarden, Pastore, Konings, & Kaback, 1991). As long as transport activity is measurable, a reduction in activity is not a grave concern as measurable activity demonstrates that the transporter still recognizes its substrates and is able to sample all the conformations necessary for transport. If complete cysteine removal ablates transport activity, other residues can be substituted for the native cysteines. This selection process can be guided by determining which amino acids are tolerated in the equivalent amino acid position to the cysteine in orthologous proteins using multiple sequence alignments (Radchenko et al., 2016).

3. THE MANY USES OF CYSTEINE

Over the years, researchers probing transporter mechanism have unearthed a remarkable number of applications that exploit the useful properties of cysteine. A selection of these applications is presented here.

3.1 Probing Transporter Oligomeric State

Structural analysis has revealed many transporters to form oligomeric complexes, and there are now several examples of dimeric and trimeric transporters in the literature (Dutzler, Campbell, Cadene, Chait, & MacKinnon, 2002; Mancusso, Gregorio, Liu, & Wang, 2012; Mindell, Maduke, Miller, & Grigorieff, 2001; Ressl, van Scheltinga, Vonrhein, Ott, & Ziegler, 2009; Yernool, Boudker, Jin, & Gouaux, 2004). However, with crystal structures (of proteins solubilized in nonnative detergent environments), there is the possibility that the protein:protein interactions observed are physiologically irrelevant and are caused by the crystallization process itself. These doubts can be assuaged if the interface is shown to form in solution or in the lipid bilayer. To demonstrate this, cysteine residues can be introduced into positions across the interface that are predicted to be in close proximity from the crystal structure. Due to the strict distance and orientation dependence of disulfide bond formation, if these cysteines form interprotomer disulfides, then the case for the physiological relevance of the interface is strengthened (Johnson, Cheong, & Lee, 2012). Successful disulfide formation is easily verified by analyzing the treated protein on sodium dodecyl sulfate polyacrylamide gel electrophoresis (SDS-PAGE) gels. Most oligomeric transporters migrate at the monomeric molecular weight on a denaturing SDS-PAGE gel (actually, membrane proteins tend to run a little faster than expected on SDS-PAGE, presumably due to incomplete unfolding); cross-linking across the oligomeric interface would result in higher molecular weight bands corresponding to multiples of the protomer molecular weight.

3.2 Testing Predicted Conformational States Using Intramolecular Cross-Linking

Transporters are highly dynamic proteins and sample many conformational states during the transport cycle. Knowing the structure of these different conformational states and how the transporter moves between them is key to fully understanding the mechanism. Ideally, one would obtain high-resolution crystal structures of the protein in its various conformations with and without its full complement of substrates. However, this is obviously a very difficult task. In the absence of multiple crystal structures, a worthwhile approach that can yield mechanistic insight is to use existing structures to predict alternative conformations, which are then tested using cysteine cross-linking. Disulfide bond formation between cysteines is highly dependent on the distance and orientation of the two participating side chains. Recently, structure-guided cross-linking studies have taken advantage of this property to gain insight into the mechanism of various transporter families, none more so than transporters that utilize an elevator-type mechanism (Coincon et al., 2016; Mulligan et al., 2016; Reyes, Ginter, & Boudker, 2009; Reyes, Oh, & Boudker, 2013; Verdon & Boudker, 2012). The exemplar of this approach is the structural and thermodynamic characterization of the excitatory amino acid transporter (EAAT) homolog, Glt_{Ph} from the archaeon *Pyrococcus horikoshii*. EAATs are essential for the robust clearance of neurotransmitter from the synaptic cleft, and malfunction of these transporters is associated with several disease states, such as ALS (Kanai et al., 2013). A major leap forward in our understanding of these transporters came with publication of the structure of Glt_{Ph} (Yernool et al., 2004). Much insight was gleaned from the initial structure of Glt_{Ph}, but only one conformational state was captured, and it did not reveal how the substrate was translocated across the membrane. To gain more mechanistic insight, Boudker and coworkers took advantage of a previous cysteine cross-linking study on the human Glt_{Ph} homolog, EAAT1, which revealed a pair of cysteines that formed a cross-link in EAAT1, but, when mapped onto the Glt_{Ph} structure, were too distant to cross-link (Ryan, Mitrovic, & Vandenberg, 2004). Structural characterization of the cross-link-trapped version of Glt_{Ph} revealed a radical conformational change in the transporter that moved the entire substrate-binding site through the membrane ~18 Å (Reyes et al., 2009). This work showcased the first example of an elevator-type transporter. Subsequently, cysteine cross-linking was used to trap Glt_{Ph} in the IFS and the OFS to facilitate thorough thermodynamic analysis (Reyes et al., 2013); was used to trap Glt_{Ph} in intermediate conformations between the IFS and OFS (Verdon & Boudker, 2012); and has been used to probe the contributions of peripheral loops to the transport cycle (Mulligan & Mindell, 2013). More recently, intramolecular cysteine cross-linking was instrumental in demonstrating that two other transporter families, completely unrelated to the EAATs, also employ elevator-type mechanisms (Coincon et al., 2016; Mulligan et al., 2016).

3.3 Monitoring Conformational Change

The specificity and precision of cysteine labeling have regularly been exploited to introduce biophysical probes into particular regions of transporters and channels to report on conformational dynamics. Two powerful biophysical approaches that have been particularly useful in measuring conformational dynamics of transport cycles are electron paramagnetic resonance (EPR) spectroscopy and fluorescence spectroscopy, which includes single-molecule Förster resonance energy transfer (smFRET) (Akyuz, Altman, Blanchard, & Boudker, 2013; Akyuz et al., 2015; Claxton, Kazmier, Mishra, & Mchaourab, 2015; Erkens, Hänelt, Goudsmits, Slotboom, & van Oijen, 2013; Georgieva, Borbat, Ginter, Freed, & Boudker, 2013; Hänelt, Wunnicke, Bordignon, Steinhoff, & Slotboom, 2013; Kazmier, Sharma, Islam, Roux, & Mchaourab, 2014; Martens et al., 2016). These solution-based methods can be used to report on local environmental changes of a single probe as well as distance changes between two partner probes specifically conjugated to different sites on the protein. These data can be used to assess how protein domain movement is modulated by interaction with substrate and/or lipid, the kinetics of conformational changes, and the rate-limiting steps of the transport cycle.

3.4 Topological Mapping of Transporters

Prior to the explosion of transporter structures in the early 2000s, the substituted cysteine accessibility method (SCAM) was an important tool used to reveal structural detail and topologically map transporters and channels (Akabas, 2015; Akabas, Stauffer, Xu, & Karlin, 1992; Bogdanov, Zhang, Xie, & Dowhan, 2005). SCAM remains a powerful tool for transporters with no known structure, especially when combined with computational modeling approaches (Butler, Davis, Bari, Nicholson, & Ruiz, 2013; Riegelhaupt, Frame, & Akabas, 2010). In this method, single-cysteine residues are inserted into the protein at regular intervals, which can be guided by structural models. The protein is expressed and while the protein is still in the lipid bilayer, the accessibility of each cysteine is assessed. There are now many variants of the SCAM procedure, but in general, to assess cysteine accessibility, each cysteine mutant is treated with an easily detectable, membrane-impermeable thiol-reactive reagent, such as the fluorescent probe, fluorescein-5-maleimide (F5M) (Poelarends & Konings, 2002), or biotin-linked reagents such as 3-(*N*-maleimidylpropionyl) biocytin (MPB) that can be tightly bound by a ~50 kDa streptavidin derivative and detected by a shift in the protein band on an SDS-PAGE gel (Matos, Fann, Yan, & Maloney, 1996). Only cysteine residues that are solvent exposed will react with the probe; thus, the amino acid position can be assigned as extracellular accessible, cytoplasmic accessible, or transmembrane, the latter of which will likely be inaccessible/nonreactive to the probes. To assign these positions, SCAM can be performed on whole cells, where only cysteines exposed to the extracellular environment will be accessible (Bogdanov et al., 2005). Alternatively, membrane vesicles can be prepared from the cells that are inside-out (ISO), where the normally cytoplasmic cysteines are exposed to the externally applied reaction buffer, or right side-out (RSO) that should mimic the accessibility seen for whole cells (Bogdanov et al., 2005). Proteins reconstituted into liposomes can also be probed using this method (see Section 3.5). Strategies used to decide on the placement of cysteine residues are detailed in Section 6.1. One of the more prodigious displays of the SCAM approach was in the characterization of lactose permease (LacY) (Frillingos, Sahin-Tóth, Wu, & Kaback, 1998). In a Herculean effort, Kaback and coworkers mutated every amino acid residue in LacY to cysteine (one at a time; 401 single-cysteine mutants in total); using this system Kaback pioneered the use of many cysteine-modification strategies and used them to reveal mechanistic aspects of transport that were later supported by the crystal structure of LacY (Abramson et al., 2003).

For proteins of unknown structure, the SCAM approach is an excellent tool to map solvent accessible loops and the boundaries of transmembrane helices (Bogdanov et al., 2005; Karlin & Akabas, 1998). In addition, for structurally characterized transporters, this method can be used to probe substrate-dependent conformational changes. Ideally, the accessibility of each cysteine should be determined by measuring the rates of modification. However, cysteine accessibility has also been measured by assessing the extent of reaction after a fixed length of time (Jacobs, Zhang, Campbell, & Rudnick, 2007; Tavoulari, Forrest, & Rudnick, 2009; Tavoulari et al., 2016).

3.5 Determining and Selecting Transporter Orientation in Proteoliposomes

The gold standard for probing the functional mechanism of transporters is to reconstitute the purified transporter into proteoliposomes. Proteoliposomes give unrivaled control over the internal and external solutions across the membrane, which allows for tightly controlled application of electrochemical gradients. Proteoliposomes are the system of choice to unambiguously determine fundamental attributes of transporters, such as substrate stoichiometry and coupling ion preference (Fitzgerald, Mulligan, & Mindell, 2017; Mulligan et al., 2014, 2009). However, one drawback to using proteoliposomes is that the transporters are often inserted into the membrane randomly, resulting in a mixture of ISO and RSO protein (Jin, Nair, & van Veen, 2014; Mulligan et al., 2014; Ryan, Compton, & Mindell, 2009). This orientational heterogeneity has several technical ramifications; the transport rates may differ between the orientations, which in a mixed orientation system makes detailed kinetic analysis impossible; and binding and transport analysis of inhibitors/substrates that only recognize one façade of the protein becomes murky. Unfortunately, the proportion of ISO and RSO transporters in proteoliposomes is impossible to know ab initio and there are currently no tried-and-true methods of manipulating the reconstitution procedure to favor one orientation. Cysteine labeling is both a convenient method of determining transporter orientation and, using approaches recently developed, can also be used to isolate particular orientations. In a similar approach to the SCAM studies (see Section 3.4), the proportion of ISO and RSO transporters in the membrane can be assessed by introducing a single cysteine into a region of the protein that is only accessible from one side of the membrane. After reconstitution, the proteoliposomes are incubated with a membrane-impermeable reporter reagent, such as a thiol-reactive fluorophore, that will react only with externally exposed cysteine residues (Ryan et al., 2009). Alternatively, the externally exposed cysteines could be blocked by a membrane-impermeable thiol-reactive reagent, the membrane solubilized, and the luminal cysteines labeled with thiol-reactive reporter (Mulligan et al., 2014). Either way, by comparing the amount of labeled protein, quantified by SDS-PAGE, to a fully labeled sample (where the membranes were solubilized prior to reaction with the reporter reagent), one can approximate the ratio of the two orientations.

More often than not, reconstitution results in a mixture of transporter orientations. However, using a similar process to the one described earlier, it is possible to functionally isolate the transporters in one of the orientations (Tsai et al., 2012). In this approach, a single cysteine is inserted into a position on the transporter that is only accessible from one side of the membrane and completely inactivates the protein upon modification by a membrane-impermeable reagent. In this way, only the population of transporters with the cysteine exposed to the external solution is inactivated, thus functionally isolating the population with the opposite orientation. This approach has been very useful in characterizing the virtual proton pumps; arginine/agmatine exchanger (AdiC) and the glutamate/γ-aminobutyric acid exchanger (GadC) from enteric bacteria; and the EAAT homolog Glt_{Ph} (McIlwain, Vandenberg, & Ryan, 2016; Tsai, McCarthy, & Miller, 2013; Tsai & Miller, 2013).

3.6 Monitoring Transporter: Ligand Interactions

Transporters undergo conformational changes during the binding and transport of substrates that alters the environment of specific regions of the protein; for example, the substrate-binding site may go from a polar, solvent-exposed environment to a buried, hydrophobic one. These environmental changes can be monitored using environment-sensitive fluorophores conjugated to cysteine residues and exploited to probe ligand-binding properties and local conformational changes (Rahman et al., 2016; Veenstra, Lanza, Hirayama, Turk, & Wright, 2004).

Another cysteine-based method of monitoring transporter:ligand interactions is a repurposed high-throughput microscale thermal stability assay originally designed to screen conditions in which the protein of interest is most stable (Alexandrov, Mileni, Chien, Hanson, & Stevens, 2008). As protein stability positively correlates with the protein's propensity to yield useful crystals for structural analysis (Sonoda et al., 2011), this method was originally designed to expedite precrystallization buffer screening. However, the thermostability of a protein also increases upon binding ligands (Allison et al., 2015; Celej, Montich, & Fidelio, 2003), and for transporters this also includes interactions with lipids (Crichton et al., 2015), resulting in a tool that can quickly screen large libraries of compounds for interactions with the transporter of interest. This particular approach relies on native (or introduced), buried cysteine residues becoming exposed as the protein is thermally denatured (in an quantitative-PCR machine) and reacting with the cysteine-reactive fluorophore *N*-[4-(7-diethylamino-4-methyl-3-coumarinyl)phenyl]maleimide (CPM), which is only fluorescent once conjugated to cysteine (Alexandrov et al., 2008). The CPM-based approach is particularly useful for transporters, and other integral membrane proteins, as the dyes used in other variants of the fluorescent thermal stability assays, such as SYPRO Orange, bind to hydrophobic moieties (Yeh, McMillan, & Stowell, 2006). This characteristic is ideal for soluble protein where the hydrophobic core is exposed as the protein is denatured, but is not as readily applied to transporters because the hydrophobic nature of the transporter leads to insuperable background fluorescence levels.

4. CYSTEINE-REACTIVE REAGENTS USED FOR CROSS-LINKING AND LABELING

The unique nature of cysteine makes it reactive to a number of different reagents. Each member of this diverse group of labeling and cross-linking moieties has advantages and disadvantages depending on the goal of the experiment and downstream uses of the labeled/cross-linked protein.

4.1 Methanethiosulfonate Reagents

Methanethiosulfonate (MTS) reagents comprise a very large and varied suite of cysteine-reactive compounds that have been used extensively to probe the structure and function of transporters and channels (Akabas et al., 1992; Jacobs et al., 2007; Karlin & Akabas, 1998; Tavoulari et al., 2009). The MTS moiety reacts preferentially with the thiolate ion of cysteine producing a mixed disulfide during the reaction (Fig. 2B). Labeling and cross-linking using MTS reagents offer a number of benefits; the reactive MTS group is small and so will likely have minimal impact on the structure and function of the transporter of interest (Fig. 2B); the reaction is fast (at least compared to maleimides) under mild conditions; most MTS reagents are hydrophilic so will only target solvent-exposed cysteines; and the labeling reaction can be reversed or blocked by addition of common reducing agents, for example, dithiothreitol (DTT), tris(2-carboxyethyl)phosphine (TCEP), or β-mercaptoethanol (BME). These characteristics make MTS labeling particularly suited to the study of transporters and channels using electrophysiological measurements where the effects of the posttranslational modification can be observed in real time (Akabas, Cheung, & Guinamard, 1997; Holmgren, Liu, Xu, & Yellen, 1996).

A large suite of MTS reagents have been developed that allow proteins to be labeled with a variety of functional groups, including fluorescent probes like fluorescein, Cy3 and Cy5; charged groups (MTSEA and MTSET), neutral bulky groups, and spin labels for use in EPR spectroscopy. Due to their hydrophilicity, most MTS reagents are membrane impermeable, allowing for selective extracellular labeling. MTSEA, on the other hand, is membrane permeable, which facilitates convenient intracellular labeling without having to disrupt the membrane (Holmgren et al., 1996). The MTS moiety has also been utilized in homobifunctional cross-linkers of varying linker length, which have been used to characterize transporter dynamics (Mulligan & Mindell, 2013; Zhou, Guan, Freites, & Kaback, 2008).

MTS reagents hydrolyze in water and have a relatively short half-life, the rate of which depends on the functional group attached to the MTS moiety; for example, MTSEA and MTSES have $t_{1/2}$s of 12 and 370 min, respectively, under the same conditions (Karlin & Akabas, 1998). Therefore, MTS reagents should be freshly made just prior to the experiment. Care should also be taken when using homobifunctional MTS reagents for cross-linking studies. If the rate of reaction is too high, then each cysteine will be labeled with an MTS cross-linker, rather than a single cross-linker connecting them. To mitigate this effect, the concentration of the cross-linker could be reduced or the pH could be lowered to increase the proportion of thiol (−SH), which has a slower reaction rate than the thiolate ion ($-S^-$) with MTS reagents (Mulligan & Mindell, 2013).

4.2 Maleimides

Maleimide-based labels and cross-linkers react with cysteine to form a thioester bond, which is reported to be resistant to reversal by reducing agents (Fig. 2C). This oft-repeated boon of maleimide chemistry is one of the main advantages maleimides have over MTS reagents, especially if subsequent reactions utilize reducing agents. However, evidence in the literature suggests that the reducing agent TCEP can react with maleimides (Shafer,

Inman, & Lees, 2000), and in our hands, maleimide-based cross-linking is completely prevented by the presence of TCEP (C. Mulligan & J. A. Mindell, unpublished data). Like MTS reagents, a suite of maleimide reagents accommodating an assortment of functional groups has been developed to modify cysteines. However, maleimides have some disadvantages when compared to MTS reagents: the reactive moiety is bulkier (Fig. 2C), the reaction rate is considerably slower, and the pH has to be carefully monitored as the maleimide reacts faster with the thiolate ion ($-S^-$) so a pH >7 is ideal; however, at >8 maleimides will react with deprotonated primary amines in addition to cysteines.

4.3 Oxidizing Agents

Several oxidizing agents have been used to catalyze the formation of disulfides in mechanistic studies of transporters (Fig. 2D), including I_2, sodium tetrathionate, and H_2O_2 (Dobrowolski & Lolkema, 2010; Monette, Somasekharan, & Forbush, 2014; Nguitragool & Miller, 2007; Wu & Kaback, 1996). However, the most commonly used oxidizing agent used for the study of transporters is copper phenanthroline (CuPhen) (Brocke, Bendahan, Grunewald, & Kanner, 2002; Crisman, Qu, Kanner, & Forrest, 2009; Groeneveld & Slotboom, 2007; Mulligan et al., 2016). CuPhen is thought to catalyze disulfide formation between cysteines by first reacting with ambient oxygen to produce oxyradicals, which then interact with the thiol groups on accessible cysteines to produce highly reactive sulfur radicals. Collision between two similarly radicalized cysteines results in a disulfide bond (Careaga & Falke, 1992). Due to the small size of the oxyradical, this method can catalyze cross-link formation between cysteines inaccessible to bulkier cross-linkers. In addition, the participating cysteine residues need to be very close together to form a CuPhen-catalyzed disulfide; therefore, cross-linking with CuPhen may have a slightly higher resolution for mapping interacting residues than other cross-linkers. This higher resolution could, however, lead to a lack of cross-linking for slightly misaligned cysteine side chains that would be detected by a cross-linker with a longer linker length. Aside from catalyzing a disulfide, treating cysteine mutants with CuPhen may also produce higher oxidation states of the individual cysteines rendering them inert (Lo Conte & Carroll, 2013). However, an attractive benefit of using CuPhen for cysteine cross-linking is that the cross-linking reaction is readily quenched by addition of excess the metal chelator, EDTA.

4.4 Hg^{2+}-Induced Cross-Linking

The heavy metal, Hg^{2+}, has a high affinity for cysteinyl thiolate groups and is able to cross-link two cysteines by directly coordinating both thiolate groups or by inducing disulfide bond formation (Fig. 2E; Arnon & Shapira, 1969). Researchers have taken advantage of this property and used Hg^{2+} cross-linking to study a number of transporters (Basilio et al., 2014; Mulligan et al., 2016; Reyes et al., 2009, 2013; Soskine, Steiner-Mordoch, & Schuldiner, 2002; Verdon & Boudker, 2012).

Using Hg^{2+} to induce cross-link formation has a number of advantages over other cross-linkers. Relative to other cross-linkers, Hg^{2+}-induced cross-link formation is fast, and due to its diminutive size, Hg^{2+} is more penetrative and is able to reach the nooks and crannies of the transporter inaccessible to bulkier cross-linking agents. For example, the EAAT homolog Glt_{Ph} can be trapped in its OFS by cross-linking the cysteine residues C66 and C300 with Hg^{2+} (Reyes et al., 2013). In our hands, the *only* method of cross-linking this cysteine pair was with Hg^{2+} treatment; attempts to cross-link these cysteines with the oxidizing agent CuPhen or the homobifunctional methanethiosulfonate, MTS-1-MTS, proved unsuccessful (C. Mulligan & J. A. Mindell, unpublished data; Mulligan & Mindell, 2013).

Despite being a very effective cross-linking agent, using Hg^{2+} for cysteine cross-linking has some foibles that may preclude its use in certain experiments. First, when a Hg^{2+}-induced cross-link forms, it is not known whether Hg^{2+} is coordinated by the two sulfhydryl groups or induces disulfide bond formation. Binding of Hg^{2+} to cross-linkable Glt_{Ph} cysteine mutants, measured by isothermal titration calorimetry (ITC), suggests a single Hg^{2+} ion binds per double-cysteine mutant (in Glt_{Ph}'s case, three ions per trimer) (Reyes et al., 2013; Verdon & Boudker, 2012). However, in X-ray structures of Hg^{2+} cross-linked transporters where the Hg^{2+} ion is visible between the cysteine side chains, the ion binding has low occupancy (estimated to be 30% for Glt_{Ph}) (Basilio et al., 2014; Reyes et al., 2009). This observation suggests that Hg^{2+} can form the initial cross-link between cysteines, but the ion is lost over time as the cysteines are oxidized to disulfides (Reyes et al., 2009). Second, Hg^{2+} binds tightly to single cysteines (Arnon & Shapira, 1969), which can affect the function of single-cysteine control proteins (C. Mulligan & J. A. Mindell, unpublished data). In addition, Hg^{2+} treatment can also reduce the activity of cysteine-free transporters, suggesting that Hg^{2+} interacts with other regions of the protein as well (Basilio et al., 2014). Finally, the Hg^{2+}-induced cross-linking reaction is not easy to quench, which can affect the trustworthiness of SDS-PAGE-based cross-link screening. If not properly quenched, cysteines not normally vicinal in the folded protein can cross-link when the protein is sodium dodecyl sulfate (SDS) denatured. The metal-chelating agent, EDTA, does not prevent Hg-induced cross-link formation, so cannot be used to quench the reaction (C. Mulligan & J. A. Mindell, unpublished observations). Excess Hg^{2+} can be removed from a reaction by buffer exchange. However, single-cysteine residues hold Hg^{2+} ions tightly, so the cysteine is "primed" to quickly react with any cysteine it comes in contact with once denatured (Mulligan & Mindell, 2013). In our hands, quenching the Hg^{2+}-induced cross-linking reaction by adding excess alkylating agent (methyl-MTS (MMTS), *N*-ethyl maleimide (NEM)) to protect unreacted cysteines also proved ineffective.

5. METHODS OF DETECTING CROSS-LINK FORMATION

One of the more challenging aspects of using intermolecular cross-linking to inform mechanism is unambiguously determining that a cross-link is actually formed. Whereas the labeling of a single cysteine usually results in a readily detectable physicochemical change in the protein, such as a shift in the molecular weight of the protein or an expansion of the protein's fluorescence emission profile, unequivocally ascertaining cross-link formation can be surprisingly challenging. In most cases, multiple lines of evidence are necessary to show convincingly that a disulfide or cross-link has formed. The detection methods can be split into two groups: direct and indirect detection methods.

5.1 Indirect Detection Methods

5.1.1 SDS-PAGE Band Shift

The simplest method of detecting intramolecular cross-link formation is to look for altered electrophoretic migration rate of the transporter when analyzed with denaturing SDS-PAGE gel (Fig. 3A). This approach makes assignment of cross-link formation easy and can also be utilized as a handy readout of the conformational states sampled, which can be exploited to probe the role of substrates, protein domains, and loops on conformational dynamics (Mulligan & Mindell, 2013). However, only two transporters to date have displayed this migration rate shift upon cross-link formation (Radchenko et al., 2016; Reyes et al., 2009; Verdon & Boudker, 2012). Therefore, this handy cross-link reporting tool is unlikely to be a feature

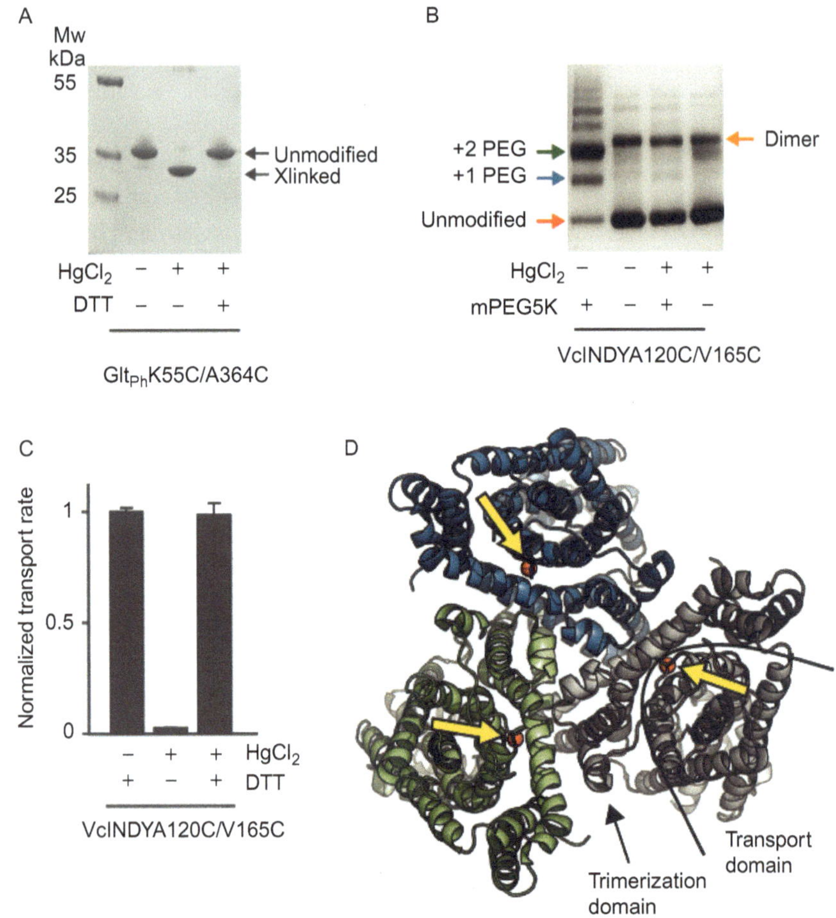

Fig. 3 Methods of determining cross-link formation. (A) Band-shift assay. SDS-PAGE gel showing the electrophoretic migration of Glt_{Ph}K55C/A364C cysteine mutant in the presence and absence of $HgCl_2$ and DTT treatment. Cross-linking Glt_{Ph}K55C/A364C with Hg^{2+} results in increased electrophoretic migration. Treatment of cross-linked protein with DTT returns the protein to the migration rate of untreated protein. (B) Cysteine-protection assay. SDS-PAGE showing the electrophoretic migration of VcINDYA120C/V165C cysteine mutant before and after treatment with $HgCl_2$, mPEG5K, or both. Treatment with mPEG5K alone results in singly (+1 PEG) and doubly PEGylated (+2 PEG) protein. Pretreatment with Hg^{2+} or treatment with Hg^{2+} alone results in the same migration pattern as untreated protein. The protein bands corresponding to unmodified monomer and dimer of VcINDY are indicated with *arrows*. (C) Transport activity assay. Normalized rate of [^{3}H]-succinate transport into VcINDYA120C/V165C-containing proteoliposomes after treatment with $HgCl_2$, DTT, or both. Cross-linking the protein with Hg^{2+} abolishes activity. Activity returns upon treatment of the Hg^{2+} cross-linked proteoliposomes with DTT. (D) *Top-down view* of the crystal structure (PDB ID: 3KBC)

of most transporters. Indeed, most transporters are indistinguishable in their cross-linked and noncross-linked forms on SDS-PAGE (Basilio et al., 2014; Coincon et al., 2016; Khantwal et al., 2016; Mulligan et al., 2016). For transporters where this approach is applicable, caution should be taken in drawing conclusions based on SDS-PAGE migration data alone because this approach can produce to false positives (Verdon & Boudker, 2012). Several double-cysteine mutants in Glt_{Ph} resulted in band shifts upon treatment with both $HgCl_2$ and CuPhen. However, subsequent crystallization of the cross-link-treated protein revealed no cross-link. Hg^{2+} cross-linkable mutants bind Hg^{2+} in a 1:1 ratio with protein when analyzed by ITC—indicative of Hg^{2+} coordination by the two sulfhydryl groups (Verdon & Boudker, 2012). Binding analysis of the false positive Glt_{Ph} mutants revealed a Hg^{2+}:protein ratio of 2:1, indicating each cysteine coordinated a Hg^{2+} ion without forming a cross-link (Verdon & Boudker, 2012). Why treatment of these particular double-cysteine mutants caused them to migrate faster on SDS-PAGE when treated with cross-linker remains a mystery, but researchers should be mindful of the possibility of false positives.

The key to SDS-PAGE-based analysis of cross-link formation is thoroughly quenching the reaction before preparing the sample for analysis. Quenching the cross-linking reaction prevents aberrant in-gel cross-link formation, while the protein is in a SDS-induced quasidenatured state. There are many ways in which these reactions can be quenched; (1) label all unreacted cysteine with a thiol-reactive reagent such as NEM or MMTS. In our experience, this is the most effective quenching method and simply adding excess MMTS (we use 100 mM for 10 μM protein) to the SDS-PAGE sample buffer is sufficient to label all unreacted cysteines (Mulligan et al., 2016; Mulligan & Mindell, 2013). (2) The cross-linking agent could be removed using buffer exchange (we routinely use Zeba Spin Desalting columns, ThermoFisher), although treating unreacted cysteines in

of Glt_{Ph}K55C/A364C cysteine mutant cross-linked with Hg^{2+} ion (*red spheres*, indicated with *yellow arrows*). The Glt_{Ph} trimer is shown with each protomer *colored* differently (*blue*, *green*, and *white*). For the *white* protomer, the transport domain is indicated, which houses all the substrate-binding residues, and the trimerization domain, which forms all the contacts at the oligomeric interface; the Hg^{2+} ion is located between these two domains. *Panels (B) and (C) are adapted from figures originally published in Mulligan, C., Fenollar-Ferrer, C., Fitzgerald, G. A., Vergara-Jaque, A., Kaufmann, D., Li, Y., et al. (2016). The bacterial dicarboxylate transporter VcINDY uses a two-domain elevator-type mechanism.* Nature Structural & Molecular Biology, *23(3), 256–263. http://doi.org/10.1038/nsmb.3166.*

conjunction is recommended to prevent aberrant reactions downstream. (3) Excess cross-linker could be mopped up to prevent further cross-linking. A CuPhen-induced cross-linking reaction can be effectively quenched by addition of the chelating agent EDTA (Groeneveld & Slotboom, 2007). To remove excess MTS or maleimide-containing cross-linkers, a thiol-containing compound, such as cysteine, could be added to quench the reaction. However, care should be taken here because excess cysteine can break disulfides by a cysteine exchange mechanism, which will result in no apparent cross-linking (C. Mulligan & J. A. Mindell, unpublished data; Gilbert, 1995). (4) Cross-linking reagents predominantly react with the ionized thiolate form of cysteine; therefore, cross-linking reactions can be quenched by dramatically reducing the pH. However, this may render the transporter inactive, potentially precluding subsequent functional studies on the treated protein. The rate of reaction with the quencher must exceed the rate of disulfide bond or cross-link formation. Therefore, MTS would be preferable over maleimide-based quenchers as MTS reagents have considerably faster reaction rates ($1 \times 10^5 M^{-1} s^{-1}$ for MTS compared to $1.5 \times 10^3 M^{-1} s^{-1}$ for maleimides; Li et al., 2002; Stauffer & Karlin, 1994). Acid quenching the reaction is basically diffusion limited so may be a good option if feasible within the subsequent experimental framework (Gilbert, 1995). It should be noted that Hg-induced cross-linking is very hard to quench using the methods described earlier as detailed in Section 4.4.

5.1.2 Protection From Modification

Another method of indirectly monitoring cross-link formation is to enumerate the free thiols available before and after treatment with cross-linker. Free thiols can be counted by treating the transporter with a thiol-reactive reporter such as methoxypolyethylene glycol maleimide (mPEG5K) that adds 5 kDa mass to the protein when it is conjugated to a cysteine, which is easily discernable on a polyacrylamide gel (Fig. 3B; Basilio et al., 2014; Coincon et al., 2016; Mulligan et al., 2016). In the absence of cross-linker, both cysteines should be available to label and, upon labeling, will add 10 kDa mass to the protein (5 kDa per cysteine) which can be visualized using SDS-PAGE analysis. If the cysteines are not readily accessible to modification in the folded protein, SDS can be added to partially unfold the protein and allow the reporter unfettered access to the cysteines. In practice, and depending the relative accessibility of each cysteine, all three variants of the transporters will be discernable on the gel: unmodified, singly PEGylated, and doubly PEGylated protein (Basilio et al., 2014; Mulligan et al.,

2016). If the protein is fully cross-linked, then neither cysteine will be available to react with the reporter and the mass of the protein remains at the unmodified level (Fig. 3B).

When visualizing the treated protein on polyacrylamide gels, it is essential to select an appropriate acrylamide percentage to ensure sufficient separation of the unmodified and PEGylated protein. If the transporter is oligomeric, the majority of the unmodified protein will migrate as a monomer on a denaturing acrylamide gel. However, there is the possibility that a small population of higher oligomeric states of the protein may also be visible on the gel and may migrate at a similar rate to the PEGylated monomer, potentially confounding the results (Fig. 3B; Mulligan et al., 2016). Other PEG-maleimide variants that add a larger mass to the protein (e.g., 10 kDa per cysteine modification) may be used to alleviate this issue, if required.

Fluorescent thiol-specific reagents, such as F5M, can also be used to reveal free thiols to indirectly assess cross-link formation (Radchenko et al., 2016; Reyes et al., 2009). In the absence of cross-linker, both cysteines should react with the dye, producing a highly fluorescent conjugate of the transporter. The extent of fluorescent labeling can also assessed using acrylamide gels and a gel documentation system with an appropriate filter. If the cysteines are involved in cross-linking, then the fluorophore will not label the protein and it will appear nonfluorescent during analysis.

An advantage of using PEG-maleimide over a fluorescent reporter is that the extent of labeling can easily be determined using mPEG5K because any unlabeled protein will remain at the unmodified level in the gel. The fluorescence-based approach does not give the same readout; under a given set of labeling conditions, it is not easy to tell what percentage of the protein is labeled.

This protection assay is an excellent screening approach. However, a lack of labeling may result from factors other than cross-link formation. For example, the cysteine may become oxidized by treatment with the cross-linking agent, especially CuPhen, rendering the cysteine inert. Extra care must be taken when cross-linking with MTS and maleimide-based cross-linkers because, if the concentration of cross-linker is too high, instead of cross-linking the two cysteines together, both cysteines will react with separate cross-linking compounds (Mulligan & Mindell, 2013). This will prevent cross-linking from occurring between the cysteines, but will also protect the cysteines from modification with mPEG5K or the fluorescent reporter, also giving rise to false positives. Due to the possibility of false positives, this method should be accompanied by complimentary approaches to ascertain cross-link formation.

5.1.3 Proteolysis Coupled With SDS-PAGE Analysis

If a cross-link-induced band shift is not apparent, then another acrylamide gel-based method of cross-link detection is to introduce a specific protease cleavage site in a loop between the two cysteine residues predicted to cross-link. If a cross-link is formed between the cysteines, then cleaving the protein will result in the protein migrating similar to full-length protein. Conversely, if cross-links do not form, then the protein will migrate as two fragments (Mulligan & Mindell, 2013; Yang et al., 2005; Zhou et al., 2008). Due to the very different band profiles produced, this approach is able to clearly distinguish between cross-linked from noncross-linked cysteine pairs.

In addition to robust quenching covered in Section 5.1.1, other issues need to be considered to successfully apply this approach. First, a protease cleavage site needs to be introduced into the protein that does not interfere with transport function. The protease cleavage site can be introduced as a cassette or residues in the protein can be mutated to formulate the cleavage site. However, the shortest, specific protease site is four amino acids for the protease Factor Xa (–ID/EGR–), so a sufficiently long loop should be selected that can accommodate the site. The protease site has to be accessible and fully cleavable. In most cases, the protein will not be used for functional studies after cleavage, so harsh conditions (increased temperature and time) may be used as long as they do not affect the integrity of the cross-link.

A variation of this approach is to express and purify a split transporter where the protein is expressed in two separate fragments, with each fragment containing a cysteine (Adler & Bibi, 2004; Bibi & Kaback, 1990; Dobrowolski & Lolkema, 2010; Yu, Kono, McKee, & Oprian, 1995). Here, the uncross-linked protein will run as two fragments on an acrylamide gel, but when cross-linked, protein will run as a single, higher molecular weight band close to the molecular weight expected for wild-type protein. Obviously, the prime requirement for this approach is that the protein can tolerate such manipulations. However, if a split transporter is possible, it offers a convenient method of cross-link assessment.

Cross-link formation can also be assessed using limited proteolysis, which utilizes a protease that recognizes multiple, naturally occurring cleavage sites throughout the protein, such as trypsin (cleaves at Arg and Lys residues) or chymotrypsin (cleaves at hydrophobic amino acids like Tyr, Trp, and Phe).

In this approach, the protein of interest is treated with cross-linker and then fragmented with the protease of choice (we recommend screening multiple proteases as each will produce a different fragment profile). The peptide fragments are separated on SDS-PAGE and compared to the fragment profile of noncross-linked protein (Reyes et al., 2009). Depending on the position of the accessible protease cleavage sites in relation to the cross-link, the presence of a cross-link can be identified by the appearance of an extra higher molecular weight band in the cross-linked fragment profile at the cost of two smaller molecular weight bands. The fragments can be excised and identified using peptide mass fingerprinting. Alternatively, cross-link trapping the transporter in one conformation may protect a protease site that is normally accessible if the transporter is allowed to fluctuate between its conformational states.

5.1.4 Activity Measurements

Measuring the transporter's functional activity is an effective method of assessing intramolecular cross-link formation (Fig. 3C). Transporters are highly dynamic and their function relies on their ability to readily sample many conformations. Intramolecular cross-links used to probe mechanism are designed to staple regions of the protein together that may only be vicinal in certain conformations during the transport cycle. If the conformational changes to and from this conformation are essential for the transport activity, then trapping the protein in this conformation will impair transport activity. Indeed, not only can it be a means of assessing cross-link formation, but it also reveals whether the conformational changes predicted are essential for transport.

Unless the transporter in question transports naturally fluorescent substrates or a substrate for which there are sensitive and specific dyes, the method of choice for measuring transport activity is to monitor the flux of a radiolabeled substrates using liquid scintillation. Transport activity can be measured in several biological samples with varying degrees of processing required; the three most common are whole cells, membrane vesicles, and proteoliposomes. Monitoring transport into whole cells requires minimal processing, but the specific activity of the transporter in question is contaminated by the activity of native transporters, and it is impossible to control the internal solution, making the application of specific gradients difficult. Membrane vesicles can be produced where the internal solution can be readily modulated, but this system is still contaminated with native transporters. Proteoliposomes are the system of choice as the protein of interest is reconstituted in isolation, and the experimenter has complete control over both internal and external solutions. However, as discussed in Section 3.5, the protein is often inserted in a mixture of ISO and RSO orientations.

Whereas whole-cell transport assays are in theory less labor intensive and faster to prepare than membrane vesicles and proteoliposomes, other issues such as the properties of the cross-linker being used and the position of the cysteines in the protein need to be considered. If the cysteines are accessible only from the cytoplasmic face of the protein, then whole-cell assay cannot be used unless a membrane permeable cross-linker is employed. As the internal solution of membrane vesicles and proteoliposomes can be easily modified, the choice of cross-linker is less constrained in these systems. In addition, cross-link formation would need to be verified by some other means (see previous sections), which would require at least partial purification of the transporter, and once purified protein is obtained, the effort required to produce proteoliposomes is considerably reduced.

Regardless of the experimental system used, to obtain a reference for transport activity, a control experiment should be included where no cross-link is induced. As disulfides can spontaneously form by air oxidation, precautions should be taken to prevent aberrant cross-linking, the most effective of which is to add a reducing agent, such as DTT, TCEP of BME to the reaction buffers. The reducing agents themselves can have effects on the transport activity. For example, transport of succinate by wild-type and cysteine-free VcINDY was potentiated by the presence of DTT, giving rise to an abnormally high basal transport rate (Mulligan et al., 2016). In addition, low concentrations (0.5 m*M*) of TCEP prevented cysteine oxidation and did not affect transport activity of VcINDY. However, a moderate increase in TCEP concentration (up to 5 m*M*) completely inhibited succinate transport, presumably due to TCEP, which shares some structural features with succinate, competing for a shared binding site (C. Mulligan & J. A. Mindell, unpublished observations). Although the mechanisms by which these reducing agents are affecting transport remain unknown and are likely specific to VcINDY, an assessment of their effects on the activity of the transporter of interest is prudent.

As it still poses a considerable technical challenge to express and purify eukaryotic membrane proteins, most cross-linking studies in these systems are performed in whole cells and protein activity is used as the readout for cross-link formation. Transport of radiolabeled substrate into whole cells can be monitored (Crisman et al., 2009; Silverstein, Crisman, Forrest, & Kanner, 2013), or if the protein of interest an ion channel or an electrogenic transporter then electrophysiology methods can be used to monitor activity (Ryan et al., 2004). Due to their tractability, relative ease of use, and available structural information, bacterial transporters often act as models for entire transporter families. However, it is important to verify that findings in bacterial proteins translate to their eukaryotic cousins. In vivo cross-linking coupled to activity measurements is a straightforward, relatively scalable method of model validation.

5.2 Direct Detection Methods

5.2.1 Obtain a Crystal Structure of the Cross-Linked Protein

The most direct evidence of cross-link formation is to obtain a high-resolution structure of the cross-linked protein (Fig. 3D). More importantly than confirming cross-link formation, obtaining a high-resolution structure of the cross-linked protein will provide invaluable information about the cross-linked conformer that is not attainable with any of the other methods mentioned here. Clearly, obtaining high-resolution structures is a difficult task. Much success has been achieved in obtaining alternative conformations by elucidating the structures of homologous proteins (Coincon et al., 2016; Jensen, Guskov, Rempel, Hänelt, & Slotboom, 2013; Quistgaard, Löw, Guettou, & Nordlund, 2016). However, there is no guarantee of success, and even if a structure is obtained, there is a strong possibility that the structure of the homolog will be in the same conformation as the original protein of interest. Cross-linking coupled to structural studies offers a more directed approach to obtaining structural information of multiple conformations of the same transporter. There are now several crystal structures of cross-linked transporters, including Glt_{Ph} and NapA that revealed very large conformational changes between the cross-linked and unrestrained structures (Coincon et al., 2016; Reyes et al., 2009); and CLC, whose structure barely changed when cross-linked, indicative of subtle conformational changes (Basilio et al., 2014; Khantwal et al., 2016).

5.2.2 Use of Mass Spectroscopy to Unambiguously Identify the Cross-Linked Peptides

A powerful approach to unambiguously determine the presence of a cross-link is to use mass spectrometry. In this approach, the protein of interest harboring a cross-linkable cysteine pair is treated with cross-linker under conditions that will achieve full cross-linking. The cross-linked protein is digested using a specific protease, which is usually trypsin, but other proteases with different recognition sites, such as chymotrypsin and pepsin, are available and often required. The peptides produced from the protein digestion are separated using liquid chromatography, which is directly linked to a mass spectrometer. The identity of each eluted peptide is identified using tandem mass spectrometry (MS/MS) (Mulligan et al., 2016). A control sample is also analyzed in which protein is maintained in reducing agent to prevent cross-link formation. Comparison of the peptide profile of cross-linked vs uncross-linked protein should reveal a cross-linked peptide in the former, which is not present in the latter. Reciprocally, the two peptides harboring the cysteine residues should be apparent in the uncross-linked sample, but missing in the cross-linked sample.

While this approach can unambiguously identify cross-linked peptides, it is not without limitations. One major issue with this approach is that peptides covering the entire protein may not be detectable due to insufficient protease cleavage sites. This means that cross-links in certain positions will either not be visible at all or not produce high-confidence data (Mulligan et al., 2016; Nguitragool & Miller, 2007). As this reduced coverage is caused by a paucity of protease cleavage sites, different proteases or combinations of proteases can be used to try and address this issue.

6. STRATEGIES USED TO DECIDE THE POSITION OF CYSTEINE RESIDUES

A major challenge to probing the mechanism of transporters using cysteine labeling and cross-linking is deciding where to introduce the cysteine residues where they will be tolerated by the protein and probe the desired parameters. It is advisable to avoid replacing functionally important amino acids, especially those around the substrate-binding site (if this information is available). If no functional data are available that suggest which residues are functionally relevant, then a surrogate for this is to avoid highly conserved residues as they are likely to have key structural and/or functional roles. Any cysteine mutants created should be assessed for activity to ensure that the cysteine(s) introduced do not compromise stability/function. The decision-making strategies employed depend on the currently available structural information for the protein of interest and what specific questions are being asked.

6.1 Topological Mapping of a Protein With No Known Structure

To topologically map a transporter using SCAM (see Section 3.4), it helps to have some structural information to guide placement of the cysteine residues. While it is possible to perform SCAM studies "blind," where no prior topological models or data are used, computational methods are available that can predict transmembrane helices and solvent-exposed loops. A topology model of the protein of interest can be made using the hydropathy profile of the amino acid sequence (for example, using TMHMM; Krogh, Larsson, von Heijne, & Sonnhammer, 2001). Topology models can act as a starting point to guide the positioning of the single-cysteine residues. Alternatively, a homology model can be created based on a homolog of known structure. With advances in computational structural biology it is possible to identify distantly related homologs (10%–15% identity) with high confidence, thus increasing the possibility of finding a structural homolog (Stamm, Staritzbichler, Khafizov, & Forrest, 2013). While the homology model produced can act as a guide for the positioning of cysteines in SCAM studies, the results of the SCAM studies can be used to validate the model and further refine it.

6.2 Validation of Predicted Conformational Changes Using Intramolecular Cross-Links

Crystal structures of transporters offer a mere glimpse of one of the many possible conformers of the protein. Knowledge of other conformations sampled during transport is essential, and cross-linking can be a means of testing alternative conformations (as detailed in Section 3.2). To avoid extraneous experiments, alternative conformations should be predicted first, which will guide the placement of cysteines in order to experimentally test the hypothesis.

The crudest way of predicting of an alternative conformation is to manually maneuver domains of the protein in the original crystal structure to different positions that satisfy the requirements of the conformation, i.e., if the OFS is being predicted, the substrate-binding site needs to be accessible to the extracellular side of the membrane. This "eye-balling" approach can lead to some success, but it can be highly inaccurate and only feasible for small transporters with a limited scope of possible conformational change. As such, this approach requires several pairs of cysteines to be screened to identify cross-linking pairs (Lu, Radchenko, Symersky, Nie, & Guo, 2013; Radchenko et al., 2016).

The most elegant approach to predicting alternative conformations using existing crystal structures is to use the symmetry of internal structural repeats found in transporter structures (Forrest, 2015; Vergara-Jaque, Fenollar-Ferrer, Kaufmann, & Forrest, 2015). Several transporter structures are composed of repeating units that are structurally related to each other, but share only limited sequence identity (Forrest, 2013, 2015). Although structurally related, these inverted repeats are not identical; they are pseudosymmetrical. Forrest and coworkers discovered that, if the structures of these internal repeats are swapped, an alternative conformation could be predicted (Vergara-Jaque et al., 2015). This repeat-swap method has been successfully applied to several transporter families of varying structural and mechanistic complexity (Crisman et al., 2009; Fowler et al., 2015; Mulligan et al., 2016; Radestock & Forrest, 2011; Schushan et al., 2012; Yaffe, Radestock, Shuster, Forrest, & Schuldiner, 2013).

To validate these models using cysteine cross-linking, cysteine residues need to be introduced into the protein that are sufficiently close enough in the predicted structure to form a cross-link ($\leq$7 Å, for a disulfide or Hg^{2+}-induced cross-link) and distant enough in the template structure that they cannot cross-link ($\geq$10 Å). If a cross-link can form, then this would suggest that the transporter can sample the predicted conformation.

6.3 Probing Conformational Dynamics With Biophysical Probes

Biophysical methods, such as EPR analysis and smFRET, are invaluable to our understanding of transporter mechanism because they are able to report on the conformational ensemble of the protein in solution. The choice of probe and the proper placement of the modifiable cysteines are key to maximizing the information derived from these methods. These choices hinge on the magnitude and direction of the conformational changes the transporter of interest undergoes during the transport cycle. The EPR-based approach, double electron electron resonance, can measure distances between 8 and 70 Å (Claxton et al., 2015). For smFRET, the energy transfer between compatible fluorophores is highly distance dependent and varies depending on the fluorophores used, for example, the combination of cyanine dyes, Cy3 and Cy5, has a FRET efficiency (R0, the distance at which energy transfer is at 50% efficiency) of ~60 Å, whereas the combination of Alexa555 and Alexa647 has an R0 of ~50 Å (Roy, Hohng, & Ha, 2008; Zhao et al., 2010). Therefore, the positioning of the cysteines has to take into consideration the detectable distance and resolution of the technique as well as the predicted conformational change of the transporter.

Ideally, at least two conformations for the transporter of interest should be known or predicted. Cysteine residues are selected for modification that have a measurable distance change when switching between these conformations. The position of these introduced cysteines needs to be accessible to, in some cases, bulky probes, and the modification with these probes cannot compromise activity. As transporter domains can move in three dimensions during the transport cycle, to obtain an accurate measure of the trajectory of movement, modified residue positions can be triangulated by separately pairing a modified position with at least two other positions (Claxton et al., 2015).

7. DETAILED PROTOCOL FOR CROSS-LINKING A TRANSPORTER

The following procedure was used to introduce cross-links into the Na^+/succinate transporter, VcINDY to test the prediction that this transporter utilizes an elevator-type transport mechanism (Mulligan et al., 2016). The X-ray structure of VcINDY revealed a homodimeric transporter, with both protomers captured in an IFS (Mancusso et al., 2012). Using symmetry considerations (see Section 6.2), we predicted an OFS that proposed an elevator-like conformational change within each protomer of ~15 Å with a ~43 degree rotation of the domain housing the substrate-binding site (Mulligan et al., 2016). Using cysteine cross-linking coupled to mass spectrometry, PEG protection assays, and functional studies, we tested this prediction.

7.1 Selecting Cysteine Positions

Using IFS crystal structure and the OFS repeat-swapped model, we selected positions distant in the IFS (>10 Å), but close enough in the predicted OFS that, if the protein sampled this conformation, a disulfide bond could form (<7 Å).

7.2 Sample Preparation

1. Remove native cysteines from protein and measure functional effects.

   Wild-type VcINDY (VcINDYwt) contains three cysteine residues that were mutagenized to serine using QuikChange site-directed mutagenesis kit (Agilent Technologies). Reconstitution of this cysteine-free

VcINDY (VcINDYcysless) and functional analysis revealed that VcINDYcysless had fivefold lower activity than VcINDYwt. Attempts to mitigate these deleterious functional effects by using combinations of serines and alanines were unsuccessful. However, aside from this modest reduction in activity, we observed no profound functional differences between VcINDYwt and VcINDYcysless; and therefore, we judged VcINDYcysless to be a satisfactory template.

2. Introduce cross-linkable cysteine residues.

   The amino acid positions identified for mutagenesis were substituted for cysteine using QuikChange site-directed mutagenesis kit, and the pET-based plasmid harboring each mutant allele (inframe with an N-terminal His tag for purification) was moved into the expression strain, BL21-AI (ThermoFisher).

3. Express, purify, and reconstitute cysteine mutants.

   Each strain was cultured (3 L), expression was induced, and membranes collected and resuspended in Lysis Buffer (50 m*M* Tris, pH 8, 400 m*M* NaCl, 10% glycerol, 0.5 m*M* TCEP). His-tagged protein was purified using immobilized metal affinity chromatography (IMAC) followed by size-exclusion chromatography (SEC). All IMAC purification buffers contained 0.5 m*M* TCEP. SEC buffer contained 1 m*M* DTT. The stability of each mutant was assessed using the monodispersity of the SEC profile as a reporter and the activity of the protein when reconstituted into liposomes. After purification, the protein is stored at −80°C until needed. We have stored protein in these conditions for several months with no apparent loss in transport activity or cysteine reactivity.

7.3 Inducing Cross-Link Formation

1. Purified protein is thawed on ice.
2. Reducing agent is removed, and the pH is lowered from 8 to 7 by exchanging the protein into Conjugation Buffer (50 m*M* Tris, pH 7, 100 m*M* NaCl, 5% glycerol, 3 m*M* *n*-decyl-β-D-maltoside (DM). The protein is kept at RT from now on.
3. Protein is diluted to 10 μ*M* using Conjugation Buffer.

   Note: The absolute amount of protein required depends on the subsequent use of the protein; for a PEGylation assay (detailed later), 5 μg of protein per gel sample is needed; for reconstitution of VcINDY, a minimum of 25 μg is needed per treatment. The cross-linking reaction can be scaled up as needed.

4. For CuPhen-induced disulfide bond formation;
   a. A stock of a 2:1 ratio of Cu:Phen is freshly made by mixing solutions of 500 m*M* 1,10-phenanthroline (dissolved in ethanol) and 250 m*M* $CuSO_4$.
   b. CuPhen is added to the protein solution to a final concentration of 100–500 μ*M* phenanthroline. Adjust the CuPhen stock concentration (dilute with dH_2O) depending on the final concentration required. Avoid adding more than 10% of the total volume so that the detergent concentration still exceeds it's critical micelle concentration.
   c. The cross-linking reaction is incubated at RT for 45 min.

   Note: The incubation time, incubation temperature, final concentration of CuPhen, and the ratio of Cu to Phen should all be determined empirically.
5. For Hg-induced cross-link formation;
   a. $HgCl_2$ is added to the protein solution to a final concentration of 50 μ*M*.
   b. Cross-linking reaction is incubated at RT for 45 min.

   Note: Incubation time and concentration of $HgCl_2$ should be determined empirically.
6. Excess cross-linker is removed by exchanging the reaction solution into Conjugation Buffer devoid of cross-linker.

Following the cross-linking reaction the protein can be analyzed by mass spectrometry, entered into a PEGylation assay, or reconstituted to measure transport activity. For each of these analyses, it is important to include a noncross-linked control. Reducing agent (1 m*M* DTT) may be added to the control reaction to prevent unwanted spontaneous cross-link formation by air oxidation.

7.4 PEGylation Assay

1. A fresh 20 m*M* stock of mPEG5K (Sigma Aldrich) is made by dissolving in dH_2O and vortexing vigorously.
2. 2 m*M* (final concentration) mPEG5K is added to 10 μ*M* cross-linked protein and the noncross-linked control.
3. SDS is added to the reaction to a final concentration 0.5% (v/v).

Note: We observe no difference between adding the SDS and mPEG5K separately or simultaneously as a premixture (which is useful when performing time courses with multiple samples).

4. The reaction is incubated at RT for 3 h.
5. SDS-PAGE sample buffer is added the reaction. The samples are ready for SDS-PAGE analysis.

REFERENCES

Abramson, J., Smirnova, I., Kasho, V., Verner, G., Kaback, H. R., & Iwata, S. (2003). Structure and mechanism of the lactose permease of Escherichia coli. *Science*, *301*(5633), 610–615. http://dx.doi.org/10.1126/science.1088196.

Accardi, A., & Miller, C. (2004). Secondary active transport mediated by a prokaryotic homologue of ClC Cl- channels. *Nature*, *427*(6977), 803–807. http://dx.doi.org/10.1038/nature02314.

Adler, J., & Bibi, E. (2004). Promiscuity in the geometry of electrostatic interactions between the E. coli Mdr transporter MdfA and cationic substrates. *Journal of Biological Chemistry*, *280*(4), 2721–2729. http://dx.doi.org/10.1074/jbc.M412332200.

Akabas, M. H. (2015). Cysteine modification: Probing channel structure, function and conformational change. In C. Ahern & S. Pless (Eds.), *Novel chemical tools to study ion channel biology. Advances in experimental medicine and biology: Vol. 869*. New York, NY: Springer New York. http://dx.doi.org/10.1007/978-1-4939-2845-3_3.

Akabas, M. H., Cheung, M., & Guinamard, R. (1997). Probing the structural and functional domains of the CFTR chloride channel. *Journal of Bioenergetics and Biomembranes*, *29*(5), 453–463.

Akabas, M. H., Stauffer, D. A., Xu, M., & Karlin, A. (1992). Acetylcholine receptor channel structure probed in cysteine-substitution mutants. *Science*, *258*(5080), 307–310.

Akyuz, N., Altman, R. B., Blanchard, S. C., & Boudker, O. (2013). Transport dynamics in a glutamate transporter homologue. *Nature*, *502*(7469), 114–118. http://dx.doi.org/10.1038/nature12265.

Akyuz, N., Georgieva, E. R., Zhou, Z., Stolzenberg, S., Cuendet, M. A., Khelashvili, G., et al. (2015). Transport domain unlocking sets the uptake rate of an aspartate transporter. *Nature*, *518*(7537), 68–73. http://dx.doi.org/10.1038/nature14158.

Alexandrov, A. I., Mileni, M., Chien, E. Y. T., Hanson, M. A., & Stevens, R. C. (2008). Microscale fluorescent thermal stability assay for membrane proteins. *Structure (London, England: 1993)*, *16*(3), 351–359. http://dx.doi.org/10.1016/j.str.2008.02.004.

Allison, T. M., Reading, E., Liko, I., Baldwin, A. J., Laganowsky, A., & Robinson, C. V. (2015). Quantifying the stabilizing effects of protein-ligand interactions in the gas phase. *Nature Communications*, *6*, 8551. http://dx.doi.org/10.1038/ncomms9551.

Arnon, R., & Shapira, E. (1969). Crystalline papain derivative containing in intramolecular mercury bridge. *Journal of Biological Chemistry*, *244*(3), 1033–1038.

Basilio, D., Noack, K., Picollo, A., & Accardi, A. (2014). Conformational changes required for $H^+/Cl-$ exchange mediated by a CLC transporter. *Nature Structural & Molecular Biology*, *21*(5), 456–463. http://dx.doi.org/10.1038/nsmb.2814.

Bibi, E., & Kaback, H. R. (1990). In vivo expression of the lacY gene in two segments leads to functional lac permease. *Proceedings of the National Academy of Sciences of the United States of America*, *87*(11), 4325–4329.

Bogdanov, M., Zhang, W., Xie, J., & Dowhan, W. (2005). Transmembrane protein topology mapping by the substituted cysteine accessibility method (SCAMTM): Application to lipid-specific membrane protein topogenesis. *Methods, 36*(2), 148–171. http://dx.doi.org/10.1016/j.ymeth.2004.11.002.

Brocke, L., Bendahan, A., Grunewald, M., & Kanner, B. I. (2002). Proximity of two oppositely oriented reentrant loops in the glutamate transporter GLT-1 identified by paired cysteine mutagenesis. *Journal of Biological Chemistry, 277*(6), 3985–3992. http://dx.doi.org/10.1074/jbc.M107735200.

Butler, E. K., Davis, R. M., Bari, V., Nicholson, P. A., & Ruiz, N. (2013). Structure-function analysis of MurJ reveals a solvent-exposed cavity containing residues essential for peptidoglycan biogenesis in Escherichia coli. *Journal of Bacteriology, 195*(20), 4639–4649. http://dx.doi.org/10.1128/JB.00731-13.

Careaga, C. L., & Falke, J. J. (1992). Structure and dynamics of Escherichia coli chemosensory receptors. Engineered sulfhydryl studies. *Biophysical Journal, 62*(1), 209–216. discussion 217–9. http://dx.doi.org/10.1016/S0006-3495(92)81806-4.

Celej, M. S., Montich, G. G., & Fidelio, G. D. (2003). Protein stability induced by ligand binding correlates with changes in protein flexibility. *Protein Science: A Publication of the Protein Society, 12*(7), 1496–1506. http://dx.doi.org/10.1110/ps.0240003.

Claxton, D. P., Kazmier, K., Mishra, S., & Mchaourab, H. S. (2015). Navigating membrane protein structure, dynamics, and energy landscapes using spin labeling and EPR spectroscopy. *Methods in Enzymology, 564*, 349–387. http://dx.doi.org/10.1016/bs.mie.2015.07.026.

Coincon, M., Uzdavinys, P., Nji, E., Dotson, D. L., Winkelmann, I., Abdul-Hussein, S., et al. (2016). Crystal structures reveal the molecular basis of ion translocation in sodium/proton antiporters. *Nature Structural & Molecular Biology, 23*(3), 248–255. http://dx.doi.org/10.1038/nsmb.3164.

Crichton, P. G., Lee, Y., Ruprecht, J. J., Cerson, E., Thangaratnarajah, C., King, M. S., et al. (2015). Trends in thermostability provide information on the nature of substrate, inhibitor, and lipid interactions with mitochondrial carriers. *The Journal of Biological Chemistry, 290*(13), 8206–8217. http://dx.doi.org/10.1074/jbc.M114.616607.

Crisman, T. J., Qu, S., Kanner, B. I., & Forrest, L. R. (2009). Inward-facing conformation of glutamate transporters as revealed by their inverted-topology structural repeats. *Proceedings of the National Academy of Sciences of the United States of America, 106*(49), 20752–20757. http://dx.doi.org/10.1073/pnas.0908570106.

Dobrowolski, A., & Lolkema, J. S. (2010). Cross-linking of trans reentrant loops in the Na^+−citrate transporter CitS of Klebsiella pneumoniae. *Biochemistry, 49*(21), 4509–4515. http://dx.doi.org/10.1021/bi100336s.

Drew, D., & Boudker, O. (2016). Shared molecular mechanisms of membrane transporters. *Annual Review of Biochemistry, 85*(1), 543–572. http://dx.doi.org/10.1146/annurev-biochem-060815-014520.

Dutzler, R., Campbell, E. B., Cadene, M., Chait, B. T., & MacKinnon, R. (2002). X-ray structure of a ClC chloride channel at 3.0|[thinsp]||[angst]| reveals the molecular basis of anion selectivity. *Nature, 415*(6869), 287–294. http://dx.doi.org/10.1038/415287a.

Eilers, M., Patel, A. B., Liu, W., & Smith, S. O. (2002). Comparison of helix interactions in membrane and soluble alpha-bundle proteins. *Biophysical Journal, 82*(5), 2720–2736. http://dx.doi.org/10.1016/S0006-3495(02)75613-0.

Erkens, G. B., Hänelt, I., Goudsmits, J. M. H., Slotboom, D.-J., & van Oijen, A. M. (2013). Unsynchronised subunit motion in single trimeric sodium-coupled aspartate transporters. *Nature, 502*(7469), 119–123. http://dx.doi.org/10.1038/nature12538.

Fitzgerald, G. A., Mulligan, C., & Mindell, J. A. (2017). A general method for determining secondary active transporter substrate stoichiometry. *eLife, 6*. e21016. http://dx.doi.org/10.7554/eLife.21016.

Forrest, L. R. (2013). Structural biology. (Pseudo-)symmetrical transport. *Science, 339*(6118), 399–401. http://dx.doi.org/10.1126/science.1228465.

Forrest, L. R. (2015). Structural symmetry in membrane proteins. *Annual Review of Biophysics*, *44*(1), 311–337. http://dx.doi.org/10.1146/annurev-biophys-051013-023008.
Forrest, L. R., Krämer, R., & Ziegler, C. (2011). The structural basis of secondary active transport mechanisms. *Biochimica et Biophysica Acta*, *1807*(2), 167–188. http://dx.doi.org/10.1016/j.bbabio.2010.10.014.
Fowler, P. W., Orwick-Rydmark, M., Radestock, S., Solcan, N., Dijkman, P. M., Lyons, J. A., et al. (2015). Gating topology of the proton-coupled oligopeptide symporters. *Structure (London, England: 1993)*, *23*(2), 290–301. http://dx.doi.org/10.1016/j.str.2014.12.012.
Frillingos, S., Sahin-Tóth, M., Wu, J., & Kaback, H. R. (1998). Cys-scanning mutagenesis: A novel approach to structure function relationships in polytopic membrane proteins. *FASEB Journal: Official Publication of the Federation of American Societies for Experimental Biology*, *12*(13), 1281–1299.
Georgieva, E. R., Borbat, P. P., Ginter, C., Freed, J. H., & Boudker, O. (2013). Conformational ensemble of the sodium-coupled aspartate transporter. *Nature Publishing Group*, *20*(2), 215–221. http://dx.doi.org/10.1038/nsmb.2494.
Giladi, M., Shor, R., Lisnyansky, M., & Khananshvili, D. (2016). Structure-functional basis of ion transport in sodium-calcium exchanger (NCX) proteins. *International Journal of Molecular Sciences*, *17*(11), 1949. http://dx.doi.org/10.3390/ijms17111949.
Gilbert, H. F. (1995). *[2]* Thiol/disulfide exchange equilibria and disulfide bond stability. In *Biothiols part A monothiols and dithiols, protein thiols, and thiyl radicals: Vol. 251* (pp. 8–28). Elsevier. http://dx.doi.org/10.1016/0076-6879(95)51107-5.
Groeneveld, M., & Slotboom, D.-J. (2007). Rigidity of the subunit interfaces of the trimeric glutamate transporter GltT during translocation. *Journal of Molecular Biology*, *372*(3), 565–570. http://dx.doi.org/10.1016/j.jmb.2007.06.067.
Hänelt, I., Wunnicke, D., Bordignon, E., Steinhoff, H.-J., & Slotboom, D.-J. (2013). Conformational heterogeneity of the aspartate transporter Glt(Ph). *Nature Publishing Group*, *20*(2), 210–214. http://dx.doi.org/10.1038/nsmb.2471.
Holmgren, M., Liu, Y., Xu, Y., & Yellen, G. (1996). On the use of thiol-modifying agents to determine channel topology. *Neuropharmacology*, *35*(7), 797–804.
Houk, J., Singh, R., & Whitesides, G. M. (1987). Measurement of thiol-disulfide interchange reactions and thiol pKa values. In *Sulfur and sulfur amino acids: Vol. 143* (pp. 129–140). Elsevier. http://dx.doi.org/10.1016/0076-6879(87)43023-1.
Jacobs, M. T., Zhang, Y.-W., Campbell, S. D., & Rudnick, G. (2007). Ibogaine, a noncompetitive inhibitor of serotonin transport, acts by stabilizing the cytoplasm-facing state of the transporter. *Journal of Biological Chemistry*, *282*(40), 29441–29447. http://dx.doi.org/10.1074/jbc.M704456200.
Jardetzky, O. (1966). Simple allosteric model for membrane pumps. *Nature*, *211*(5052), 969–970.
Jensen, S., Guskov, A., Rempel, S., Hänelt, I., & Slotboom, D.-J. (2013). Crystal structure of a substrate-free aspartate transporter. *Nature Publishing Group*, *20*(10), 1224–1226. http://dx.doi.org/10.1038/nsmb.2663.
Jin, Y., Nair, A., & van Veen, H. W. (2014). Multidrug transport protein NorM from Vibrio cholerae simultaneously couples to sodium- and proton-motive force. *The Journal of Biological Chemistry*, *289*(21), 14624–14632. http://dx.doi.org/10.1074/jbc.M113.546770.
Johnson, Z. L., Cheong, C.-G., & Lee, S.-Y. (2012). Crystal structure of a concentrative nucleoside transporter from Vibrio cholerae at 2.4 Å. *Nature*, *483*(7390), 489–493. http://dx.doi.org/10.1038/nature10882.
Kanai, Y., Clémençon, B., Simonin, A., Leuenberger, M., Lochner, M., Weisstanner, M., et al. (2013). The SLC1 high-affinity glutamate and neutral amino acid transporter family. *Molecular Aspects of Medicine*, *34*(2–3), 108–120. http://dx.doi.org/10.1016/j.mam.2013.01.001.

Karlin, A., & Akabas, M. H. (1998). Substituted-cysteine accessibility method. *Methods in Enzymology*, *293*, 123–145.

Kazmier, K., Sharma, S., Islam, S. M., Roux, B., & Mchaourab, H. S. (2014). Conformational cycle and ion-coupling mechanism of the Na+/hydantoin transporter Mhp1. *Proceedings of the National Academy of Sciences of the United States of America*, *111*(41), 14752–14757. http://dx.doi.org/10.1073/pnas.1410431111.

Khantwal, C. M., Abraham, S. J., Han, W., Jiang, T., Chavan, T. S., Cheng, R. C., et al. (2016). Revealing an outward-facing open conformational state in a CLC Cl(−)/H(+) exchange transporter. *eLife*, *5*, 209. http://dx.doi.org/10.7554/eLife.11189.

Krogh, A., Larsson, B., von Heijne, G., & Sonnhammer, E. L. (2001). Predicting transmembrane protein topology with a hidden Markov model: Application to complete genomes. *Journal of Molecular Biology*, *305*(3), 567–580. http://dx.doi.org/10.1006/jmbi.2000.4315.

Li, J., Xu, Q., Cortes, D. M., Perozo, E., Laskey, A., & Karlin, A. (2002). Reactions of cysteines substituted in the amphipathic N-terminal tail of a bacterial potassium channel with hydrophilic and hydrophobic maleimides. *Proceedings of the National Academy of Sciences of the United States of America*, *99*(18), 11605–11610. http://dx.doi.org/10.1073/pnas.192439299.

Lo Conte, M., & Carroll, K. S. (2013). The redox biochemistry of protein sulfenylation and sulfinylation. *The Journal of Biological Chemistry*, *288*(37), 26480–26488. http://dx.doi.org/10.1074/jbc.R113.467738.

Locher, K. P. (2016). Mechanistic diversity in ATP-binding cassette (ABC) transporters. *Nature Structural & Molecular Biology*, *23*(6), 487–493. http://dx.doi.org/10.1038/nsmb.3216.

Lu, M., Radchenko, M., Symersky, J., Nie, R., & Guo, Y. (2013). Structural insights into H+-coupled multidrug extrusion by a MATE transporter. *Nature Structural & Molecular Biology*, *20*(11), 1310–1317. http://dx.doi.org/10.1038/nsmb.2687.

Mancusso, R., Gregorio, G. G., Liu, Q., & Wang, D. N. (2012). Structure and mechanism of a bacterial sodium-dependent dicarboxylate transporter. *Nature*, *491*(7425), 622–626. http://dx.doi.org/10.1038/nature11542.

Marino, S. M., & Gladyshev, V. N. (2010). Cysteine function governs its conservation and degeneration and restricts its utilization on protein surfaces. *Journal of Molecular Biology*, *404*(5), 902–916. http://dx.doi.org/10.1016/j.jmb.2010.09.027.

Martens, C., Stein, R. A., Masureel, M., Roth, A., Mishra, S., Dawaliby, R., et al. (2016). Lipids modulate the conformational dynamics of a secondary multidrug transporter. *Nature Publishing Group*, *23*(8), 744–751. http://dx.doi.org/10.1038/nsmb.3262.

Matos, M., Fann, M. C., Yan, R. T., & Maloney, P. C. (1996). Enzymatic and biochemical probes of residues external to the translocation pathway of UhpT, the sugar phosphate carrier of Escherichia coli. *Journal of Biological Chemistry*, *271*(31), 18571–18575.

McIlwain, B. C., Vandenberg, R. J., & Ryan, R. M. (2016). Characterization of the inward- and outward-facing substrate binding sites of the prokaryotic aspartate transporter, Glt Ph. *Biochemistry*, *55*(49), 6801–6810. http://dx.doi.org/10.1021/acs.biochem.6b00795.

Mindell, J. A., Maduke, M., Miller, C., & Grigorieff, N. (2001). Projection structure of a ClC-type chloride channel at 6.5 Å resolution. *Nature*, *409*(6817), 219–223. http://dx.doi.org/10.1038/35051631.

Monette, M. Y., Somasekharan, S., & Forbush, B. (2014). Molecular motions involved in Na-K-Cl cotransporter-mediated ion transport and transporter activation revealed by internal cross-linking between transmembrane domains 10 and 11/12. *The Journal of Biological Chemistry*, *289*(11), 7569–7579. http://dx.doi.org/10.1074/jbc.M113.542258.

Mulligan, C., Fenollar-Ferrer, C., Fitzgerald, G. A., Vergara-Jaque, A., Kaufmann, D., Li, Y., et al. (2016). The bacterial dicarboxylate transporter VcINDY uses a two-domain elevator-type mechanism. *Nature Structural & Molecular Biology*, *23*(3), 256–263. http://dx.doi.org/10.1038/nsmb.3166.

Mulligan, C., Fitzgerald, G. A., Wang, D. N., & Mindell, J. A. (2014). Functional characterization of a Na+-dependent dicarboxylate transporter from Vibrio cholerae. *The Journal of General Physiology*, *143*(6), 745–759. http://dx.doi.org/10.1085/jgp.201311141.

Mulligan, C., Geertsma, E. R., Severi, E., Kelly, D. J., Poolman, B., & Thomas, G. H. (2009). The substrate-binding protein imposes directionality on an electrochemical sodium gradient-driven TRAP transporter. *Proceedings of the National Academy of Sciences of the United States of America*, *106*(6), 1778–1783. http://dx.doi.org/10.1073/pnas.0809979106.

Mulligan, C., & Mindell, J. A. (2013). Mechanism of transport modulation by an extracellular loop in an archaeal excitatory amino acid transporter (EAAT) homolog. *The Journal of Biological Chemistry*, *288*(49), 35266–35276. http://dx.doi.org/10.1074/jbc.M113.508408.

Nguitragool, W., & Miller, C. (2007). CLC Cl /H+ transporters constrained by covalent cross-linking. *Proceedings of the National Academy of Sciences of the United States of America*, *104*(52), 20659–20665. http://dx.doi.org/10.1073/pnas.0708639104.

Poelarends, G. J., & Konings, W. N. (2002). The transmembrane domains of the ABC multidrug transporter LmrA form a cytoplasmic exposed, aqueous chamber within the membrane. *Journal of Biological Chemistry*, *277*(45), 42891–42898. http://dx.doi.org/10.1074/jbc.M206508200.

Quistgaard, E. M., Löw, C., Guettou, F., & Nordlund, P. (2016). Understanding transport by the major facilitator superfamily (MFS): Structures pave the way. *Nature Reviews. Molecular Cell Biology*, *17*(2), 123–132. http://dx.doi.org/10.1038/nrm.2015.25.

Radchenko, M., Nie, R., & Lu, M. (2016). Disulfide cross-linking of a multidrug and toxic compound extrusion transporter impacts multidrug efflux. *Journal of Biological Chemistry*, *291*(18), 9818–9826. http://dx.doi.org/10.1074/jbc.M116.715227.

Radestock, S., & Forrest, L. R. (2011). The alternating-access mechanism of MFS transporters arises from inverted-topology repeats. *Journal of Molecular Biology*, *407*(5), 698–715. http://dx.doi.org/10.1016/j.jmb.2011.02.008.

Rahman, M., Ismat, F., Jiao, L., Baldwin, J. M., Sharples, D. J., Baldwin, S. A., et al. (2016). Characterisation of the DAACS family Escherichia coli glutamate/aspartate-proton symporter GltP using computational, chemical, biochemical and biophysical methods. *The Journal of Membrane Biology*, *250*, 145–162. http://dx.doi.org/10.1007/s00232-016-9942-x.

Ressl, S., van Scheltinga, A. C. T., Vonrhein, C., Ott, V., & Ziegler, C. (2009). Molecular basis of transport and regulation in the Na+/betaine symporter BetP. *Nature*, *458*(7234), 47–52. http://dx.doi.org/10.1038/nature07819.

Reyes, N., Ginter, C., & Boudker, O. (2009). Transport mechanism of a bacterial homologue of glutamate transporters. *Nature*, *462*(7275), 880–885. http://dx.doi.org/10.1038/nature08616.

Reyes, N., Oh, S., & Boudker, O. (2013). Binding thermodynamics of a glutamate transporter homolog. *Nature Structural & Molecular Biology*, *20*(5), 634–640. http://dx.doi.org/10.1038/nsmb.2548.

Riegelhaupt, P. M., Frame, I. J., & Akabas, M. H. (2010). Transmembrane segment 11 appears to line the purine permeation pathway of the plasmodium falciparum equilibrative nucleoside transporter 1 (PfENT1). *The Journal of Biological Chemistry*, *285*(22), 17001–17010. http://dx.doi.org/10.1074/jbc.M110.115758.

Roy, R., Hohng, S., & Ha, T. (2008). A practical guide to single-molecule FRET. *Nature Methods*, *5*(6), 507–516. http://dx.doi.org/10.1038/nmeth.1208.

Ryan, R. M., Compton, E. L. R., & Mindell, J. A. (2009). Functional characterization of a Na+-dependent aspartate transporter from Pyrococcus horikoshii. *Journal of Biological Chemistry*, *284*(26), 17540–17548. http://dx.doi.org/10.1074/jbc.M109.005926.

Ryan, R. M., Mitrovic, A. D., & Vandenberg, R. J. (2004). The chloride permeation pathway of a glutamate transporter and its proximity to the glutamate translocation pathway. *Journal of Biological Chemistry, 279*(20), 20742–20751. http://dx.doi.org/10.1074/jbc.M304433200.

Schushan, M., Rimon, A., Haliloglu, T., Forrest, L. R., Padan, E., & Ben-Tal, N. (2012). A model-structure of a periplasm-facing state of the NhaA antiporter suggests the molecular underpinnings of pH-induced conformational changes. *The Journal of Biological Chemistry, 287*(22), 18249–18261. http://dx.doi.org/10.1074/jbc.M111.336446.

Shafer, D. E., Inman, J. K., & Lees, A. (2000). Reaction of tris(2-carboxyethyl)phosphine (TCEP) with maleimide and alpha-haloacyl groups: Anomalous elution of TCEP by gel filtration. *Analytical Biochemistry, 282*(1), 161–164. http://dx.doi.org/10.1006/abio.2000.4609.

Silverstein, N., Crisman, T. J., Forrest, L. R., & Kanner, B. I. (2013). Cysteine scanning mutagenesis of transmembrane helix 3 of a brain glutamate transporter reveals two conformationally sensitive positions. *The Journal of Biological Chemistry, 288*(2), 964–973. http://dx.doi.org/10.1074/jbc.M112.403576.

Slotboom, D. J. (2014). Structural and mechanistic insights into prokaryotic energy-coupling factor transporters. *Nature Reviews Microbiology, 12*(2), 79–87. http://dx.doi.org/10.1038/nrmicro3175.

Sonoda, Y., Newstead, S., Hu, N.-J., Alguel, Y., Nji, E., Beis, K., et al. (2011). Benchmarking membrane protein detergent stability for improving throughput of high-resolution X-ray structures. *Structure (London, England: 1993), 19*(1), 17–25. http://dx.doi.org/10.1016/j.str.2010.12.001.

Soskine, M., Steiner-Mordoch, S., & Schuldiner, S. (2002). Crosslinking of membrane-embedded cysteines reveals contact points in the EmrE oligomer. *Proceedings of the National Academy of Sciences of the United States of America, 99*(19), 12043–12048. http://dx.doi.org/10.1073/pnas.192392899.

Stamm, M., Staritzbichler, R., Khafizov, K., & Forrest, L. R. (2013). Alignment of helical membrane protein sequences using AlignMe. *PLoS One, 8*(3), e57731. http://dx.doi.org/10.1371/journal.pone.0057731.

Stauffer, D. A., & Karlin, A. (1994). Electrostatic potential of the acetylcholine binding sites in the nicotinic receptor probed by reactions of binding-site cysteines with charged methanethiosulfonates. *Biochemistry, 33*, 6840–6849.

Swier, L. J. Y. M., Guskov, A., & Slotboom, D. J. (2016). Structural insight in the toppling mechanism of an energy-coupling factor transporter. *Nature Communications*, 7, 11072. http://dx.doi.org/10.1038/ncomms11072.

Tavoulari, S., Forrest, L. R., & Rudnick, G. (2009). Fluoxetine (Prozac) binding to serotonin transporter is modulated by chloride and conformational changes. *Journal of Neuroscience, 29*(30), 9635–9643. http://dx.doi.org/10.1523/JNEUROSCI.0440-09.2009.

Tavoulari, S., Margheritis, E., Nagarajan, A., DeWitt, D. C., Zhang, Y.-W., Rosado, E., et al. (2016). Two Na+ sites control conformational change in a neurotransmitter transporter homolog. *The Journal of Biological Chemistry, 291*(3), 1456–1471. http://dx.doi.org/10.1074/jbc.M115.692012.

Tsai, M.-F., Fang, Y., & Miller, C. (2012). Sided functions of an arginine–agmatine antiporter oriented in liposomes. *Biochemistry, 51*(8), 1577–1585. http://dx.doi.org/10.1021/bi201897t.

Tsai, M.-F., McCarthy, P., & Miller, C. (2013). Substrate selectivity in glutamate-dependent acid resistance in enteric bacteria. *Proceedings of the National Academy of Sciences of the United States of America, 110*(15), 5898–5902. http://dx.doi.org/10.1073/pnas.1301444110.

Tsai, M.-F., & Miller, C. (2013). Substrate selectivity in arginine-dependent acid resistance in enteric bacteria. *Proceedings of the National Academy of Sciences of the United States of America, 110*(15), 5893–5897. http://dx.doi.org/10.1073/pnas.1301442110.

van Iwaarden, P. R., Pastore, J. C., Konings, W. N., & Kaback, H. R. (1991). Construction of a functional lactose permease devoid of cysteine residues. *Biochemistry*, *30*(40), 9595–9600.

Veenstra, M., Lanza, S., Hirayama, B. A., Turk, E., & Wright, E. M. (2004). Local conformational changes in the Vibrio Na+/galactose cotransporter. *Biochemistry*, *43*(12), 3620–3627. http://dx.doi.org/10.1021/bi0357210.

Verdon, G., & Boudker, O. (2012). Crystal structure of an asymmetric trimer of a bacterial glutamate transporter homolog. *Nature Structural & Molecular Biology*, *19*(3), 355–357. http://dx.doi.org/10.1038/nsmb.2233.

Vergara-Jaque, A., Fenollar-Ferrer, C., Kaufmann, D., & Forrest, L. R. (2015). Repeat-swap homology modeling of secondary active transporters: Updated protocol and prediction of elevator-type mechanisms. *Frontiers in Pharmacology*, *6*, 183. http://dx.doi.org/10.3389/fphar.2015.00183.

Wu, J., & Kaback, H. R. (1996). A general method for determining helix packing in membrane proteins in situ: Helices I and II are close to helix VII in the lactose permease of Escherichia coli. *Proceedings of the National Academy of Sciences of the United States of America*, *93*(25), 14498–14502.

Yaffe, D., Radestock, S., Shuster, Y., Forrest, L. R., & Schuldiner, S. (2013). Identification of molecular hinge points mediating alternating access in the vesicular monoamine transporter VMAT2. *Proceedings of the National Academy of Sciences of the United States of America*, *110*(15), E1332–E1341. http://dx.doi.org/10.1073/pnas.1220497110.

Yang, Q., Wang, X., Ye, L., Mentrikoski, M., Mohammadi, E., Kim, Y.-M., et al. (2005). Experimental tests of a homology model for OxlT, the oxalate transporter of Oxalobacter formigenes. *Proceedings of the National Academy of Sciences of the United States of America*, *102*(24), 8513–8518. http://dx.doi.org/10.1073/pnas.0503533102.

Yeh, A. P., McMillan, A., & Stowell, M. H. B. (2006). Rapid and simple protein-stability screens: Application to membrane proteins. *Acta Crystallographica. Section D, Biological Crystallography*, *62*(Pt. 4), 451–457. http://dx.doi.org/10.1107/S0907444906005233.

Yernool, D., Boudker, O., Jin, Y., & Gouaux, E. (2004). Structure of a glutamate transporter homologue from Pyrococcus horikoshii. *Nature*, *431*(7010), 811–818. http://dx.doi.org/10.1038/nature03018.

Yu, H., Kono, M., McKee, T. D., & Oprian, D. D. (1995). A general method for mapping tertiary contacts between amino acid residues in membrane-embedded proteins. *Biochemistry*, *34*(46), 14963–14969.

Zhao, Y., Terry, D., Shi, L., Weinstein, H., Blanchard, S. C., & Javitch, J. A. (2010). Single-molecule dynamics of gating in a neurotransmitter transporter homologue. *Nature*, *465*(7295), 188–193. http://dx.doi.org/10.1038/nature09057.

Zhou, Y., Guan, L., Freites, J. A., & Kaback, H. R. (2008). Opening and closing of the periplasmic gate in lactose permease. *Proceedings of the National Academy of Sciences of the United States of America*, *105*(10), 3774–3778. http://dx.doi.org/10.1073/pnas.0800825105.

CHAPTER EIGHT

Ion Channel Conformation and Oligomerization Assessment by Site-Directed Spin Labeling and Pulsed-EPR

Christos Pliotas[1]
University of St Andrews, St Andrews, Fife, United Kingdom
[1]Corresponding author: e-mail address: c.pliotas@st-andrews.ac.uk

Contents

Abstract

Mechanosensitive (MS) ion channels are multimeric integral membrane proteins that respond to increased lipid bilayer tension by opening their nonselective pores to release solutes and relieve increased cytoplasmic pressure. These systems undergo major conformational changes during gating and the elucidation of their mechanism requires a deep understanding of the interplay between lipids and proteins. Lipids are responsible for transmitting lateral tension to MS channels and therefore play a key role in obtaining a molecular-detail model for mechanosensation. Site-directed spin labeling combined

Methods in Enzymology, Volume 594
ISSN 0076-6879
http://dx.doi.org/10.1016/bs.mie.2017.05.013

with electron paramagnetic resonance (EPR) spectroscopy is a powerful spectroscopic tool in the study of proteins. The main bottleneck for its use relates to challenges associated with successful isolation of the protein of interest, introduction of paramagnetic labels on desired sites, and access to specialized instrumentation and expertise. The design of sophisticated experiments, which combine a variety of existing EPR methodologies to address a diversity of specific questions, require knowledge of the limitations and strengths, characteristic of each particular EPR method. This chapter is using the MS ion channels as paradigms and focuses on the application of different EPR techniques to ion channels, in order to investigate oligomerization, conformation, and the effect of lipids on their regulation. The methodology we followed, from the initial strategic selection of mutants and sample preparation, including protein purification, spin labeling, reconstitution into lipid mimics to the complete set-up of the pulsed-EPR experiments, is described in detail.

1. INTRODUCTION

Membrane proteins (MPs) are partly or fully embedded within the lipid membrane. They fold and achieve structural stability by assembling into their most energetically preferred oligomeric and conformational state. In that respect, they depend on their lifetime partners, the lipids. For many years, it was believed that the main role of the lipid bilayer was to provide structural stability to MPs with negligible influence to their regulation and function.

The pivotal moment for the understanding of the active role of the bilayer and its constituents, the lipids, was the discovery of a certain class of channels first characterized in bacteria (Sukharev, Blount, Martinac, Blattner, & Kung, 1994). These channels, named mechanosensitive (MS) channels, responded only to pressure, when reconstituted in vitro in giant liposomes and subjected to a gradient of lateral bilayer tension. Their main function was to protect the bacterial cell from rupture against severe hypo-osmotic shock (Levina et al., 1999).

Recently, we have used a combination of structural, chemical, and biophysical techniques and showed that the lipid membrane is not a passive but an active component of the regulatory mechanism for the tension sensing MS ion channel MscS (Pliotas et al., 2015; Pliotas & Naismith, 2016). Eukaryotic potassium channels have been also shown to be partly regulated by lipid bilayer tension (Brohawn, Campbell, & MacKinnon, 2014; Brohawn, Su, & MacKinnon, 2014; Pliotas & Naismith, 2016) suggesting that mechanosensation is not a property restricted to a certain class of channels, but a common mechanism for ion channel regulation.

X-ray crystallography has been the most successful scientific methodology for providing structural information at high resolution for MPs. However, it faces major technical hurdles, resulting in a significant amount of time required to resolve each structure. Moreover, the structures obtained, represent only a snapshot of their highly dynamic behavior during gating. Therefore, experimental methodologies, which can provide information on the dynamic properties of these systems and characterize their gating transitions, such as electron paramagnetic resonance (EPR) spectroscopy, are emerging as powerful methods. This technique does not only complement X-ray crystallography but also addresses specific mechanistic questions in solution and MP's native environment.

The key-discovery moment in the investigation of proteins by EPR spectroscopy came along with the introduction of site-directed spin labeling (SDSL) as a combinatory tool to EPR for studying protein dynamics (Hubbell, Cafiso, & Altenbach, 2000). Continuous wave (CW) EPR combined with SDSL constituted a powerful method for providing information regarding the local environment of the introduced paramagnetic labels. However, the methodology was not appropriate in obtaining long distances present in large protein complexes to address structure-based biological questions, where spin-to-spin intramolecular distances far exceed the 15–20 Å limit of CW EPR measurements.

Recent advances in magnetic resonance hardware made pulsed EPR possible allowing the emergence and development of the pulsed electron double resonance (PELDOR) or double electron–electron resonance (DEER) methodology for longer distance measurements in biomolecules (Jeschke, 2002; Schiemann & Prisner, 2007). These included the development of pulse sequences applicable to biological systems (Pannier, Veit, Godt, Jeschke, & Spiess, 2000; Prisner, Rohrer, & MacMillan, 2001). PELDOR has been proven to constitute a robust methodology in assigning conformational and oligomeric states in a variety of MPs, including both ion channels and transporters (Dalmas, Sompornpisut, Bezanilla, & Perozo, 2014; Endeward, Butterwick, MacKinnon, & Prisner, 2009; Georgieva, Borbat, Ginter, Freed, & Boudker, 2013; Hagelueken et al., 2009; Hanelt, Wunnicke, Bordignon, Steinhoff, & Slotboom, 2013; Kazmier et al., 2014; Martens et al., 2016; Pliotas et al., 2012; Verhalen et al., 2017; Ward et al., 2014).

We have recently applied the PELDOR methodology to investigate the MscS channel under physiological conditions in a detergent environment

(Pliotas et al., 2012) and lipid bilayer mimics, including nanodiscs and bicelles (Ward et al., 2014). In parallel, a protocol was developed to measure the extend of cysteine modification on MPs and accurately quantitate the labeling efficiency of the paramagnetic spin labels on MscS, a crucial step for the success of the project (Branigan, Pliotas, Hagelueken, & Naismith, 2013).

In this chapter, I will attempt to give an overview and describe the methodology we used to investigate MscS; a methodology applicable to many MPs. I will also explore the potential of the EPR methodology on assessing solvent accessibility of the modified protein residues, using electron spin echo envelope modulation (ESEEM) spectroscopy, on the existing samples destined for PELDOR measurements. The potential of endogenous lipid localization in respect to paramagnetic labels on MPs by EPR will be also discussed. Methods to prepare and test protein samples suitable for EPR experiments will be suggested.

2. DESIGNING THE PULSED-EPR EXPERIMENT

For the application of pulsed-EPR experiment on complex multimeric proteins, the sample preparation stages have to be cautiously designed. This is mainly to determine whether measurements are feasible and able to produce high-quality data. In the next two sections, I will discuss spin-labeling strategies and sample requirements for multimeric MS ion channels. Similar strategies could be applied to a variety of MPs and in simpler cases to soluble proteins.

2.1 Spin-Labeling Strategies and Mutant Selection Criteria

Initially, the residues that will be modified to accommodate the paramagnetic spin labels should be carefully selected. Therefore, certain considerations should be taken into account prior to the EPR experiment, protein expression, and isolation. These include the following requirements, which will be analyzed in this chapter:

- Mutations should not be disruptive to protein function and stability.
- Sites should be readily accessible to spin labels.
- Cysteine modification(s) should not interfere with any binding site(s).
- Expected distances should be within the PELDOR range.

2.1.1 Protein Stability and Function

Care should be taken on whether resulting mutations affect protein structure and function. Owing to their highly reactive nature and their ability to form disulfide bonds, cysteines often have a severe impact on protein folding and function. Cysteine mutants should be always tested for any loss of protein activity, compared to the wild-type (WT) protein, as part of the experimental design.

For MscS, single-cysteine mutants were tested by cell viability assays for their ability to protect cells against osmotic shock (Pliotas et al., 2012). For the cell viability assays, MscS was overexpressed in a bacterial strain, which lacked the genes of the tested proteins from its genome (Levina et al., 1999). Then cells were grown in a high salt environment (0.5 *M*). Cells expressing MscS were subjected to osmotic shock and judged for their survival rates against WT cells. Each individual mutant destined for EPR was able to protect the cell against osmotic downshock.

In parallel, patch-clamp electrophysiology was also employed to test the gating ability of mutated channels in spheroplasts formed from a strain deficient in MscS channels (Pliotas et al., 2012). Therefore, pore conductance should only arise from the overexpressed cysteine mutants tested each time. MscS mutants presented slight variable channel properties and pressure gating profiles, as expected, but they were functional (Pliotas et al., 2012).

Despite EPR being widely considered a nondestructive experimental method, one should also take into account the effect of the attached spin label and in general the cysteine modification on spin-labeled (SL) proteins. Especially, in cases where selected residues are known to affect binding sites, and for MPs in particular, lipid-binding sites (Pliotas et al., 2015; Pliotas & Naismith, 2016). The severe effect that different sulfhydryl substituents have on single-cysteine MscL mutants, particularly disrupting potential lipid-binding sites, was recently demonstrated using a high-throughput study (Iscla, Wray, Eaton, & Blount, 2015).

2.1.2 Label Accessibility, Cysteine Exposure, and X-Ray Limitation Awareness

EPR spectroscopy is a method, which entirely relies on the presence of paramagnetic species to produce a detectable signal, exploiting the absence of unpaired electrons in natural proteins. An essential requirement for EPR is that the selected sites should be readily accessible to labels in order to achieve the desired level of labeling efficiency.

When structural X-ray, NMR, or cryo-EM models exist, a computational in silico scanning is highly recommended to indicate exposed residues by spin labels. By attaching a spin label through a *plug-in* in PyMOL (*MttslWizard*) (Hagelueken, Ward, Naismith, & Schiemann, 2012), or a Matlab interface routine (MMM) (Polyhach, Bordignon, & Jeschke, 2011), the available spin label conformations and accessibility for each individual site can be evaluated. An appearing low number of conformers would be an indication of inaccessible sites and thus challenging to label.

Such computational approaches rely solely on existing structural models and for MPs in particular, these models do not often take into account the presence of unresolved endogenous lipid and/or detergent molecules in the models. These endogenous, most often annular lipids are transferred from the parental membrane after extraction, while detergent is essential to keep MPs in solution. Both of these types of molecules are often disordered and are not resolved in the generally lower resolution of MP structures, hindering label accessibility, and limiting label conformations.

Such a case was demonstrated when a significantly higher resolution of MscS was reported, including resolved lipid acyl chains (Pliotas et al., 2015). Those were present but not resolved in the previously reported lower resolution MscS X-ray structures (Bass, Strop, Barclay, & Rees, 2002; Wang et al., 2008), accounting for their inherent highly mobile nature and heterogeneity. By performing an in silico attachment of the S-(2,2,5,5-tetramethyl-2,5-dihydro-1H-pyrrol-3-yl)methyl methanesulfonothioate) (MTSL) label and searching for conformers on sites that lipid acyl chains are in close proximity, would be expected to cause a severe discrepancy to predicted distance distributions.

In order to demonstrate the severe lipid effect to in silico labeling of MPs, the following example is presented here. We used *MttslWizard* with identical search parameters for both cases (MTSL, thorough search, and loose). We performed in silico labeling in the presence and absence of lipid acyl chains modeled into the X-ray structure. In order to perform this computational investigation, we selected an MscS residue that is located in close proximity to the resolved acyl chains as evidenced in the X-ray structure. In addition, the fluorescence intensity of this residue, when mutated to tryptophan, is strongly quenched by the presence of exogenously added brominated lipids (Pliotas et al., 2015).

MscS A119 is a residue located within a hydrophobic region of MscS, which has surprisingly shown a strong fluorescence signal quenching by brominated lipids, equally to an exposed residue facing the lipid bilayer when mutated to tryptophan. Therefore, A119 is expected to be located in very close proximity to lipids, in the heart of the transmembrane (TM) annular lipid path (Pliotas et al., 2015).

In the high-resolution MscS structure, A119 is indeed facing toward the resolved lipid acyl chains (PDB 5AJI, Fig. 1A). In silico labeling of A119C with MTSL resulted in a group of MTSL conformers pointing away from the acyl chains (Fig. 1B). When acyl chains were removed from the structure and thus not considered by *MtsslWizard*, an additional group of conformers was now present pointing toward the region, the acyl chains used to occupy (Fig. 1C). This resulted in two very different distance distributions for the aforementioned cases. Notably, a number of new distance peaks appeared in the absence of lipids (Fig. 1D). Therefore, the presence of lipid-like acyl chains, either originating from exogenously added detergents or endogenous lipids, would limit label conformers resulting in changes in the distance distributions.

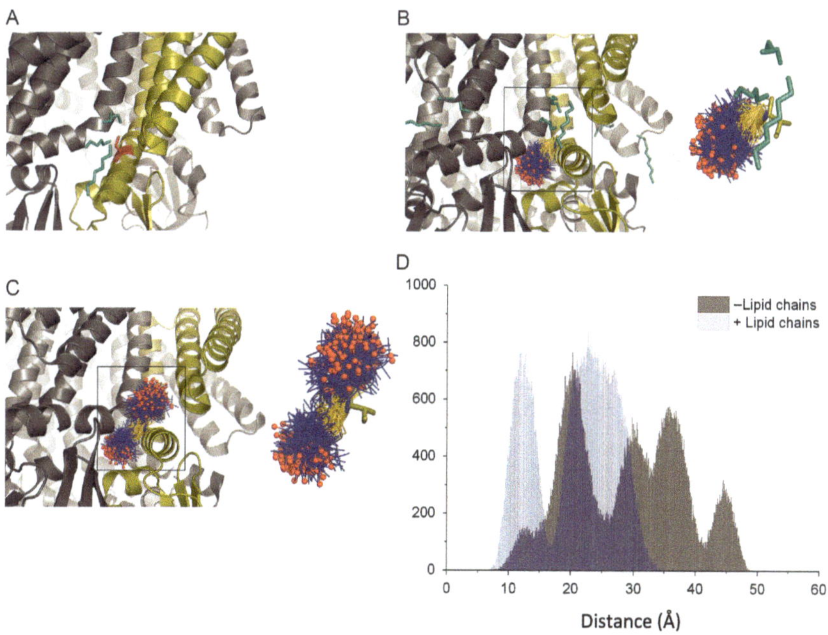

Fig. 1 Effect of lipid acyl chains to spin label conformations and PELDOR distance distribution. (A) MscS A119 is in close proximity to resolved lipid acyl chains of the MscS high-resolution X-ray structure (PDB 5AJI). (B) Available conformers of MscS A119C in the presence of lipids. *MttslWizard* was used for in silico labeling. (C) Available conformers in the absence of lipids. (D) Resulting distance distributions in the absence (*gray*) and presence (*light blue*) lipids.

Furthermore, lipid-like chains are also expected to mask cysteines by making them inaccessible to spin labels, resulting in a poorer labeling efficiency for a given site. Therefore, one should consider their existence in the protein samples even if they have not been resolved in the X-ray crystal structure.

To overcome this effect, one could simulate lipids using molecular dynamic simulations on the TM region of the MP of interest and investigate which sites may be affected by their presence, before costly experiments are performed. Basic coarse-grained self-assembly molecular dynamic simulations of MPs within lipid bilayers could be performed on a freely accessible platform (MemProtMD; Stansfeld et al., 2015), providing a useful insight. However, it is advisable that information regarding endogenous lipids, related to their type, number, and binding site should be obtained using appropriate spectroscopic techniques prior, or in parallel, to computational analysis.

To experimentally address the accessibility of native or engineered cysteines and the quantification of that extend, a similar protocol that we have developed during MscS investigation could be used. This protocol requiring small amounts of protein is fast, accurate, and equally applicable to both soluble and TM portions of MP residues (Branigan et al., 2013). It may be used as a reliable screening method to overcome cysteine accessibility issues before the large-scale preparations, required for the EPR experiments.

2.1.3 Distance Measurement Restrictions

Distance restrictions set by intrinsic label properties, hardware performance, and sample limitations, should also be considered and are discussed in this section. The shortest possible measured distance by PELDOR, mostly imposed by the pulse lengths, is ~15–20 Å. Several parameters determine the upper distance limit, but the most limiting factor is the relatively short relaxation time of the commonly used nitroxide spin labels in the EPR studies. However, it was shown that the upper limit could be significantly increased, by full protein deuteration (El Mkami & Norman, 2015; Ward et al., 2010).

Quantitative interpretation of distance measurements becomes even more challenging for higher-order multimeric proteins where more than two spins (one distance) are present. The data analysis is therefore affected by the so-called multispin effects (Edwards et al., 2014; Giannoulis, Ward, Branigan, Naismith, & Bode, 2013; Pliotas et al., 2012; Valera et al., 2016). These effects could be suppressed by substoichiometric labeling strategies. When they are combined with protein deuteration, distances of around 160 Å for multimeric soluble proteins could be measured (Schmidt, Walti, Baber, Hustedt, & Clore, 2016).

2.1.4 TM vs Cytosolic Mutant Selection

Mutant selection has proven to be far more challenging for TM than cytosolic or periplasmic domains of MPs. Whereas almost any soluble part yielded high-quality data, the rate of TM mutants in respect to producing visual oscillation in the raw data, labeling efficiency, and modulation depth, was much lower. Most TM mutants generated quality CW spectra, presenting variable label mobility. However, all cytosolic and a handful of TM mutants (out of over 25 purified and SL) resulted in defined interpretable PELDOR distributions (Pliotas et al., 2012; Ward et al., 2014). The reasons behind this are not well understood, but assumptions could be made.

The presence of multiple oligomeric and/or conformational MP states or endogenous lipids and detergents may have a severe impact on distance measurements. The use of Q-band EPR spectrometers equipped with high power TWT (traveling wave tube) has significantly increased the sensitivity and allowed longer time windows to be recorded compared to X-band spectrometers. In addition, low temperatures and deuteration of the buffers and proteins increased the resulting relaxation times, allowing much longer distances to be reliably measured.

The use of an X-band spectrometer on a seven-spin system, such as MscS (Pliotas et al., 2012; Ward et al., 2014) has allowed us to reliably measure distances between 15 and 70 Å, with optimal distance range between 20 and 60 Å. However, we were severely limited by the short timing window and the significantly long averaging time required for achieving a good signal-to-noise ratio (~16 h per sample). An example of PELDOR time trace and its resulting distance distribution analyzed by *DeerAnalysis* (Jeschke et al., 2006) for MscS D67R1 are presented in Fig. 2. We have recently been using a Q-band spectrometer on a variety of multimeric MPs, which has significantly improved all aspects of measuring performance. This includes a great enhancement of the signal-to-noise ratio (sensitivity), the length of time windows (longer distances), and low averaging (time acquisition reduction).

2.2 Sample Requirements

During the preparation of the SL MP samples, there is a significant number of parameters that should be considered. These include protein purity, monodispersity, stability, and the extend of cysteine modification. Furthermore, optimum protein–spin concentration and sample volume, which depends

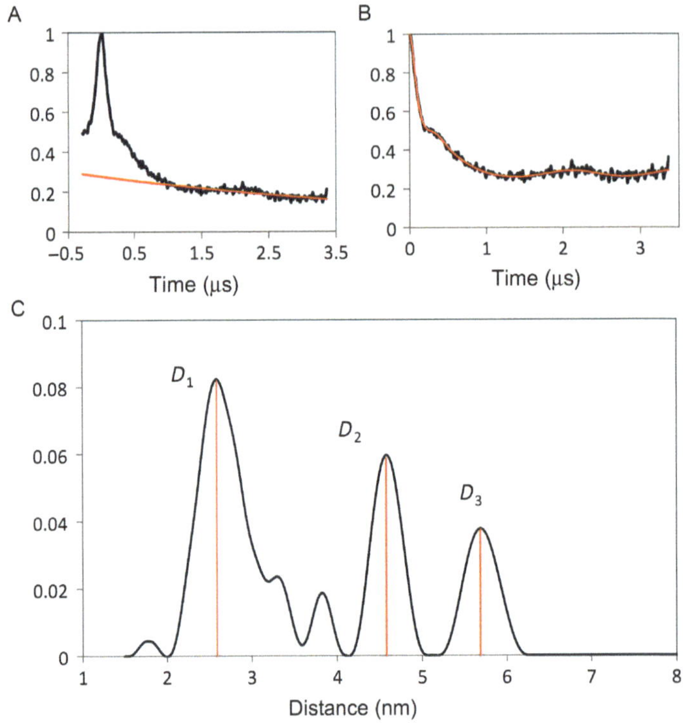

Fig. 2 Distance distribution of MscS D67R1 using X-band PELDOR (DEER) spectroscopy. (A) Raw PELDOR time trace from D67R1. (B) Background corrected data trace. (C) Resulting distance distributions using *DeerAnalysis* (Jeschke et al., 2006).

on the type of the EPR cavity used, should also be considered. Cryoprotection, essential for low temperature pulsed-EPR measurements, is also critical. In this section, I will discuss and briefly describe the sample preparation and requirements to obtain a sample suitable for EPR measurements.

2.2.1 Protein Purity, Dispersity, and Buffer Requirements

For our MS channel preparations, single-cysteine mutants are expressed and purified by Ni-NTA affinity chromatography, then SL and finally spin label excess and imidazole are being removed by size-exclusion chromatography (SEC). The protocol is detailed in the following sections. All mutants investigated, eluted as a single homogeneous peak of same volume in SEC, and presented no signs of aggregation or dissociation (Pliotas et al., 2012).

As buffer system we tend to use the conditions, where the protein is most stable in, in terms of salt, detergent, and additives, judged by stability assays (Sonoda et al., 2011). For MscS, the best buffer conditions were found to be

50 mM of phosphate pH 7.5, 300 mM NaCl, and 0.05% dodecyl-β-D-maltopyranoside (DDM). Nevertheless, for pulsed-EPR we exchanged the buffer system. In cases protein was stable in the new buffer, including a range of pHs and detergents, measurements were not affected yielding quality data. All MscS samples used in the EPR studies have been subjected to gel filtration and dynamic light scattering for further assessment of monodispersity and size evaluation.

2.2.2 Protein and Spin Concentration Limits

EPR is a sensitive technique but signal detection sensitivity limits depend on the different hardware used. For measurements on MscS (Pliotas et al., 2012; Ward et al., 2014), we used an X-band spectrometer and all the data were recorded at 50 K. Total sample loading requirements for the 4 mm or 5 mm EPR tubes is a minimum of 100 μL. Highly reproducible results were obtained by loading 120 μL of total sample volume. Fifty percent of this final sample volume was deuterated ethylene glycol (d-EG), added for cryoprotection. Therefore, a two times protein concentrated sample of 60 μL in volume was required for each experiment. A spin concentration of SL MscS around ~200 μM was largely enough to generate sufficient echo signal allowing a reliably set-up for pulsed-EPR experiments.

Protein sample requirements were standard for the majority of MscS mutants, which presented a 80% labeling efficiency, as previously described (Branigan et al., 2013). Final monomer protein concentration was around 200 μM accounting for an approximate number of spin concentration between 160 and 200 μM, and around 25 μM of MscS heptamers.

Under these conditions with a Bruker X-band spectrometer, overnight measurements from 16 to 24 h were required to produce a sufficient signal-to-noise ratio. In case of labeling efficiency lower than 40% for individual sites, the protein concentration was increased accordingly to reach the required spin concentration. Recently, we have started using a Bruker pulsed Q-band spectrometer, for MPs and multimeric ion channels, performing distance measurements. For this set-up sample requirements are much lower, both in volume and spin concentration. By comparison, we use 3 mm tubes, which require half of the total sample volume (i.e., 60 μL) and similar spin concentration to X-band. For a comparable signal-to-noise ratio to be achieved, on the same amount of sample at X- and Q-bands, only few averages are required at Q-band, about 15–20 min acquisition time, while at X-band, 16–24 h is needed. These great advantages allow longer time windows to be recorded and longer distances to be more reliably measured.

3. SAMPLE PREPARATION FOR EPR EXPERIMENTS

Perhaps the most crucial step for a successful EPR experiment is the preparation of the SL protein samples. MP samples are particularly challenging to prepare, as it is also the case for X-ray crystallography. In this section, I will describe the process we followed for MscS, a methodology that is suitable to a wide range of MPs, with minor alterations and adjustments.

3.1 Site-Directed Mutagenesis

MscS and other MS channels such as MscL benefit from the fact they are cysteine free in the native WT form. But in most cases where native cysteines exist, a procedure described previously for assessing protein stability and function (Section 2.1.1), should be followed. For MscS (Pliotas et al., 2012), we selected seven mutants on the basis they could discriminate among the different gating models previously proposed. We based the design of our primers and site-directed mutagenesis approach on a previously published protocol (Liu & Naismith, 2008).

3.2 Protein Purification

The following detailed protocol includes purification and spin-labeling steps for MscS mutants, to be used in both CW and pulsed-EPR studies.

3.2.1 Buffers Used

Lysis Buffer	**Per 100 mL**
50 m*M* Na-phosphate	10 mL of 500 m*M*
300 m*M* NaCl	15 mL of 2 *M*

Solubilization Buffer	**Per 50 mL**
1.5% detergent (DDM)	0.75 g
50 m*M* Na-phosphate buffer pH 7.5	5 mL of 500 m*M*
300 m*M* NaCl	7.5 mL of 2 *M*
10% glycerol	10 mL of 50%
50 m*M* imidazole	2.5 mL of 1 *M*
Complete EDTA-free protease inhibitor (Roche)	1 tablet

2× Purification Buffer (Degassed)	Per 25 mL
100 m*M* Na-phosphate buffer pH 7.5	5 mL of 500 m*M*
600 m*M* NaCl	7.5 mL of 2 *M*
20% glycerol	10 mL of 50%
0.1% DDM	0.025 g

Wash Buffer	Per 50 mL
1 × Purification buffer	25 mL of 2 ×
50 m*M* imidazole	2.5 mL of 1 *M*

Elution Buffer First	Per 1 mL
1 × Purification buffer	0.5 mL of 2 ×
50 m*M* imidazole	50 μL of 1 *M*

Elution Buffer Second	Per 6 mL
1 × Purification buffer	3 mL of 2 ×
300 m*M* imidazole	1.8 mL of 1 *M*

Gel Filtration Buffer	Per 500 mL
50 m*M* Na-phosphate buffer pH 7.5	50 mL of 500 m*M*
300 m*M* NaCl	75 mL of 2 *M*
0.05% DDM	0.25 g

3.2.2 O/N Preparation

1. Innoculate 10 mL LB (sterile) with appropriate antibiotic.
2. Working under sterile conditions, choose a single colony from the Petri dish containing the freshly transformed plasmid.
3. Place the falcon tube into a beaker and leave it shaking at 37°C o/n.

3.2.3 Cell Growth

Materials: 120 mL LB (autoclaved 500 mL flask), 500 mL LB (2 L flask).

Note: Work under sterile conditions.

1. Pipette 4 mL from o/n's solution into the 120 mL LB preculture (use 500 mL flask).
2. Grow for ~1.5 h at 37°C (shaken 200 shaking speed) until OD_{600} is ~0.5 to 0.6.
3. After reaching the required OD_{600} levels, transfer 60 mL of the preculture into the 500 mL LB (use 2 L flask).
4. Grow main culture of 500 mL at 37°C and 200 rpm shaking speed till OD_{600} ~0.8 (~2 h).
5. Cool down to 25°C and induce with 1 m*M* IPTG.
6. Grow the culture at 25°C for 4 more hours shaken at the same speed.
7. Harvest the cells in Sorval flasks by centrifugation at 5000 × *g* for 15 min at 4°C.
8. Weigh the *Sorval* flask with the pellet after the supernatant is removed.
9. Collect the pellets in flacon tubes (use 1 falcon tube per liter of culture).
10. Weigh the empty *Sorval* flasks again and calculate the pellet mass.
11. Keep the pellets at −80°C freezer until use.

3.2.4 First Purification Day

12. Resuspend the frozen pellets in 20 mL lysis buffer.
13. Centrifuge at 4000 × *g*, 4°C, for 20 min.
14. Transfer 30 mL of the supernatant in each of preweighted *Beckman* tubes.
15. Ultracentrifugate at 100,000 × *g*, 4°C, for 1 h.
16. Remove supernatant, weigh tubes with pellet, and calculate the total mass of membrane.
17. Resuspend each pellet in 4 mL of solubilization buffer.
18. Combine the suspensions in homogenizer and adjust the volume: for each g of membrane use ~25 mL solubilization buffer.

    Note: Solubilization buffer volume could vary depending on protein expression (in membranes), which should be checked for each individual preparation and adjust accordingly. The volume ratio we recommended is optimal for the specific protein construct and expression strain we used for these specific studies on MscS.
19. Solubilize for a further hour in cold room with gentle shake.
20. Centrifuge at 4000 × *g*, 4°C, for 10 min.
21. Pass the sample through the equilibrated Ni^{2+}–NTA (0.4 mL resin =0.8 mL suspension).
22. Wash with 10 mL of wash buffer.

3.3 Cysteine Modification by Nitroxide Spin Labels

The following protocol was used in order to modify single-cysteine mutants for MscS. It is a generally applicable protocol and with minor modifications could be applied to a wide variety of membrane and soluble proteins (or soluble portions of MPs).

3.3.1 *Second Purification Day and SDSL*

23. Dissolve MTSL spin label in wash buffer (10 × the expected concentration of the eluted protein) in 1.2 mL final volume.

Note: Reducing agent such as DDT or TCEP may be added at this step.

24. Add 1 mL of fresh MTSL of 10 × excess of protein.

25. Transfer the column to room temperature (RT) for 20 min and then transfer it to 4°C for another 40 min.

26. Wash with 10 mL of wash buffer.

27. Elute with first elution buffer (50 m*M* imidazole), 1 mL of elution fraction (E_0).

28. Elute with the second elution buffer (300 m*M* imidazole), 1 mL fractions of purified protein (E_1 to E_4).

29. Measure UV/vis spectra for E_0 to E_4 with a UV–VIS spectrometer.

30. Subject the eluted protein to gel filtration, using a *Superdex200* column.

31. Collect fractions for loading onto the tubes for later EPR use.

3.4 Labeling Efficiency

Accurate quantification of cysteine modification (spin-labeling efficiency) is essential for the success of the EPR experiments. Low efficiency translates into low signal-to-noise ratio requiring higher number of spectra averages and longer acquisition times.

Efficiency is heavily dependent on the chosen site. Even if the initial strategy for optimal residue selection described in previous sections is followed, there may still be:

(1) Residues of exceptional interest one needs to test with low predicted label accessibility.

(2) Practical limitations of the selection strategy, due to the absence of resolved bound lipid or detergent in the X-ray MP structural models. Their presence will significantly decrease the accessibility to sites predicted to be accessible.

(3) No existing structural models present in the PDB database, a very common case for MPs, since they only represent 1% of the PDB database entries.

(4) Low resolution of the available MP X-ray structures, having an impact on the in silico label modeling accuracy.

(5) A structural model only in one conformation of the protein, or irrelevant to its solution or native structure, which might directly affect the accessibility of the residues under study.

Therefore, it is recommended to test the efficiency of all mutants prior to the EPR experiments and large-scale preparations, to save time, effort, and resources. Appropriate testing would also allow fine-tuning and optimization of the conditions used for spin labeling.

There is a variety of methods suggested to measure the efficiency, with each one presenting certain advantages and disadvantages. The analysis of each one of these methods is out of the scope of this chapter and I will only describe here a protocol developed and optimized during our MscS project. This protocol, requiring very little amount of protein and standard equipment, is fast and provides highly accurate quantification of labeling efficiency (Branigan et al., 2013).

Accuracy is extremely important for cases such as multimeric ion channels, since fixed values of labeling efficiency are required, in order to overcome the multispin effects (Edwards et al., 2014; Giannoulis et al., 2013; Pliotas et al., 2012; Valera et al., 2016). Substoichiometric (or sparse) labeling has been shown to improve significantly the resolution of the measured distance distributions as long as signal intensity is adequate for quality spectra acquisition. However, the optimal labeling values are expected to differ for systems with different number of subunits (Schmidt et al., 2016). In such cases accurate quantification is essential and the proposed protocol, described in detail in our recent study, could offer the accuracy required (Branigan et al., 2013).

An alternative for measuring labeling efficiency is common to CW EPR methodology. Briefly, CW spectra of known protein concentration and volume samples are recorded and then doubly integrated. The integral represents the total number of spins present in the sample and is compared to paramagnetic standards of known concentration. Measurements are performed under the same conditions, to calculate the total labeling efficiency. This method requires all EPR experimental parameters to be identical for reliable conclusions, including identical positioning of the sample, temperature control, etc., and thus being prone to significant measurement errors.

3.5 Protein in Solution

In this section, I am going to describe the final steps of preparation for a protein that has been purified, SL and has been subjected to SEC. These steps include loading onto an EPR tube and flash freezing of the sample. The procedure described is similar to the one used for the studies on MscS channel in solution (Pliotas et al., 2012) with minor modifications.

Prior to EPR tube loading:

1. Keep the protein in gel filtration buffer or exchanged to D_2O buffer containing 100 m*M* TES pH 7.4, 100 m*M* NaCl, and 0.05% DDM.
2. Concentrate protein to around 400 μ*M* (~12 mg/mL for MscS monomers and mutants presenting high labeling efficiency, >80%).

Note: For mutants presenting a labeling efficiency lower than 60%, the protein concentration was further increased in order to exceed the 300 μ*M* of spin concentration.

3. Dilute protein samples by 50% by adding d-EG as cryoprotectant.
4. Transfer aliquots of 120 μL (60 μL protein sample plus 60 μL of d-EG) into 4 mm EPR tubes for X-band or 70 μL (35 μL protein plus 35 μL d-EG) into 3 mm tubes for Q-band.
5. Flash-freeze EPR tubes by immersing them into a freezing mixture (methylcyclohexan:isopentane = 1:4) cooled to −80°C with liquid nitrogen.

Note: For special cases and in order to improve the signal-to-noise ratio (SNR), samples were degassed by several pump-thaw cycles and kept under nitrogen.

3.6 Reconstitution in Lipid Mimics

A very important and powerful aspect of the PELDOR methodology is that conformation and oligomeric assembly of MPs could be assessed within lipid bilayer mimics, such as nanodiscs, bicelles, and liposomes (Dastvan et al., 2010; Joseph, Korkhov, Yulikov, Jeschke, & Bordignon, 2014; Ward et al., 2014). Recently, we have demonstrated a high-resolution method by which liposomes and nanodiscs which carry reconstituted MscL channel could be accurately quantified and visualized (or measured) with increased contrast by Transmission Electron Microscopy (TEM) (Asadi et al., 2017;

Hacker et al., 2016). In this section, I will describe the sample preparation protocol we followed to study MscS in nanodiscs and bicelles by EPR (Ward et al., 2014), a method applicable to a variety of MPs with minor modifications.

After MscS is purified and SL in detergent solution (Sections 3.2 and 3.3), it could be reconstituted in either empty nanodiscs or bicelles to perform PELDOR distance measurements within lipid environment. This allows the assessment of MP helical arrangement in native-like environment.

3.6.1 Nanodiscs

Note: MSP1D1E3 proteins are expressed in *Escherichia coli* cells and purified to homogeneity as described previously (Denisov, Baas, Grinkova, & Sligar, 2007; Denisov, Grinkova, Lazarides, & Sligar, 2004).

For MscS experiments 1,2-dimyristoyl-sn-glycero-3-phosphocholine (DMPC) lipids, purchased from Avanti Polar Lipids were used. We have also used in similar type of experiments *E. coli Polar Extract* (Avanti 100600) with equal efficiency regarding nanodisc formation.

1. Use MscS monomer to MSP1D1E3 protein ratio of 1:2.
2. Add sodium cholate (0.5 *M* stock) directly to MSP1D1E3 protein to a final 0.11 *M* cholate concentration.
3. Dissolve DMPC in buffer containing 50 m*M* sodium phosphate pH 7.5, 300 m*M* NaCl and 0.1 *M* sodium cholate by subsequent rounds of vortexing and sonication.
4. Add the MSP1D1E3 protein to give a molar ratio of 150:1 (DMPC: MSP1D1E3). At this stage empty nanodiscs are formed.
5. Add 1.5 mg of purified MscS in 0.05% DDM (~5 × CMC), 50 m*M* sodium phosphate pH 7.5, 300 m*M* NaCl to the MSP1D1E3-lipid mixture (empty nanodiscs).
6. Add buffer containing 50 m*M* sodium phosphate pH 7.5, 300 m*M* NaCl and 0.1 *M* sodium cholate to the mixture such that the concentration of DDM in the sample is diluted to a concentration below the critical micelle concentration of DDM (~0.0087%).
7. Incubate the reconstitution mixture at RT (21°C) for 2 h.
8. Dialyze the sample mixture three times 2 h against 3 L PBS buffer at 21°C, the buffer to remove detergent and allow for the nanodiscs with reconstituted MscS to be formed.
9. Exchange sample buffer after dialysis into D_2O buffer (as described in Section 3.5).

10. Concentrate the reconstituted mixture to ~30 mg/mL (total protein concentration including both MscS and MSP1D1E3), which could give more that sufficient EPR signal for X-band EPR of moderately labeled mutants.

Note: In case a Q-band spectrometer is used the final concentration could be reduced to half or even a third (i.e., 10 mg/mL total protein concentration).

3.6.2 Bicelles

1. Dissolve 10% (w/v) CHAPSO in D_2O buffer containing 100 m*M* TES pH 7.4 and 100 m*M* NaCl.
2. Make a 40% (w/v) stock solution by mixing DMPC lipids with the 10% (w/v) CHAPSO solution.
3. Perform 5 × cycles, which include 5 min ice cooling, 2 min vortexing, 5 min sonication, and 5 min subsequent centrifugation until the mixture becomes a clear liquid-like solution (~16°C) or a viscous gel (~4°C) (Poulos, Morgan, Zimmer, & Faham, 2015).

Note: DHPC could also be used instead of CHAPSO but it is more hygroscopic and difficult to work with.

4. Add the 40% (w/v) bicelle solution to ~15 mg/mL purified SL MscS in buffer containing 100 m*M* TES pH 7.4, 100 m*M* NaCl and 0.05% DDM and obtain a homogeneous solution after mixing
5. Add d-EG to a final concentration of 25% (v/v) (different from the rest of detergent and/or nanodisc samples, where 50% d-EG was used instead).

Note: Final spin/MscS-monomer concentration of ~200 μ*M* is sufficient for X-band measurements (60 μ*M* may be used for Q-band) and final bicelle concentration is ~8% (w/v).

4. THE EPR METHODOLOGY TOOLBOX

Different EPR methodologies could be used to SL MPs in order to monitor the mobility and local environment of the spin probes, measure intramolecular distances and determine their accessibility to solvent (via deuterium and proton). In addition, EPR could also be used to identify the location of either endogenous or exogenous membrane lipids in respect to label position and measure the distance between them.

In this section, after briefly describing the CW EPR methodology used for initial protein sample and labeling efficiency testing, I will focus on the PELDOR (or DEER) spectroscopy, a pulsed-EPR method we applied to

MscS, predicting its oligomeric and conformational state, applicable to the majority of MPs. In addition, I will describe another pulsed-EPR method, known as ESEEM, which requires the same experimental equipment, much shorter measuring times and uses the same samples to measure solvent accessibility of spin labels (via proton and deuterium). Furthermore, I will give examples from the literature, present the limitations and describe the main tools and equipment needed for each method.

4.1 CW EPR

CW EPR could be used for an initial assessment of the labeling efficiency, as described in Section 3.4. Despite this methodology presents certain drawbacks when compared to the thermofluorescent accurate quantification (Branigan et al., 2013), it still constitutes an alternative quick method for assessing the extend of spin-labeling efficiency.

In addition, it could provide important information regarding the probe. The local environment could have a direct effect on the mobility of the spin label, while its accessibility to certain hydrophilic or hydrophobic reagents could also be assessed. Occasionally, spin labels follow protein side-chain changes and CW spectra differences arising from these changes are attributed to protein conformational changes (Hubbell et al., 2000; Perozo, Cortes, Sompornpisut, Kloda, & Martinac, 2002; Roosild et al., 2010; Vasquez, Sotomayor, Cordero-Morales, Schulten, & Perozo, 2008).

However, prominent changes on CW EPR spectra of MPs could be caused by detergent exchange and are not directly linked to conformational changes (Jimenez, Freed, & Cafiso, 2011; Pliotas et al., 2012). In other reported cases, CW EPR spectra showed no difference, between two distinct protein conformational states identified by PELDOR measurements (Glaenzer, Peter, Thomas, & Hagelueken, 2017), suggesting the environment and the resulting mobility of the spin label has not changed between the two different protein states. We have previously observed that detergent exchange of MscS D67R1 resulted in local changes of the immediate environment of the label translating to significant changes in the CW spectra, when switching from DDM to Fos-Choline-14, used for MscS extraction and solubilization (Fig. 3A and B). Nevertheless, PELDOR time traces and distance distributions remained unchanged, suggesting same MscS conformation for both DDM and Fos-Choline-14 (Pliotas et al., 2012).

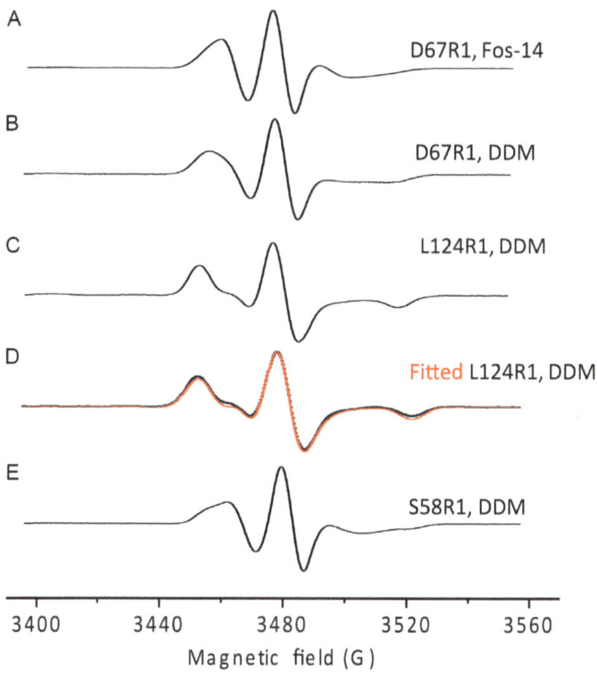

Fig. 3 X-band CW EPR spectra of MscS mutants of varying mobilities. (A) CW spectrum of MscS D67R1 extracted, solubilized, and purified in Fos-Choline-14. (B) CW spectrum of MscS D67R1 extracted, solubilized, and purified in DDM shows difference in spin label mobility and overall shape compared to (A), but no change in MscS state (both were in the open conformation) was observed by PELDOR. (C) CW spectrum of MscS L124R1 in DDM, which was the most rigid spectrum we observed compared to the ones tested. (D) Fitted CW spectrum of MscS L124R1 in DDM, using *EasySpin 4.0* (Stoll & Schweiger, 2006), yielding a correlation time of ~70 ns. (E) CW spectrum of MscS S58R1 in DDM which was one of the most mobile MscS spectra we observed.

PELDOR is directly linked to the state of the protein, while for CW EPR this is not always the case. Therefore, when CW EPR spectra are taken as experimental evidence to support conformational changes (Bavi et al., 2016; Dimitrova et al., 2016; Perozo et al., 2002; Vasquez et al., 2008), further experiments providing independent structural restraints should be performed before safe conclusions could be made. PELDOR could offer ~Å resolution, as evidenced by a comparison of the high-resolution MscS D67R1 X-ray structure distances, in which all seven TM spin labels have been resolved, with the PELDOR mean distances (Table 1 and Fig. 4). PELDOR is in complete agreement with X-ray

Table 1 Comparison of PELDOR (DEER) Distance Measurements With X-Ray Spin Labeled MscS D67R1 X-Ray Structure

	MscS D67R1 X-Ray Structure (Å)[a]	PELDOR Mean Distance (Å)[b]	PELDOR Mean Distance Ratio (D_x/D_1)	PELDOR Mean Distance Ratio (D_x/D_2)
D_1	25.9 ± 0.2	25.9	1	0.56 (7-mer)
D_2	46.7 ± 0.2	45.8	1.8 (7-mer)	1
D_3	58.3 ± 0.1	56.7	2.2 (~7-mer)	1.24 (7-mer)

[a]Values adapted from Pliotas, C., Dahl, A. C., Rasmussen, T., Mahendran, K. R., Smith, T. K., Marius, P., et al. (2015). The role of lipids in mechanosensation. *Nature Structural & Molecular Biology, 22*(12), 991–998. doi:10.1038/nsmb.3120.
[b]Values adapted from Pliotas, C., Ward, R., Branigan, E., Rasmussen, A., Hagelueken, G., Huang, H., et al. (2012). Conformational state of the MscS mechanosensitive channel in solution revealed by pulsed electron–electron double resonance (PELDOR) spectroscopy. *Proceedings of the National Academy of Sciences of the United States, 109*(40), E2675–E2682. doi:10.1073/pnas.1202286109.
Ratios between D_1, D_2, and D_3 distances of MscS heptamer derived from PELDOR measurements predicting the oligomeric state of the protein.

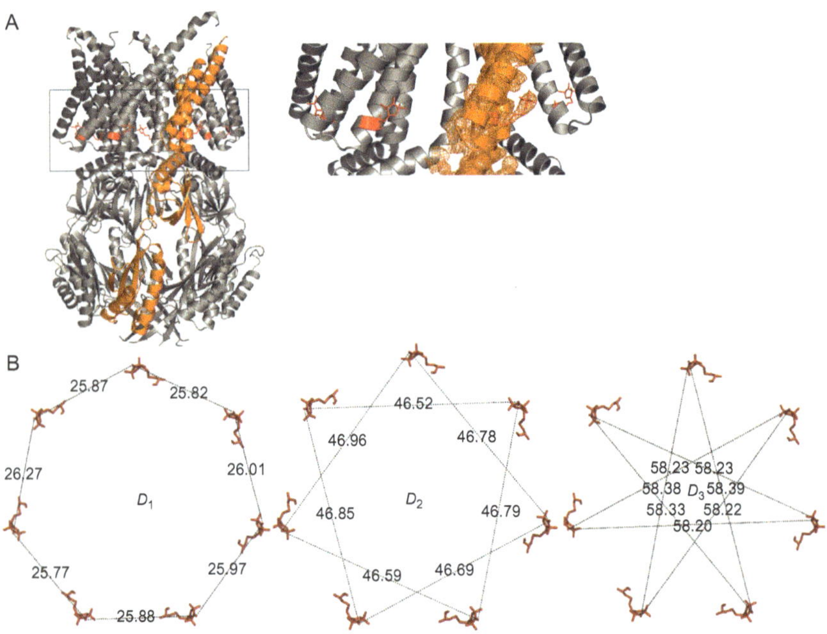

Fig. 4 Spin-to-spin distances between the seven resolved MTSL labels on the MscS D67R1 high-resolution structure. (A) Location of D67R1 (*red sticks*) in the hydrophobic pocket of MscS in the TM region within TM3 and TM3 helices. Zoom in of one of the labels and $2F_0$–F_c electron density (*golden mesh*), contoured at 1.5σ (*inset*). (B) D_1, D_2, and D_3 spin-to-spin distances of the MTSL label measured on the high-resolution MscS D67R1 X-ray structure (PDB 5AJI) (Pliotas et al., 2015).

providing Å or even sub-Å resolution, especially for the first distance (D_1), which was exclusively used to draw conclusions in our previous study (Pliotas et al., 2015, 2012).

Both the oligomeric and conformational states of the protein were accurately predicted by PELDOR. The ratios among the three mean distances were consistent with a heptameric assembly (Table 1, Figs. 2 and 5D) (Pliotas et al., 2012). D_1, D_2, and D_3 distances were in complete agreement with the X-ray structure-derived distances, suggesting MscS in solution is in the same conformation, that is the open state (Table 1, Figs. 2 and 4) (Pliotas et al., 2015). PELDOR (or DEER) spectroscopy has been proven a very reliable tool in predicting the number of MscS subunits and its structural state and should be the preferred method of choice also among other MPs, when the equipment and expertise are available.

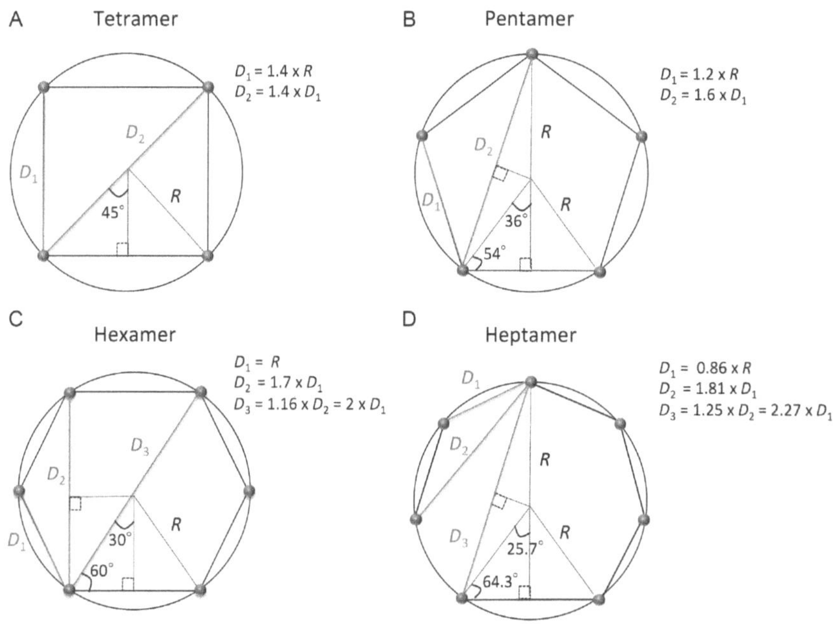

Fig. 5 Oligomerization assessment by PELDOR (DEER) spectroscopy. The ratios among different measured distances could lead to number of subunit identification for a channel with unknown oligomeric state and symmetry. Relation between D_1 and D_2 distances for a symmetric (A) tetramer, (B) pentamer and D_1, D_2, and D_3 for a symmetric, (C) hexamer, and (D) heptamer.

4.1.1 Spin Label Mobility and Local Environment Probing

In this section, I will describe the steps we employed to record RT CW EPR spectra for a number of MscS SL mutants. That includes the EPR parameters, reagents, and experimental set-up.

Protocol: SL MscS samples after size-exclusion chromatography were concentrated to around 100 μ*M* (MscS monomer concentration), judged by UV absorption at 280 nm.

4.1.1.1 CW EPR Experiment

All CW EPR experiments (Pliotas et al., 2012; Ward et al., 2014) were performed on a Bruker X-band CW spectrometer with either TM4103 or SHQ, Bruker resonators, at RT.

1. Load 50 μL of the SL protein sample on a bottom sealed glass EPR tube, using a Gilson pipette.
2. Place the sample tube in the resonator.
3. Adjust the bias (at 50 dB), and then the frequency, signal phase, and the iris while power is gradually being increased from 60 db to a large value 10 dB.
4. Record spectrum using the following main EPR parameters, no. of averages 10–20, power 10 mW, and modulation amplitude 1–2G.

Note: The *X*- and *Y*-axes data values from the spectra were then imported and fitted using the slow-motional regime option (e.g., chili function) of the *EasySpin 4.0* routine (Stoll & Schweiger, 2006). An example of the fitting using EasySpin is given in Fig. 3D for L124R1 in DDM. The fitting parameters of the CW spectra allowed an estimation of the correlation time for each site varying from a couple of ns (5.4 ns for D67R1 and 3.3 ns for S58R1 in DDM, Fig. 3B and E), to ~100 ns (70 ns for L124R1 in DDM, Fig. 3C and D) to be obtained (Pliotas et al., 2012).

In this study (Pliotas et al., 2012), we only commented on the mobility of the label as a result of its surrounding environment and we did not assign any changes of spectra mobility to conformational changes or state of the protein. However, we consistently observed restricted spin label mobility (i.e., close to rigid EPR limit and 100 ns correlation time) for exposed sites, which presented lower labeling efficiency. Interestingly, most of these sites were predicted to be totally solvent accessible in the X-ray model structures.

Based on our experience from large datasets of both published and unpublished work we could speculate that this unexpected label restriction for could be attributed to bound endogenous lipid and/or detergent molecules located in a very close proximity to modified residues. Most X-ray models of the TM regions could not predict accurate orientation positioning of the spin labels because lipids and/or detergents are poorly resolved, despite still present. This effect seems not to be an issue for non-TM MP solvent accessible sites and modeling of spin label positioning prediction for these sites is significantly more accurate.

4.2 PELDOR/DEER Spectroscopy

The pulsed-EPR methodology could be used to assess the oligomeric and conformational state, both either under physiological conditions in solution or within lipid/native environment for multimeric MPs. All this crucial information could be contained within a single quality spectrum.

4.2.1 Oligomerization Assessment

Depending on the symmetry of the specific multimer, there will be an expected number (distribution) of distances. PELDOR has successfully been applied on measuring number of present monomers in radical bearing model systems (Bode et al., 2007) while the theoretical background of these multispin system distances has been extensively investigated (Giannoulis et al., 2013). In addition, apart from the number of expected distances a mathematical relation between these distances would also exist depending on the number of subunits/monomers of the multimeric assembly. In Fig. 5, representative calculations for a hypothetical tetrameric, pentameric, hexameric and heptameric proteins are shown. As an example, the biologically relevant oligomeric state of MscL has been continuously questioned (Chang, Spencer, Lee, Barclay, & Rees, 1998; Dorwart, Wray, Brautigam, Jiang, & Blount, 2010; Liu, Gandhi, & Rees, 2009) and reports based on native mass spectrometry have shown that it could exist as tetramer, pentamer, or even hexamer, depending on a variety of conditions (Reading et al., 2015). In such cases, PELDOR could constitute an excellent tool to assess oligomeric state in a variety of physiologically relevant and native conditions (Pliotas et al., 2012; Ward et al., 2014).

Two distances are expected if MscL is a tetramer or pentamer and three in case the protein assembles into hexamers (Fig. 5). Thus, even from this first observation one could experimentally confirm or exclude the hexameric organization. To discriminate between a tetramer and pentamer one would need the ratio between the first and second expected distances, which differs significantly for these two cases (Fig. 5).

4.2.2 Conformational Characterization

PELDOR's strength for assigning the conformation of complex MPs has been demonstrated over the past years either leading or supporting other structural biology techniques toward the elucidation of complex ion channel and transporter gating mechanisms (Dalmas et al., 2014; Georgieva et al., 2013; Kazmier et al., 2014; Martens et al., 2016; Pliotas & Naismith, 2016; Pliotas et al., 2012; Verhalen et al., 2017; Zhu et al., 2016). In the particular case of MscS, PELDOR has been shown to be in agreement and complement well with X-ray crystallography, highlighted by the SL high-resolution MscS X-ray structure (Table 1 and Fig. 4) (Pliotas et al., 2015).

4.2.3 Distance Measurements by PELDOR/DEER, Set-Up, and Running

After the protein sample has been assessed for stability, purity, spin-labeling efficiency, and total spin concentration, it was loaded to a PELDOR tube following the procedure described in Section 3.5.

The procedure we followed to record X-band spectra for MscS was based on the method described earlier (Schiemann, 2009). More recently, we have performed Q-band measurements on several ion channels including MS channels and potassium transporters, using a similar protocol, including alterations for improvement coming from our experience is described here.

Q-band has become almost essential nowadays since they are the preferred platforms for PELDOR measurements, due to the great advantages they offer, mainly the increased sensitivity and acquisition time (for a given SNR 1/2h is required to o/n measurements required for X-band for comparable SNR).

These advantages combined, offer longer time windows for signal acquisition, increase the reliability of measured distances and importantly allow acquisition for very long distances to be recorded (~160 Å). The following protocol could be equally applied to both X-band and Q-band measurements with the appropriate adjustments, with few of them highlighted in the following protocol.

For MscS PELDOR experiments (Pliotas et al., 2012; Ward et al., 2014), an X-band Bruker ELEXSYS E580 spectrometer with a dielectric ring resonator and a Bruker 400U second microwave source unit were used. Measurements were made at 50 K with an overcoupled resonator. The four-pulse, dead-time free PELDOR sequence was used (Pannier et al., 2000). Pump pulse frequency was positioned at the center of the nitroxide spectrum. Despite these measurements were performed at X-band the following protocol could be equally applied to Q-band measurements (e.g., we currently use a 150 W Bruker ELEXSYS E580 Q-band spectrometer), with minor appropriate adjustments, which are highlighted in the following protocol.

4.2.3.1 Switching on the Spectrometer

1. Turn on the heat exchanger, the power for the console and the power supply for the magnet.
2. *Software*: Start the Xepr program and then go to Acquisition and click Connect to spectrometer.

4.2.3.2 Tuning the Microwave Cavity and Bridge

3. Open the Microwave Bridge tuning dialog box and switch to tune mode.
4. Set the microwave attenuator to 30 dB and scan for the cavity dip (the resonance frequency of the empty cavity).
5. Insert your sample and this will shift the cavity dip far away from the center, therefore you will need to bring it back to the center again.

Note: Depending on which type of experiment, the cavity needs to be either critically coupled for CW experiments or overcoupled for pulse measurements. Detailed instructions on how to adjust the coupling in both cases can be found in ELEXYS E580 user's manual provided by Bruker.

4.2.3.3 Acquiring Spectra

In the case of pulsed experiments, once the cavity is overcoupled switch to Pulse mode in the Bridge Configuration tab on the FT Bridge panel.

1. Create a new experiment by clicking the New Experiment button in the monitoring panel.
2. Click the Pulse tab and select Advanced button and hit the Create button.

Note: In pulse experiment, you do have either option to run from tables or from pulse PulseSPEL program depending on the type of experiment that you would like to perform. Usually, one uses tables for simple experiments that do not require phase cycling whereas PulseSpel program is more suitable for more sophisticated sequences that require phase cycling to remove the unwanted echoes and receiver offsets.

Before performing any pulse experiments, a safety test is necessary to avoid damaging the pulse bridge. We recommend following the safety instructions provided in the ELEXSYS E580 user's manual prior to run any pulse experiment.

The first experiment to start with in pulse measurements is the echo-detected field swept EPR, which produces an EPR spectrum analogous to the CW EPR spectrum. The experiment is performed by sweeping the magnetic field and integrating the echo intensity, generated by a two-pulse sequence echo. Most of the experiments presented in this chapter have been performed using the PulseSPEL program.

4.2.3.4 Field Sweep

1. Set the magnetic field around 3460G (X-band) for a nitroxide spin label. Activate the run PulseSPEL button in the acquisition panel.

Note: Frequency is round 9.8GHz for X-band and 34GHz for Q-band.

2. Load the PulseSPEL program and its corresponding variables definitions. Each kind of experiment (Field sweep, T_2, DEER) has its own PulseSPEL program.
3. Compile both programs.
4. Set the acquisition trigger position by adjusting d_0 in PulseSPEL program (d_0 = 360 ns in our case). This is adjusted in a way that the acquisition should start just before the echo.
5. Set the length of the acquisition trigger pg to cover the whole echo (pg = 160 ns in our case).
6. Set the sweep width to 120.00G.

Note: A PulseSPEL program can contain several individual experiments. Therefore one should select the experiment that he/she aims for.

7. Select the Field sweep in the experiment box with no phase cycling.
8. Press the Run button and save the spectrum.

Note: Detailed steps can be found in the ELYXSYS E580 user's manual.

4.2.3.5 T_2 Relaxation Time Experiment

This experiment is performed at fixed magnetic field position, and the intensity of the echo is monitored by incrementing the position of the second pulse.

Note: Once the field sweep echo spectrum is saved check the magnetic field value corresponding to the highest point in the spectrum and set it as center field at which you will run your T_2 experiment. The PulseSPEL program used for the field sweep echo, also includes the T_2 experiment.

1. Change the experiment in the experiment box to 2P_ESE_decay.

Note: Since we are monitoring the height of the echo, we only need to integrate the area around its maximum, which corresponds to the length of the long pulse used in the pulse sequence.

2. Set the integrator gate length to pg = 32 ns (our pi pulse length).
3. Set the delay between the two pulses $d_1 = 200$ ns.
4. Set the acquisition trigger position $d_0 = 418$ ns or 420 ns to cover only 32 ns area symmetric around the highest point of the echo.

Note: To save all the optimized variable values in PulseSPEL program, one needs to click run button after each modification.

5. Start SpecJet in order to check whether the signal is truncated.
6. Select 2-step phase cycling.

Note: This 2-step phase cycling is needed to suppress any receiver offsets.

7. Click run and save the data.

Note: Processing options are provided by Bruker software to get a good estimation of T_2.

4.2.3.6 The PELDOR Experiment

PELDOR is a two-microwave frequency pulse EPR technique that allows to measure distances between spin labels. A successful PELDOR (DEER) measurement relies on a number of setup experiments that needs to be conducted properly.

1. Activate the run PulseSPEL button in the acquisition panel.
2. Load the PELDOR PulseSPEL program and its corresponding variables definitions.
3. Compile both programs.

In the PELDOR experiment, one has to consider that the modulation depth of the PELDOR time trace mostly depends on the amount of spins excited by the pump pulse. Bruker has provided an optional MPFU (microwave pulse forming unit) channels in the pulse bridge that allow achieving this goal. In all our PELDOR experiments, we made the use of these MPFU channels. More details can be found in the ELEXSYS E580 user's manual.

Note: Prior to run a PELDOR experiment many adjustments are required and PELDOR PulseSPEL program provides all the steps to do so. Although one could still run a PELDOR experiment from tables, we recommend the use of PulseSPEL to acquire a data with DC offsets free, thereby making the analysis much easier.

PELDOR experiment is performed at fixed magnetic field position and two distinct microwave frequencies. We set the pump pulse at the highest point in the field sweep and the detection is set at least 80 MHz offset.

4. Set the magnetic field position to the highest point in the FS spectrum.
5. Load in the experiment box $\langle +x \rangle$ setup and in the phase cycling $\langle +x \rangle$ as well.
6. Set the Microwave attenuation to full power (i.e., 0 dB) and adjust the $\langle +x \rangle$ amplitude and the phase in a way to achieve a maximum echo.
7. Real part (i.e., green line) has to be positive.

Note: Once the power and the phase are adjusted, next load the ELDOR setup experiment in the experiment box to optimize the pump pulse length.

8. Load in Phase cycling box pi_frq2.

Note: Since it's a nutation like experiment, we do need to set the ELDOR frequency to the same value of the observer frequency.

9. Set the ELDOR attenuation to 0 dB.

Note: Peak should be flipped in SpecJet.

10. Adjust the pump pulse length in order to get a maximum inverted echo.
11. Set the pump pulse length p_2 to the value found in the previous step.
12. Stop the SpecJet.
13. Set high power channel to 60 dB.
14. Switch TWT to standby mode.
15. Set high power channel to 0 dB.

Note: Once the observed frequency is shifted by the required offset, you do need to go back and check the amplitude and the phase for the $\langle +x \rangle$ as both are affected by such frequency shift.

16. Load in the experiment and phase cycling boxes $\langle +x \rangle$ setup and $\langle +x \rangle$, respectively.
17. Play it.
18. Optimize $\langle +x \rangle$ using the $\langle +x \rangle$ MPFU control amplitude and signal phase.
19. Real part must be positive.

Note: As we are using 2-steps phase cycling in the PELDOR experiment, we do need to adjust the amplitude and the phase of $\langle -x \rangle$ channel. This is done in way to get the same amplitude like $\langle +x \rangle$ but opposite phase.

20. Change experiment to $\langle -x \rangle$ setup and phase cycling to $\langle -x \rangle$ as well.
21. Play it.
22. Optimize $\langle -x \rangle$ using the $\langle -x \rangle$ MPFU control amplitude and signal phase.

Note: Real part must be negative.

Relaxation time T_2 plays a key role in determining the optimum dipolar evolution time (d_2 in PulseSPEL program). A standing DEER experiment is needed to optimize such delay d_2.

23. Set experiment to Standing DEER.
24. Set phase cycling Standing DEER (only observer).
25. Set, in the first instance, the delay $d_2 = 1\ \mu s$ to phase the signal properly and set the right acquisition trigger position d_0. At short d_2, the signal of the echo is usually strong enough to phase it properly.
26. Optimize the signal using the main signal phase.
27. Check for optimal delay d_2 parameter that allows you to get a decent signal and also to cover your expected distance. This will depend mainly on T_2 relaxation time. The stronger the echo at your optimum d_2, the shorter is the averaging time. We do need to calculate the dimension of X-axis, which is a combination of the delays d_1, d_2, d_3 and the step size of the pump pulse d_{30} to avoid pulse overlapping.
28. Set d_{30} to the desired value.
29. Change $\dim_3$ coordinately according to the formula described in the program.
30. Compile the program.
31. Set experiment to DEER.
32. Phase cycling to DEER run.
33. Adjust number of scans (n).
34. Run it.

4.2.4 Exogenous SL Lipid Localization

Exogenous SL lipids could be directly added into nonlabeled MP/detergent samples and in case there is affinity for them, bind to the protein. I will briefly describe here how samples could be made and distances between bound SL lipids could be measured by PELDOR using a similar set up to the one described in previous sections, to allow exogenous lipid localization. The latter could work in a similar way to global positioning system where the intercept of three spheres in space should define a single point. This has not been demonstrated for lipids yet but it has been successfully demonstrated for the localization of metal ions in simpler biomolecules (Abdullin, Florin, Hagelueken, & Schiemann, 2015).

Lipid localization faces certain challenges such as:

a) The affinity for specific exogenous lipids of the MP of interest.
b) Efficient binding competition of exogenous SL lipids with either endogenous lipids ones or bound detergent used for extraction.
c) Symmetry of lipid binding to MPs.

d) Total number of bound lipids should be kept low per MP, otherwise a great number of distances should appear making data interpretation extremely challenging.
e) Considerations of the lipid properties and effectiveness after the modification with spin label compared to the native one.

Endogenous lipids and detergent present in purified MscS have been previously exchanged to exogenously added *Lyso*-PC (Pliotas et al., 2015). A similar protocol described here could be used for efficient exchange with SL lipids instead.

4.2.4.1 SL-Lipid Treatment of Purified MscS

Procedure

1. Dissolve SL lipids of choice in 0.05% DDM, 50 m*M* sodium phosphate pH 7.5, and 300 m*M* NaCl.
2. Add 30% molar ratio (SL lipid/WT MscS monomer) in a microcentrifuge tube.
3. Place the tube on a rocker at RT for 5 min and subsequently sonicate for a further 5 min.
4. Keep the MP target protein-containing tube and SL lipids on ice for several minutes.
5. Repeat Steps 3 and 4 two more times.
6. Subject the sample to gel filtration to remove the excess of nonbound SL lipids, in a buffer containing 50 m*M* sodium phosphate pH 7.5, 300 m*M* NaCl, and 0.05% DDM.

Note: Free SL lipids in solution will contribute to the total background EPR signal, which is unwanted.

7. Collect the fractions of the single homogeneous peak corresponding to the heptamer of WT MscS or equally to the MP or interest.
8. Concentrate the fraction using 100 kDa cut-off viva spin concentrators and subject the concentrated sample to further lipid mass spectrometry analysis.

4.3 Electron Spin Echo Envelope Modulation (ESEEM)

ESEEM spectroscopy is a hyperfine pulsed-EPR technique which is not widely used in the case of biomolecules but it has shown great sensitivity in detecting proton and deuterium coupling among chemical molecules and radical forms (Deligiannakis & Rutherford, 2001). It can be also used to probe local secondary structure of MPs (Liu et al., 2016, 2015) and native environment on the KcsA potassium channel including phospholipid headgroups (Cieslak, Focia, & Gross, 2010).

In this section, I will describe a methodology by which it could be possible to assess the accessibility of protein spin label(s) to deuterium (represents the vast majority of solvent in our PELDOR protein samples) and protons (arise from surrounding residues and/or acyl chains coming from either detergents or endogenous lipids and thus very informative for localization and binding of lipids or lipid-like molecules). Deuterium, proton and ^{31}P accessibility for a given SL residue could be estimated via their corresponding hyperfine couplings in performing a single experiment, which significantly reduces the averaging time. Protein sample preparation is identical to PELDOR samples and the same samples could be used and measurements could be performed at 50 K or equally at 77 K (e.g., using liquid N_2).

ESEEM measurements are much faster compared to PELDOR measurements varying with spin concentration, while the experimental set-up is quite simpler. In this way, more information could be obtained from the same set of samples with no extra sample preparation needed, reducing both time and cost of experiments.

4.3.1 Protein, Lipid, and Solvent Accessibility

Proton accessibility could pin point hydrophobic or buried sites in the TM domain of MPs. Residues in close proximity of the spin label or surrounded by acyl chains from lipid-like molecules both endogenous and/or exogenous, could be monitored by deuterium and proton accessibility ESEEM. Phospholipid headgroups may also be monitored using the same spectra and experimental set-up, as has been previously demonstrated for the KcsA channel (Cieslak et al., 2010).

Since the same PELDOR samples could be directly remeasured, as long as they have been safely stored in liquid nitrogen without protein state and stability to be affected, an identical protocol, described in this section, could be followed.

I will describe here the procedure, by which 2p- and 3p-ESEEM spectra could be recorded in order to measure proton, deuterium, and lipid accessibilities in a single experiment. I will then present ESEEM unpublished data on MscS and explain how these data could be interpreted to lead to conclusions, regarding its largely hydrophobic surface landscape.

The limitation of this methodology when applied to MPs is that proton accessibility could not discriminate between proton contribution originating from acyl chains or neighboring protein residues. Nevertheless, ^{31}P, which is only present in phospholipid headgroups could be detected by ESEEM. Therefore, lipid headgroup localization may be achieved in respect to a reference spin label in case there is strong coupling between (a) spin(s) and a phosphorous nucleus.

4.3.1.1 The ESEEM Experiment

For MscS ESEEM experiments presented in Fig. 6, a X-band Bruker ELEXSYS E580 spectrometer with a dielectric ring resonator and a Bruker 400U second microwave source unit was used. Measurements were done at 50 K but we have also performed measurements at 77 K, using liquid N_2, producing similar quality ESEEM spectra.

Sample loading

1. Insert the tube through the small sample holder and the latter is screwed up to the long rod.
2. Insert the sample in the cavity and resume the liquid N_2 transfer by switching on the vacuum pump.

Note: When transferring the sample in to the cavity at low temperature, a careful handling is needed to avoid ice forming inside the cavity. We strongly recommend referring to the manual, which describes the procedure required for changing samples at low temperatures.

Note: Similar tuning procedure of the resonator cavity is required for most pulsed experiments; the main difference resides in the type of the experiment that you would like to perform which requires a specific pulse sequence and therefore specific settings.

4.3.1.2 Run the ESEEM Experiment

Prior to run an ESEEM experiment, a field sweep EPR spectrum is recorded. Usually this is part of the experiments set in the PulseSPEL program of 2p- or 3p-ESEEM. Once the spectrum is recorded one needs to check the magnetic field position at which he will set-up his ESEEM experiment. Usually this coincides with the highest point in the FS spectrum, which corresponds to the highest sensitivity point.

Once the desired value of the magnetic field is set, one needs to select and optimize the right echo by adjusting the amplitude and the phase. In the case of the simple 2p-ESEEM sequence a single echo is generated, while in the 3p-ESEEM four echoes are generated after the last pulse and the targeted

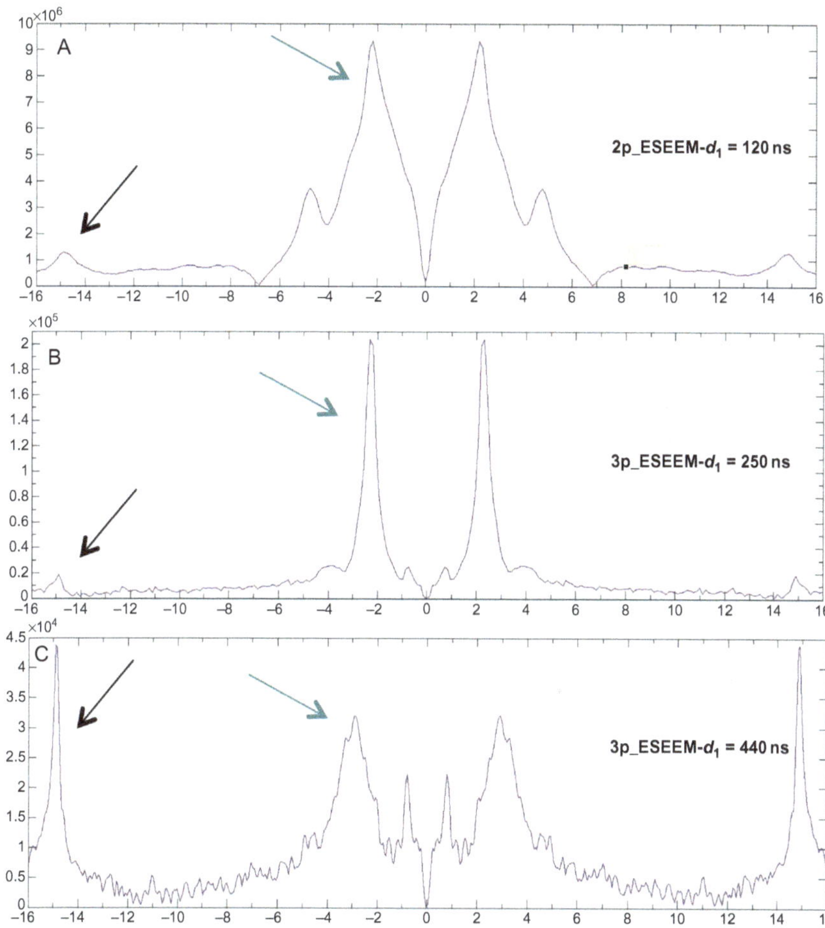

Fig. 6 Deuterium, proton, and ^{31}P accessibility measured by X-band ESEEM Fourier transform of 2p- and 3p-ESEEM spectra of spin-labeled D67R1 MscS. (A) 2p-ESEEM with $d_1 = 120$ ns. (B) 3p-ESEEM with $d_1 = 250$ ns. (C) 3p-ESEEM with $d_1 = 440$ ns. Spectra were recorded with different d_1 could suppress either the proton (*black arrows*) at 14.62 MHz or deuterium (*cyan arrows*) at 2.2 MHz signal contribution, resulting in better resolution. For this particular spin-labeled mutant we did not observe any signal for ^{31}P, expected at 5.96 MHz, possibly due to the small detection limit of ^{31}P–spin interaction of ~5 Å.

echo in this case is the stimulated echo, the one appearing just after the last pulse. In both cases, the acquisition trigger position should be set just before the desired echo. A phase cycling is needed particularly for the 3p-ESEEM to avoid the unwanted echoes. Both experiments are acquired via the PulseSPEL program. The definition of all the delays and pulse lengths is described in the ELYXSYS E580 user's manual.

Note: Echo should be symmetrical for either 2p- or 3p-ESEEM.

3. Run a FieldSweep and record the maximum intensity.
4. Adjust the Centre Field accordingly.
5. Go to Acquisition and check Run from PulseSPEL.
6. Open PulseSPEL.
7. To run 2p-ESEEM, choose the 2p_decay program and run, for 3p-ESEEM select 3p-ESEEM.
8. At setup, choose the 4-phase cycling program.

Note: Through the PulseSPEL Variable tab you can change the program's variables. For the tau1 type "d_1=," for the number of scans "n=" and for the number of points per shot "h=."

9. Adjust the $d_1 \approx 454$ ns to suppress the Deuterium and ≈ 142 ns to suppress the Proton contribution. The actual values may vary slightly depending on the Centre Field.

Note: Every time you adjust tau1 check that the signal is not truncated.

ACKNOWLEDGMENTS

C.P. is a Royal Society of Edinburgh (RSE) Personal Research Fellow, funded by the Scottish Government. C.P. would also like to acknowledge additional support by a Tenovus Scotland grant award (T15/41). Finally, C.P. would like to acknowledge Charalampos Kapsalis and Bela Bode for helpful discussions, Hassane El Mkami, Juan Carlos Penedo, and Christophe Rouillon for critical suggestions on the manuscript and the following, to work described in this chapter, Richard Ward, Olav Schiemann, and James H. Naismith.

REFERENCES

Abdullin, D., Florin, N., Hagelueken, G., & Schiemann, O. (2015). EPR-based approach for the localization of paramagnetic metal ions in biomolecules. *Angewandte Chemie (International Ed. in English)*, *54*(6), 1827–1831. http://dx.doi.org/10.1002/anie.201410396.

Asadi, J., Ferguson, S., Raja, H., Hacker, C., Marius, P., Ward, R., et al. (2017). Enhanced imaging of lipid rich nanoparticles embedded in methylcellulose films for transmission electron microscopy using mixtures of heavy metals. *Micron*, *99*, 40–48. http://dx.doi.org/10.1016/j.micron.2017.03.019.

Bass, R. B., Strop, P., Barclay, M., & Rees, D. C. (2002). Crystal structure of *Escherichia coli* MscS, a voltage-modulated and mechanosensitive channel. *Science*, *298*(5598), 1582–1587. http://dx.doi.org/10.1126/science.1077945.

Bavi, N., Cortes, D. M., Cox, C. D., Rohde, P. R., Liu, W., Deitmer, J. W., et al. (2016). The role of MscL amphipathic N terminus indicates a blueprint for bilayer-mediated gating of mechanosensitive channels. *Nature Communications*, 7, 11984. http://dx.doi.org/10.1038/ncomms11984.

Bode, B. E., Margraf, D., Plackmeyer, J., Durner, G., Prisner, T. F., & Schiemann, O. (2007). Counting the monomers in nanometer-sized oligomers by pulsed electron-electron double resonance. *Journal of the American Chemical Society*, *129*(21), 6736–6745. http://dx.doi.org/10.1021/ja065787t.

Branigan, E., Pliotas, C., Hagelueken, G., & Naismith, J. H. (2013). Quantification of free cysteines in membrane and soluble proteins using a fluorescent dye and thermal unfolding. *Nature Protocols*, *8*(11), 2090–2097. http://dx.doi.org/10.1038/nprot.2013.128.

Brohawn, S. G., Campbell, E. B., & MacKinnon, R. (2014). Physical mechanism for gating and mechanosensitivity of the human TRAAK K+ channel. *Nature*, *516*(7529), 126–130. http://dx.doi.org/10.1038/nature14013.

Brohawn, S. G., Su, Z., & MacKinnon, R. (2014). Mechanosensitivity is mediated directly by the lipid membrane in TRAAK and TREK1 K+ channels. *Proceedings of the National Academy of Sciences of the United States of America*, *111*(9), 3614–3619. http://dx.doi.org/10.1073/pnas.1320768111.

Chang, G., Spencer, R. H., Lee, A. T., Barclay, M. T., & Rees, D. C. (1998). Structure of the MscL homolog from mycobacterium tuberculosis: A gated mechanosensitive ion channel. *Science*, *282*(5397), 2220–2226.

Cieslak, J. A., Focia, P. J., & Gross, A. (2010). Electron spin-echo envelope modulation (ESEEM) reveals water and phosphate interactions with the KcsA potassium channel. *Biochemistry*, *49*(7), 1486–1494. http://dx.doi.org/10.1021/bi9016523.

Dalmas, O., Sompornpisut, P., Bezanilla, F., & Perozo, E. (2014). Molecular mechanism of Mg2+-dependent gating in CorA. *Nature Communications*, *5*, 3590. http://dx.doi.org/10.1038/ncomms4590.

Dastvan, R., Bode, B. E., Karuppiah, M. P., Marko, A., Lyubenova, S., Schwalbe, H., et al. (2010). Optimization of transversal relaxation of nitroxides for pulsed electron–electron double resonance spectroscopy in phospholipid membranes. *The Journal of Physical Chemistry. B*, *114*(42), 13507–13516. http://dx.doi.org/10.1021/jp1060039.

Deligiannakis, Y., & Rutherford, A. W. (2001). Electron spin echo envelope modulation spectroscopy in photosystem I. *Biochimica et Biophysica Acta*, *1507*(1–3), 226–246.

Denisov, I. G., Baas, B. J., Grinkova, Y. V., & Sligar, S. G. (2007). Cooperativity in cytochrome P450 3A4: Linkages in substrate binding, spin state, uncoupling, and product formation. *The Journal of Biological Chemistry*, *282*(10), 7066–7076. http://dx.doi.org/10.1074/jbc.M609589200.

Denisov, I. G., Grinkova, Y. V., Lazarides, A. A., & Sligar, S. G. (2004). Directed self-assembly of monodisperse phospholipid bilayer nanodiscs with controlled size. *Journal of the American Chemical Society*, *126*(11), 3477–3487. http://dx.doi.org/10.1021/ja0393574.

Dimitrova, A., Walko, M., Hashemi Shabestari, M., Kumar, P., Huber, M., & Kocer, A. (2016). In situ, reversible gating of a mechanosensitive ion channel through protein–lipid interactions. *Frontiers in Physiology*, 7, 409. http://dx.doi.org/10.3389/fphys.2016.00409.

Dorwart, M. R., Wray, R., Brautigam, C. A., Jiang, Y., & Blount, P. (2010). *S. aureus* MscL is a pentamer in vivo but of variable stoichiometries in vitro: Implications for detergent-solubilized membrane proteins. *PLoS Biology*, *8*(12), e1000555. http://dx.doi.org/10.1371/journal.pbio.1000555.

Edwards, D. T., Huber, T., Hussain, S., Stone, K. M., Kinnebrew, M., Kaminker, I., et al. (2014). Determining the oligomeric structure of proteorhodopsin by Gd3+-based pulsed dipolar spectroscopy of multiple distances. *Structure*, *22*(11), 1677–1686. http://dx.doi.org/10.1016/j.str.2014.09.008.

El Mkami, H., & Norman, D. G. (2015). EPR distance measurements in deuterated proteins. *Methods in Enzymology*, *564*, 125–152. http://dx.doi.org/10.1016/bs.mie.2015.05.027.

Endeward, B., Butterwick, J. A., MacKinnon, R., & Prisner, T. F. (2009). Pulsed electron–electron double-resonance determination of spin-label distances and orientations on the tetrameric potassium ion channel KcsA. *Journal of the American Chemical Society*, *131*(42), 15246–15250. http://dx.doi.org/10.1021/ja904808n.

Georgieva, E. R., Borbat, P. P., Ginter, C., Freed, J. H., & Boudker, O. (2013). Conformational ensemble of the sodium-coupled aspartate transporter. *Nature Structural & Molecular Biology*, *20*(2), 215–221. http://dx.doi.org/10.1038/nsmb.2494.

Giannoulis, A., Ward, R., Branigan, E., Naismith, J. H., & Bode, B. E. (2013). PELDOR in rotationally symmetric homo-oligomers. *Molecular Physics*, *111*(18–19), 2845–2854. http://dx.doi.org/10.1080/00268976.2013.798697.
Glaenzer, J., Peter, M. F., Thomas, G. H., & Hagelueken, G. (2017). PELDOR spectroscopy reveals two defined states of a sialic acid TRAP transporter SBP in solution. *Biophysical Journal*, *112*(1), 109–120. http://dx.doi.org/10.1016/j.bpj.2016.12.010.
Hacker, C., Asadi, J., Pliotas, C., Ferguson, S., Sherry, L., Marius, P., et al. (2016). Nanoparticle suspensions enclosed in methylcellulose: A new approach for quantifying nanoparticles in transmission electron microscopy. *Scientific Reports*, *6*, 25275. http://dx.doi.org/10.1038/srep25275.
Hagelueken, G., Ingledew, W. J., Huang, H., Petrovic-Stojanovska, B., Whitfield, C., ElMkami, H., et al. (2009). PELDOR spectroscopy distance fingerprinting of the octameric outer-membrane protein Wza from *Escherichia coli*. *Angewandte Chemie (International Ed. in English)*, *48*(16), 2904–2906. http://dx.doi.org/10.1002/anie.200805758.
Hagelueken, G., Ward, R., Naismith, J. H., & Schiemann, O. (2012). MtsslWizard: In silico spin-labeling and generation of distance distributions in PyMOL. *Applied Magnetic Resonance*, *42*(3), 377–391. http://dx.doi.org/10.1007/s00723-012-0314-0.
Hanelt, I., Wunnicke, D., Bordignon, E., Steinhoff, H. J., & Slotboom, D. J. (2013). Conformational heterogeneity of the aspartate transporter Glt(Ph). *Nature Structural & Molecular Biology*, *20*(2), 210–214. http://dx.doi.org/10.1038/nsmb.2471.
Hubbell, W. L., Cafiso, D. S., & Altenbach, C. (2000). Identifying conformational changes with site-directed spin labeling. *Nature Structural Biology*, *7*(9), 735–739. http://dx.doi.org/10.1038/78956.
Iscla, I., Wray, R., Eaton, C., & Blount, P. (2015). Scanning MscL channels with targeted post-translational modifications for functional alterations. *PLoS One*, *10*(9), e0137994. http://dx.doi.org/10.1371/journal.pone.0137994.
Jeschke, G. (2002). Distance measurements in the nanometer range by pulse EPR. *Chemphyschem*, *3*(11), 927–932. http://dx.doi.org/10.1002/1439-7641(20021115)3:11<927::aid-cphc927>3.0.co;2-q.
Jeschke, G., Chechik, V., Ionita, P., Godt, A., Zimmermann, H., Banham, J., et al. (2006). DeerAnalysis2006—A comprehensive software package for analyzing pulsed ELDOR data. *Applied Magnetic Resonance*, *30*(3), 473–498. http://dx.doi.org/10.1007/bf03166213.
Jimenez, R. H., Freed, D. M., & Cafiso, D. S. (2011). Lipid and membrane mimetic environments modulate spin label side chain configuration in the outer membrane protein A. *The Journal of Physical Chemistry. B*, *115*(49), 14822–14830. http://dx.doi.org/10.1021/jp207420d.
Joseph, B., Korkhov, V. M., Yulikov, M., Jeschke, G., & Bordignon, E. (2014). Conformational cycle of the vitamin B12 ABC importer in liposomes detected by double electron–electron resonance (DEER). *The Journal of Biological Chemistry*, *289*(6), 3176–3185. http://dx.doi.org/10.1074/jbc.M113.512178.
Kazmier, K., Sharma, S., Quick, M., Islam, S. M., Roux, B., Weinstein, H., et al. (2014). Conformational dynamics of ligand-dependent alternating access in LeuT. *Nature Structural & Molecular Biology*, *21*(5), 472–479. http://dx.doi.org/10.1038/nsmb.2816.
Levina, N., Totemeyer, S., Stokes, N. R., Louis, P., Jones, M. A., & Booth, I. R. (1999). Protection of *Escherichia coli* cells against extreme turgor by activation of MscS and MscL mechanosensitive channels: Identification of genes required for MscS activity. *The EMBO Journal*, *18*(7), 1730–1737.
Liu, Z., Gandhi, C. S., & Rees, D. C. (2009). Structure of a tetrameric MscL in an expanded intermediate state. *Nature*, *461*(7260), 120–124. http://dx.doi.org/10.1038/nature08277.
Liu, L., Hess, J., Sahu, I. D., FitzGerald, P. G., McCarrick, R. M., & Lorigan, G. A. (2016). Probing the local secondary structure of human vimentin with electron spin echo envelope modulation (ESEEM) spectroscopy. *The Journal of Physical Chemistry. B*, *120*(48), 12321–12326. http://dx.doi.org/10.1021/acs.jpcb.6b10054.

Liu, L., Mayo, D. J., Sahu, I. D., Zhou, A., Zhang, R., McCarrick, R. M., et al. (2015). Determining the secondary structure of membrane proteins and peptides via electron spin echo envelope modulation (ESEEM) spectroscopy. *Methods in Enzymology*, *564*, 289–313. http://dx.doi.org/10.1016/bs.mie.2015.06.037.

Liu, H., & Naismith, J. H. (2008). An efficient one-step site-directed deletion, insertion, single and multiple-site plasmid mutagenesis protocol. *BMC Biotechnology*, *8*, 91. http://dx.doi.org/10.1186/1472-6750-8-91.

Martens, C., Stein, R. A., Masureel, M., Roth, A., Mishra, S., Dawaliby, R., et al. (2016). Lipids modulate the conformational dynamics of a secondary multidrug transporter. *Nature Structural & Molecular Biology*, *23*(8), 744–751. http://dx.doi.org/10.1038/nsmb.3262.

Pannier, M., Veit, S., Godt, A., Jeschke, G., & Spiess, H. W. (2000). Dead-time free measurement of dipole-dipole interactions between electron spins. *Journal of Magnetic Resonance*, *142*(2), 331–340. http://dx.doi.org/10.1006/jmre.1999.1944.

Perozo, E., Cortes, D. M., Sompornpisut, P., Kloda, A., & Martinac, B. (2002). Open channel structure of MscL and the gating mechanism of mechanosensitive channels. *Nature*, *418*(6901), 942–948.

Pliotas, C., Dahl, A. C., Rasmussen, T., Mahendran, K. R., Smith, T. K., Marius, P., et al. (2015). The role of lipids in mechanosensation. *Nature Structural & Molecular Biology*, *22*(12), 991–998. http://dx.doi.org/10.1038/nsmb.3120.

Pliotas, C., & Naismith, J. H. (2016). Spectator no more, the role of the membrane in regulating ion channel function. *Current Opinion in Structural Biology*, *45*, 59–66. http://dx.doi.org/10.1016/j.sbi.2016.10.017.

Pliotas, C., Ward, R., Branigan, E., Rasmussen, A., Hagelueken, G., Huang, H., et al. (2012). Conformational state of the MscS mechanosensitive channel in solution revealed by pulsed electron-electron double resonance (PELDOR) spectroscopy. *Proceedings of the National Academy of Sciences of the United States of America*, *109*(40), E2675–E2682. http://dx.doi.org/10.1073/pnas.1202286109.

Polyhach, Y., Bordignon, E., & Jeschke, G. (2011). Rotamer libraries of spin labelled cysteines for protein studies. *Physical Chemistry Chemical Physics*, *13*(6), 2356–2366. http://dx.doi.org/10.1039/c0cp01865a.

Poulos, S., Morgan, J. L., Zimmer, J., & Faham, S. (2015). Bicelles coming of age: An empirical approach to bicelle crystallization. *Methods in Enzymology*, *557*, 393–416. http://dx.doi.org/10.1016/bs.mie.2014.12.024.

Prisner, T., Rohrer, M., & MacMillan, F. (2001). Pulsed EPR spectroscopy: Biological applications. *Annual Review of Physical Chemistry*, *52*, 279–313. http://dx.doi.org/10.1146/annurev.physchem.52.1.279.

Reading, E., Walton, T. A., Liko, I., Marty, M. T., Laganowsky, A., Rees, D. C., et al. (2015). The effect of detergent, temperature, and lipid on the oligomeric state of MscL constructs: Insights from mass spectrometry. *Chemistry & Biology*, *22*(5), 593–603. http://dx.doi.org/10.1016/j.chembiol.2015.04.016.

Roosild, T. P., Castronovo, S., Healy, J., Miller, S., Pliotas, C., Rasmussen, T., et al. (2010). Mechanism of ligand-gated potassium efflux in bacterial pathogens. *Proceedings of the National Academy of Sciences of the United States of America*, *107*(46), 19784–19789. http://dx.doi.org/10.1073/pnas.1012716107.

Schiemann, O. (2009). Mapping global folds of oligonucleotides by pulsed electron–electron double resonance. *Methods in Enzymology*, *469*, 329–351. http://dx.doi.org/10.1016/S0076-6879(09)69016-9.

Schiemann, O., & Prisner, T. F. (2007). Long-range distance determinations in biomacromolecules by EPR spectroscopy. *Quarterly Reviews of Biophysics*, *40*(1), 1–53. http://dx.doi.org/10.1017/S003358350700460X.

Schmidt, T., Walti, M. A., Baber, J. L., Hustedt, E. J., & Clore, G. M. (2016). Long distance measurements up to 160 a in the GroEL tetradecamer using Q-band DEER EPR

spectroscopy. *Angewandte Chemie (International Ed. in English)*, *55*(51), 15905–15909. http://dx.doi.org/10.1002/anie.201609617.
Sonoda, Y., Newstead, S., Hu, N. J., Alguel, Y., Nji, E., Beis, K., et al. (2011). Benchmarking membrane protein detergent stability for improving throughput of high-resolution X-ray structures. *Structure*, *19*(1), 17–25. http://dx.doi.org/10.1016/j.str.2010.12.001.
Stansfeld, P. J., Goose, J. E., Caffrey, M., Carpenter, E. P., Parker, J. L., Newstead, S., et al. (2015). MemProtMD: Automated insertion of membrane protein structures into explicit lipid membranes. *Structure*, *23*(7), 1350–1361. http://dx.doi.org/10.1016/j.str.2015.05.006.
Stoll, S., & Schweiger, A. (2006). EasySpin, a comprehensive software package for spectral simulation and analysis in EPR. *Journal of Magnetic Resonance*, *178*(1), 42–55. http://dx.doi.org/10.1016/j.jmr.2005.08.013.
Sukharev, S. I., Blount, P., Martinac, B., Blattner, F. R., & Kung, C. (1994). A large-conductance mechanosensitive channel in *E. coli* encoded by mscL alone. *Nature*, *368*(6468), 265–268. http://dx.doi.org/10.1038/368265a0.
Valera, S., Ackermann, K., Pliotas, C., Huang, H., Naismith, J. H., & Bode, B. E. (2016). Accurate extraction of nanometer distances in multimers by pulse EPR. *Chemistry*, *22*(14), 4700–4703. http://dx.doi.org/10.1002/chem.201505143.
Vasquez, V., Sotomayor, M., Cordero-Morales, J., Schulten, K., & Perozo, E. (2008). A structural mechanism for MscS gating in lipid bilayers. *Science*, *321*(5893), 1210–1214. http://dx.doi.org/10.1126/science.1159674.
Verhalen, B., Dastvan, R., Thangapandian, S., Peskova, Y., Koteiche, H. A., Nakamoto, R. K., et al. (2017). Energy transduction and alternating access of the mammalian ABC transporter P-glycoprotein. *Nature*, *543*, 738–741. http://dx.doi.org/10.1038/nature21414.
Wang, W., Black, S. S., Edwards, M. D., Miller, S., Morrison, E. L., Bartlett, W., et al. (2008). The structure of an open form of an *E. coli* mechanosensitive channel at 3.45 A resolution. *Science*, *321*(5893), 1179–1183. http://dx.doi.org/10.1126/science.1159262.
Ward, R., Bowman, A., Sozudogru, E., El-Mkami, H., Owen-Hughes, T., & Norman, D. G. (2010). EPR distance measurements in deuterated proteins. *Journal of Magnetic Resonance*, *207*(1), 164–167. http://dx.doi.org/10.1016/j.jmr.2010.08.002.
Ward, R., Pliotas, C., Branigan, E., Hacker, C., Rasmussen, A., Hagelueken, G., et al. (2014). Probing the structure of the mechanosensitive channel of small conductance in lipid bilayers with pulsed electron–electron double resonance. *Biophysical Journal*, *106*(4), 834–842. http://dx.doi.org/10.1016/j.bpj.2014.01.008.
Zhu, S., Stein, R. A., Yoshioka, C., Lee, C. H., Goehring, A., McHaourab, H. S., et al. (2016). Mechanism of NMDA receptor inhibition and activation. *Cell*, *165*(3), 704–714. http://dx.doi.org/10.1016/j.cell.2016.03.028.

CHAPTER NINE

Strategy for the Thermostabilization of an Agonist-Bound GPCR Coupled to a G Protein

Annette Strege, Byron Carpenter[1], Patricia C. Edwards, Christopher G. Tate[2]
MRC Laboratory of Molecular Biology, Cambridge, United Kingdom
[2]Corresponding author: e-mail address: cgt@mrc-lmb.cam.ac.uk

Contents

Abstract

Structure determination of G protein-coupled receptors (GPCRs) in the inactive state bound to high-affinity antagonists has been very successful through the implementation of a number of protein engineering and crystallization strategies. However, the structure determination of GPCRs in their fully active state coupled to a G protein is still very challenging. Recently, mini-G proteins were developed, which recapitulate the coupling of a full heterotrimeric G protein to a GPCR despite being less than one-third of the size. This allowed the structure determination of the agonist-bound adenosine A_{2A} receptor (A_{2A}R) coupled to mini-G_s. Although this is extremely encouraging, A_{2A}R is very stable compared with many other GPCRs, particularly when an agonist is bound. In contrast, the agonist-bound conformation of the human corticotropin-releasing factor

[1]Present address: The University of Warwick, Coventry, United Kingdom.

Methods in Enzymology, Volume 594
ISSN 0076-6879
http://dx.doi.org/10.1016/bs.mie.2017.05.014

receptor is considerably less stable, impeding the formation of good quality crystals for structure determination. We have therefore developed a novel strategy for the thermostabilization of a GPCR–mini-G protein complex. In this chapter, we will describe the theoretical and practical principles of the thermostability assay for stabilizing this complex, discuss its strengths and weaknesses, and show some typical results from the thermostabilization process.

1. INTRODUCTION

G protein-coupled receptors (GPCRs) are integral membrane proteins present in the plasma membrane of cells throughout the human body (Fredriksson & Schioth, 2005). They are the corner stone of intercellular communication with over 800 genes in the human genome encoding receptors that are activated by stimuli such as light, ions, small molecules, peptides, or protein hormones (Foord et al., 2005). GPCRs bind the activator (agonist) in the orthosteric binding pocket found in the extracellular half of the receptor (Venkatakrishnan et al., 2013). This leads to a conformational change at the cytoplasmic face of the GPCR that opens a cleft where the C-terminus of a G protein binds (Rosenbaum, Rasmussen, & Kobilka, 2009; Venkatakrishnan et al., 2016). This, in turn, results in a conformational change in the G protein (Flock et al., 2015) that ultimately leads to changes in the intracellular concentration of second messenger molecules such as Ca^{2+} and cAMP. Given their key role in signal transduction, GPCRs are the target of about 40% of commercially available small molecule drugs (Hopkins & Groom, 2002) and there is a huge potential for the development of new therapeutics with improved efficacy, potency, and reduced side effects through structure-based drug design (Congreve, Langmead, Mason, & Marshall, 2011; Lagerstrom & Schioth, 2008). The key challenge is to be able to determine the structure of any GPCR in either an inactive state bound to an antagonist or an active state bound to an agonist, depending on whether the drug should act as either an antagonist or an agonist. Ideally, structures of both conformational states should be determined to provide detailed insights into the molecular mechanism of GPCR activation. Many structures have been determined of GPCRs in the inactive state (Venkatakrishnan et al., 2013), because this is the most stable conformation of the receptor. In contrast, very few structures have been determined of GPCRs in an active conformation (Lebon, Warne, & Tate, 2012), and only two crystal structures have been determined of a GPCR coupled to a G protein, the β_2-adrenergic receptor (β_2AR) coupled to G_s

(Rasmussen et al., 2011) and the adenosine A_{2A} receptor (A_{2A}R) coupled to mini-G_s (Carpenter, Nehme, Warne, Leslie, & Tate, 2016). Both β_2AR and A_{2A}R are remarkably stable GPCRs, so the technology used to obtain diffracting three-dimensional crystals of β_2AR–G_s and A_{2A}R–mini-G_s may not be suitable for less stable receptors. Therefore, other technologies are required to stabilize GPCR–G protein complexes to enable structure determination. We have chosen to do this by further developing the methodology of conformational thermostabilization (Magnani et al., 2016; Tate, 2012).

Conformational thermostabilization employs systematic mutagenesis combined with a radioligand thermostability assay of detergent-solubilized GPCRs to stabilize them in a specific conformation. The receptors can then be crystallized using short chain detergents bound to virtually any ligand, even ligands that bind with low affinity. Conformational thermostabilization has been used successfully to stabilize the inactive, antagonist-bound state of the β_1-adrenergic receptor (Miller & Tate, 2011; Serrano-Vega, Magnani, Shibata, & Tate, 2008), A_{2A}R (Magnani, Shibata, Serrano-Vega, & Tate, 2008; Robertson et al., 2011), the glutamate receptor $mGlu_5$ (Dore et al., 2014), the glucagon receptor (Jazayeri et al., 2016), the corticotropin-releasing factor receptor 1 (CRF_1R) (Hollenstein et al., 2013), and the chemokine receptor CCR9 (Oswald et al., 2016), resulting in their structure determination (Dore et al., 2014, 2011; Hollenstein et al., 2013; Jazayeri et al., 2016; Miller-Gallacher et al., 2014; Oswald et al., 2016; Warne et al., 2011, 2008). Similarly, the agonist-bound states of A_{2A}R (Lebon, Bennett, Jazayeri & Tate, 2011) and the neurotensin receptor NTSR1 (Shibata et al., 2013, 2009) were stabilized, and their structures determined bound to the native agonist adenosine (Lebon, Warne, et al., 2011) and the active portion of the neuropeptide neurotensin (NT^{6-13}) (White et al., 2012), respectively. However, these latter two structures showed that the receptors were in an active-intermediate state and further conformational changes were required to attain the fully active state, as defined by the structure of the β_2AR–G_s complex (Rasmussen et al., 2011). Another strategy was therefore required to allow the structure determination of unstable GPCRs in their fully active state bound to G proteins. This was made possible by the development of mini-G proteins (Carpenter & Tate, 2016) that are expressed in *Escherichia coli* at levels up to 100mg per liter of culture. Mini-G proteins couple to GPCRs and increase the agonist affinity as previously reported for heterotrimeric G proteins (Carpenter & Tate, 2016; Carpenter et al., 2016). In addition, we have also observed an increase in the thermostability of agonist-bound GPCRs coupled to mini-G proteins compared to the agonist-bound GPCR alone (Carpenter & Tate, 2016; Carpenter et al., 2016).

The methodology we describe here was developed for the thermostabilization of agonist-bound human CRF_1R coupled to mini-G_s. The agonist used for these studies is a 40 amino acid residue peptide, sauvagine, which is commercially available in a radiolabeled form. Thermostability assays of ^{125}I-sauvagine-bound CRF_1R showed that it was very unstable in the mild detergent dodecylmaltoside (DDM), suggesting that crystallization would be very difficult and that stabilization was necessary. A small degree of stabilization was observed when mini-G_s was coupled to the receptor, but not enough to allow the use of short chain detergents such as octylthioglucoside that was used for the crystallization of the A_{2A}R–mini-G_s complex (Carpenter et al., 2016). We therefore devised a strategy for the thermostabilization of the ^{125}I-sauvagine-bound CRF_1R–mini-G_s complex.

2. A THERMOSTABILITY ASSAY FOR MEASURING THE STABILITY OF GPCR–G PROTEIN COMPLEXES

Thermostability assays are performed on receptors that have been directly detergent solubilized from whole cells (Magnani et al., 2016). The receptors are therefore unpurified and consequently the ligand must bind with high specificity to only the target of interest. Three different assays have been developed previously for the conformational thermostabilization of GPCRs, namely the minus format, the plus format, and the super-plus format (Magnani et al., 2016). These formats differ only in the step at which the radioligand is added to the assay. The most successful strategy for unstable receptors has been the super-plus format where the radioligand is bound to the receptor in cells, followed by detergent solubilization, and then the thermostability assay. In the plus format, the ligand is added after solubilization, but prior to heating, and in the minus format the ligand is added after heating. One reason for the success of the super-plus format over the minus and plus formats is that the ligand confers stability on the receptors when they are solubilized. The stability of a GPCR in detergent is invariably less than when it is in the membrane due to the dynamics of the detergent molecules and their ability to interpenetrate between transmembrane helices (Lee et al., 2016). In addition, binding the agonist before solubilization is essential when mutants that stabilize the active conformation are sought, because these mutations often destabilize the inactive state.

There are five essential components to consider in the development of an assay capable of measuring the amount of radioligand bound to a detergent-solubilized receptor (Magnani et al., 2016): (i) radioligand; (ii) buffer pH; (iii) ionic composition; (iv) detergent; and (v) separation strategy. The highest affinity ligand that is available should be used, ideally one with an apparent K_D of 100 nM or lower, which is important for two main reasons. First, a very high affinity ligand will stabilize the GPCR during the subsequent solubilization and heating steps (Zhang, Stevens, & Xu, 2015). Second, a high-affinity ligand is less likely to dissociate from the receptor during the separation of the free radioligand from receptor-bound radioligand on the mini gel filtration columns (spin columns). The scientific literature will generally give initial starting compositions for buffers and ionic strength that will allow the measurement of radioligand binding to the receptor in the membrane. Ensuring that these assays are reproducible on membrane-embedded receptors is recommended before starting to develop an assay for the detergent solubilized receptor. For agonist-binding assays in particular, the Na^+ ion composition needs careful optimization, because intramembrane Na^+ binding sites have been identified in the crystal structures of a variety of GPCRs (Liu et al., 2012; Miller-Gallacher et al., 2014; Zhang et al., 2012). These have been found to stabilize the inactive state of the receptor and therefore Na^+ can act as an allosteric antagonist, as observed for $A_{2A}R$ (Liu et al., 2012). Thus, a high Na^+ ion concentration may inhibit agonist binding. To find a suitable detergent, we recommend initially using the mildest detergents that are available to get the assay working, namely digitonin, GDN, or LMNG (Tate, 2010). A number of separation strategies have already been described (Magnani et al., 2016), however, this step can also require substantial optimization for ligands and/or receptors with novel properties. For the thermostabilization of the sauvagine–CRF_1R–mini-G_s complex, the novel properties of the ligand represented the main challenge for developing the thermostability assay. The 40 amino acid peptide sauvagine is much larger than any of the small molecules or short peptides used previously in other conformational thermostabilization projects. Furthermore, this ligand was found to associate with detergent micelles. Therefore, the size-based separation of the free radioligand ^{125}I-sauvagine (molecular weight 4.5 kDa, apparent molecular weight in a micelle ~50 kDa) from detergent-solubilized ^{125}I-sauvagine-bound CRF_1R (molecular weight in a micelle ~140 kDa) required screening a number of gel filtration matrices, column lengths, and spin speeds. Optimal conditions were achieved using spin columns containing

~2 mL of Sephacryl S-100, which has a higher molecular-weight cut-off than gel filtration matrices used previously in other conformational thermostabilization projects. This highlights the importance of considering the physicochemical characteristics of a given radioligand and its affinity for the target GPCR when designing a conformational thermostabilization methodology.

The super-plus mini-G format of the thermostability assay described here (Fig. 1) is based on the previously described super-plus format (Shibata et al., 2013). The main addition to the existing format is the addition of the mini-G_s protein before heating in order to promote stabilization of the fully active conformation of CRF_1R. Second, for this study we also aimed to compare the effects of introducing chemically different amino acids at the same positions by performing a RAVE scan. More specifically, we assessed mutations at 104 positions, where each position was mutated to Ala (or to Leu if it was already Ala), as described previously, and also to Glu, Arg, and Val (or to Trp if it was already Glu, Arg, or Val), resulting in a total of 416 point mutants. The method described later is specifically for the thermostabilization of the ^{125}I-sauvagine–CRF_1R–mini-G_s complex. However, further information has been included in the notes of each step to describe how this methodology can be optimized to function for the thermostabilization of any such ternary complex of interest.

The methodology described below has been subdivided into six sections (Sections 2.1–2.6). This has been done solely to facilitate the description and discussion of each distinct step in the method. In actuality, Sections 2.1–2.6 are performed sequentially in a seamless fashion and will not normally take longer than a day.

2.1 Ligand Binding

One of the features of a good thermostability assay is the use of a highly specific high-affinity ligand, which biases the conformation of the receptor in the state whose structure is required. The ligand is used in a radiolabeled form to determine the number of ligand-bound receptors in detergent solution in the thermostability assay. Once a suitable radioligand has been selected, a buffer condition must be identified to allow high specific binding to the receptor in whole cells, while keeping nonspecific binding low. The ligand of choice for stabilizing the CRF_1R–G protein complex was ^{125}I-sauvagine, a 40 amino acid peptide with a reported affinity of 100 pM for CRF_1R. After the binding assay has been applied successfully to receptors

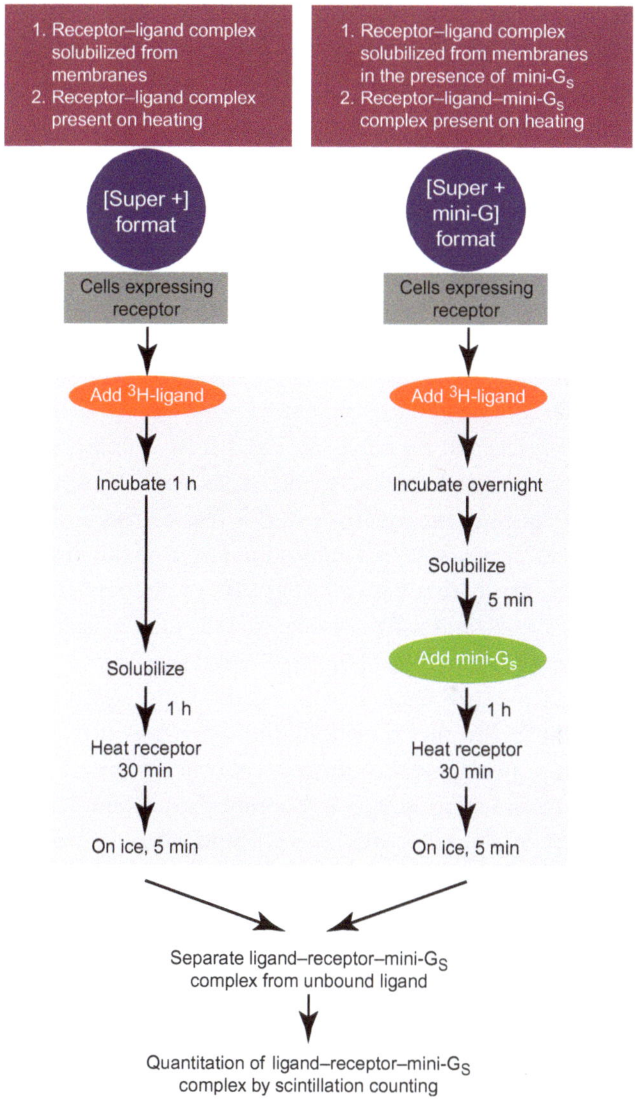

Fig. 1 Flow chart for determining the thermostability of a GPCR–mini-G protein complex. The flow chart on the left shows the preferred methodology for the isolation of mutants in either the inactive state or an active-intermediate state. On the right is the flow chart for the methodology described here for the thermostabilization of a GPCR–mini-G protein complex.

in whole cells, the assay must be altered to detect binding to the detergent-solubilized receptor. The development of such an assay is outlined in Section 2.1.3 and an in-depth discussion can be found in Magnani et al. (2016).

2.1.1 Buffers and Reagents

- Radioligand binding assay buffer (50 m*M* HEPES pH 7.5 (KOH), 20 m*M* $MgCl_2$, 0.3% BSA, 0.2 mg/mL bacitracin, 1 complete EDTA-free protease inhibitor tablet (Roche) per 50 mL buffer)
- Frozen HEK293 GnTI$^-$ cell aliquots expressing the GPCR of interest (e.g., CRF_1R) at an appropriate cell density (10 × higher than final cell density in assay, e.g., 4×10^7 cells/mL)
- High-affinity radioligand (e.g., ^{125}I-sauvagine (Hartmann Analytic)) and equivalent cold ligand (e.g., sauvagine (Sigma-Aldrich))

2.1.2 Procedure

1. Prepare a 10 × concentrated ligand stock solution at 100-fold above the apparent K_D for ligand binding. We use a final ligand concentration of 1 n*M* in the assay, and therefore need a concentrated sauvagine stock with a final concentration of 10 n*M*. To reduce the cost of this assay, we use a mixture of 2 n*M* ^{125}I-sauvagine and 8 n*M* cold sauvagine.
2. Binding reactions are set up by diluting a 10 × concentrated stock of cells expressing CRF_1R and a 10 × concentrated stock of ligand into radioligand binding assay buffer. The total reaction volume needed at this stage depends on whether a seven- or two-point assay format will be used, and whether triplicate or duplicate separation steps will be carried out. To ensure that there is sufficient volume for subsequent steps, we prepare 20% more volume than is theoretically needed. Each assay point requires 50 μL of detergent-solubilized ligand-bound receptor, so a two-point assay (two temperatures assessed) performed in triplicate will need a total of 6 × 50 μL to be loaded onto six separate spin columns, so a final total of 360 μL would be needed. At this stage, the volume required is lower, because subsequent additions of detergent and mini-G protein will increase the volume.
3. Incubate binding reactions on ice overnight.

2.1.3 Notes

In order to determine the ideal binding conditions for any receptor–ligand combination of interest, the scientific literature should be consulted to obtain a starting point for the buffer composition. Binding should then be optimized from initial experiments with the aim of maximizing specific binding of the radioligand to the receptor, while minimizing nonspecific binding of the radioligand to membranes or other proteins. Improvements in specific binding are observed as a result of screening various salts (e.g., NaCl, $MgCl_2$, and KCl), pHs, and buffers (e.g., HEPES, Tris, MES, and MOPS). Nonspecific binding can be reduced by the addition of blocking agents (e.g., BSA and bacitracin). Various concentrations of any buffer additives with beneficial effects should also be tested to maximize these effects. Furthermore, the cell density can be adjusted to optimize the signal for any ligand binding assay. The ideal cell density to use varies for different projects, since this depends on several factors, including the number of molecules of receptor expressed per cell, and the total amount of radiolabeled ligand used to set up the binding reaction. The ratio of radiolabeled: unlabeled ligand used in the assay is also dictated by the expression levels of the receptor and may require further optimization if the expression levels of the target receptor are very low. Another factor that must be optimized for each combination of radioligand and receptor is the incubation temperature and time for ligand binding. For ligands with fast on and off rates, this binding incubation time should be short and reactions should be kept on ice, whereas for ligands with slow on and off rates this binding time may need to be longer, and perhaps performed at room temperature. Further details on all these optimization steps can be found in Magnani et al. (2016).

2.2 Solubilization and Mini-G_s Binding

It is essential to find a detergent that will maintain the receptor-G protein complex in a functional state for sufficient time to measure its stability. During the initial search for assay conditions that permit a thermostability curve to be obtained, we recommend working with the mildest detergents available, namely digitonin, GDN, or LMNG. For the CRF_1R project described here, we started by using GDN. However, for assessment of the thermostabilizing effects of the point mutants, it is important to work in a detergent composition that allows easy discrimination of improvements in stability. The complex between sauvagine, wild-type CRF_1R, and mini-G_s had a very high apparent T_m in pure GDN (~40°C), making it more difficult to see improvements in thermostability arising from point mutations. We found that the addition of DDM destabilized this complex, so that its apparent T_m dropped by approximately 20°C. We therefore used a 1:3 mixture of GDN and DDM during the thermostability screen of the point mutants in order to improve the resolution of the assay.

Binding of the mini-G_s protein cannot be carried out before the cell membrane is disrupted, since it binds to the cytoplasmic face of the receptor. The addition of solubilizing concentrations of detergents disrupts membranes very quickly, and binding of mini-G_s to the ^{125}I-sauvagine–CRF_1R complex can be measured after 1 h incubation.

2.2.1 Buffers and Reagents

- Appropriate stock solutions of detergents needed for solubilization (e.g., 5% GDN (Anatrace) and 10% DDM (Glycon Biochemicals) for this project)
- Mini-G_s binding buffer (50 m*M* HEPES pH 7.5 (KOH), 20 m*M* $MgCl_2$, 0.3% BSA, 0.2 mg/mL bacitracin, 1 cOmplete EDTA-free protease inhibitor tablet per 50 mL buffer, 50 m*M* NaCl, 0.1 U/mL apyrase)
- Appropriate stock solutions of additives needed for mini-G_s binding; we use:
  - 1 M NaCl
  - 1000 U/mL apyrase
  - 250 μ*M* mini-G_s in mini-G_s binding buffer (Carpenter & Tate, 2016)

2.2.2 Procedure

1. Solubilize by adding 0.1% GDN/0.3% DDM to all samples.
2. Immediately after solubilization (within 5 min of addition of detergents), add the components required to stabilize mini-G_s and promote its binding to the receptor. The final concentrations should be 50 m*M* NaCl, 0.1 U/mL apyrase, and 25 μ*M* mini-G_s.
3. Incubate for 1 h on ice to allow the ternary complex to form.

2.2.3 Notes

A suitable detergent composition must be determined for each target and conformation of interest. Whether a harsher or milder detergent is required depends first and foremost on the stability of the GPCR of interest, and this must be tested by trial and error for each target. Furthermore, the effects of detergents on the stability of the mini-G protein must be considered as well, since this also impacts on the overall stability of the complex (Nehmé et al., 2017). The buffer additives to promote mini-G_s binding described here generally help to stabilize the GPCR-bound conformation of the mini-G_s protein, and should therefore be transferable to similar studies on other GPCRs.

In the method presented here, the binding of the agonist ^{125}I-sauvagine is performed prior to solubilization and complex formation with mini-G_s. This is done because ^{125}I-sauvagine binding to CRF_1R is very slow and is performed overnight. In addition, NaCl inhibits ^{125}I-sauvagine binding to the receptor, but NaCl is required for mini-G_s binding. However, agonist binding to many GPCRs is very fast and tolerates NaCl. Thus in most instances we add the radioligand, mini-G protein, and detergent simultaneously to form a detergent solubilized receptor–agonist–mini-Gs complex.

2.3 Heating

Heating of solubilized ^{125}I-sauvagine–CRF$_1$R–mini-G_s complex is done to measure the relative changes in thermostability arising from point mutations. To determine an accurate apparent T_m value, seven-point thermostability curves must be generated. For this, solubilized samples are split into seven aliquots and incubated at a range of temperatures between 0 and 80°C. For a high-throughput two-point assay screen, samples are split into two aliquots, where one is incubated on ice and the other just beyond the apparent T_m of the wild-type control, as determined by previous seven-point curves. Normally, we would screen all mutants in a library using a two-point assay, take the most thermostable mutants from this screen, and then use the seven-point assay to define their apparent T_m.

2.3.1 Equipment

- PCR machines or metal blocks for heating (e.g., Eppendorf Mastercycler Nexus GX2)
- Prechilled metal blocks for rapid cooling (e.g., BioCision CoolRack)

2.3.2 Procedure

1. Aliquot the reactions containing the solubilized ^{125}I-sauvagine–CRF$_1$R–mini-G_s complexes into prechilled PCR tubes prior to heating at different temperatures. For triplicates (each of which will use a volume of 50 µL), a volume of 180 µL per temperature point is recommended. Ensure that reactions are evenly resuspended by gentle pipetting prior to aliquoting.
2. Place tubes into PCR machines at the temperature(s) required for exactly 30 min. For a two-point assay, one sample is kept on ice and the other heated just beyond the apparent T_m value of the wild-type complex (here: 21°C). For seven-point curves, one sample is incubated on ice, and the others at 10, 20, 30, 40, 50, and 80°C.
3. During this incubation, spin columns should be packed with gel filtration matrix as described later.
4. At the end of the incubation time, immediately remove the tubes from the PCR machine and shock-cool on a cold metal block to stop further

denaturation of receptors. Bring into a cold room and separate as described later as quickly as possible.

2.3.3 Notes

The heating step does not require substantial optimization for different projects, as the range of temperatures between 0 and 80°C allows visualization of the unfolding process of most agonist–GPCR–mini-G complexes. The apparent T_m value obtained for each sample will vary not only depending on which GPCR is used, but also depending on the presence of stabilizing detergents and other buffer additives. It is essential to ensure that a consistent T_m value can be obtained reproducibly for the wild-type receptor complex before deciding on a temperature to use in the two-point assay format for screening point mutants.

2.4 Separation

Separation of the bound ^{125}I-sauvagine–CRF$_1$R–mini-G$_s$ complex from unbound ^{125}I-sauvagine after heating allows the quantification of the amount of intact complex after incubating at each of the temperatures assessed. This step relies on the use of small spin columns packed with a gel filtration matrix that has a suitable molecular-weight cut-off for the target of interest, so that unbound radioligand is retained on the matrix, and radioligand bound to the complex is eluted.

2.4.1 Equipment

- Small disposable filter columns (e.g., from Fisherbrand)
- Low-speed, spinning bucket centrifuge (e.g., Eppendorf centrifuge 5920R)
- Scintillation vials (e.g., Beckman SnapCap Bio-Vials (4 mL))

2.4.2 Buffers and Reagents

- Gel filtration matrix with suitable molecular-weight cut-off (for this project we use Sephacryl S-100 HR, GE Healthcare).

2.4.3 Procedure

1. The gel filtration matrix must be equilibrated in radioligand binding assay buffer and detergent (we used 0.1% GDN). The final matrix-to-buffer ratio should be 2-to-1. Three spin columns are required per temperature point. To set up spin columns, pipette 2 mL Sephacryl S-100 HR size exclusion medium into each filter column, then centrifuge

briefly (600 × *g*, 1 min) to give a final packed volume of 1.3 mL. Discard the flow through and move the spin columns onto 4 mL snap-cap scintillation vials.

2. For each temperature point, carefully pipette 3 × 50 μL onto the center of the matrix in the three spin columns, using a fresh pipette tip for each aliquot.
3. Centrifuge the columns to separate the unbound radioligand from the radioligand–GPCR–mini-G_s complex (for the ^{125}I-sauvagine–CRF_1R–mini-G_s complex, two spins were carried out: an initial loading spin at 600 × *g* for 1 min, followed by an elution spin at 910 × *g* for 5 min).
4. We recommend also aliquoting 5 μL of the unseparated reactions into scintillation vials, in order to have a measure of total radioactivity in the reaction.

2.4.4 Notes

The type of gel filtration matrix, column lengths, and spin speeds must be optimized for each complex of interest (Magnani et al., 2016). For projects working with small-molecule ligands, gel filtration matrices with a lower molecular-weight cut-off may be used, whereas for this project a higher molecular-weight cut-off was needed to retain the large peptide ligand on the column in the presence of detergents. Column lengths and spin speeds can also impact on the resolution and reproducibility of an assay, and some time should be spent optimizing these conditions for any target complex. We highly recommend initially working with triplicates, since the pipetting technique in this step requires some practice to obtain acceptable repeats.

2.5 Counting

2.5.1 Equipment

- Scintillation counter, e.g., Perkin Elmer Tri-Carb 2910TR

2.5.2 Buffers and Reagents

- Appropriate scintillant, e.g., Perkin Elmer Ultima Gold

2.5.3 Procedure

1. Add 4 mL scintillant to each of the vials containing the flow through from the spin columns.

2. Load vials onto scintillation counter and count beta emission of each vial for 1 min, using an appropriate quench curve to determine disintegrations per minute (dpm).

2.6 Data Analysis

The aim of a seven-point assay is to determine a thermostability curve for the complex of interest under the given conditions. The amount of binding of ^{125}I–labeled sauvagine to CRF_1R decreases as the proteins unfold with increasing temperature, giving rise to a sigmoidal curve. The mid-way point of this curve is defined as the apparent T_m value. When the buffer constituents and detergent composition are consistent, differences in stability between the wild-type complex and complexes containing point mutants can be detected using this assay format. Relative changes in thermostability arising from point mutations can also be observed using the two-point assay format. This format provides a method for screening large numbers of mutants in a cost-effective and time-efficient manner, allowing the identification of a subset of mutants that are more stable than the wild-type receptor.

2.6.1 Equipment

- Suitable analysis software, e.g., GraphPad Prism 7

2.6.2 Procedure

1. Import raw data from triplicate repeats, and subtract the baseline. For initial experiments and high-throughput two-point assays, untransfected cells must be included into each round of the assay to provide these baseline values. During the more detailed analysis of mutants of interest, the 80°C point of each respective thermostability curve may be used as a baseline value.
2. For two-point assays, two values are of interest for determining the usefulness of a point mutant. First, the amount of bound radioligand after incubating the complex on ice is informative, since it represents a measure of functional expression relative to the wild-type samples (providing the mutant does not change the affinity of ligand binding). Mutants that have lost their ability to bind the agonist are not of interest for thermostabilization of the active conformation of the receptor, whereas mutants that have a higher or comparable agonist affinity to the wild-type receptor are useful. Hence high agonist binding prior to heating is a desirable feature. Second, the amount of bound radioligand after

heating just beyond the apparent T_m of the wild-type sample indicates whether the apparent T_m of the point mutant is likely to be higher or lower than that of the wild-type receptor complex. This change in relative stability can be calculated as a percentage, by dividing the counts detected after heating by the counts detected on ice and multiplying them by 100.

3. For full seven-point curves, the resulting values are arranged as a function of temperature, analyzed by nonlinear regression and a sigmoidal dose–response curve is fitted through the data. This curve should be constrained so that the highest value in the dataset is defined as the top, and the lowest value is defined as the bottom. The resulting $LogEC_{50}$ value given by the software is the apparent T_m value for the sample.

3. SUMMARY AND CONCLUSION

For the thermostabilization of any target, we recommend performing all the two-point assays in triplicates, and we advise that all incubation times are strictly adhered to in order to minimize day-to-day variability. In order to apply the methodology described here to the thermostabilization of any agonist–GPCR–mini-G protein complex, it is essential to consider the unique physicochemical properties of each receptor and agonist of interest. The choice of agonist has implications for both the ligand binding and separation steps of the thermostability assay (Magnani et al., 2016), while the choice of receptor impacts on which mini-G protein can be used during the thermostabilization of the complex (Nehmé et al., 2017). The characteristics of both the receptor and its mini-G protein in turn affect the choice of detergent, since both proteins will have different detergent sensitivities.

There will usually be several detergent compositions that could be suitable for the assessment of point mutations of any given target complex. However, it is important to ensure that a detergent composition is selected that yields reproducible apparent T_m values in a temperature range of 20–30°C. Fig. 2 illustrates how adding increasing concentrations of the more destabilizing detergent DDM to the more stabilizing detergent GDN can gradually reduce the apparent T_m of ^{125}I-sauvagine-bound CRF_1R. Initially, CRF_1R had an apparent T_m of ~40°C in 0.1% GDN (data not shown), which dropped to just under 30°C in a 0.1% GDN/0.1% DDM, and was further reduced to ~20°C when using 0.1% GDN/0.3% DDM. We generally recommend to aim for a detergent composition that generates an apparent T_m value of ~20°C for the wild-type protein.

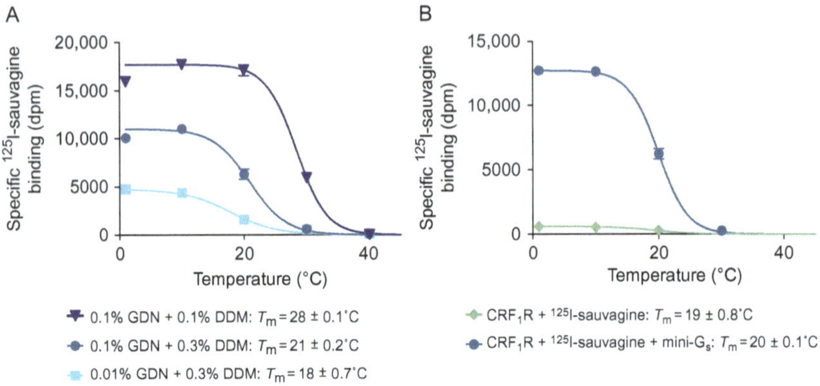

Fig. 2 Thermostability curves of the wild-type ^{125}I-sauvagine–CRF$_1$R–mini-G$_s$ complex, illustrating how to find an ideal detergent composition for assessment of a library of point mutants. (A) Increasing concentrations of DDM reduces the apparent T_m of the ^{125}I-sauvagine–CRF$_1$R–mini-G$_s$ ternary complex. (B) The stability of ^{125}I-sauvagine-bound CRF$_1$R in the absence and presence of mini-G$_s$ in the detergent composition chosen for assessment of the mutant library in this study. Accurate assessment of changes in thermostability in 0.1% GDN/0.3% DDM is only possible in the presence of mini-G$_s$.

Choosing a detergent composition in which the apparent T_m value of the wild-type receptor is much lower than 20°C can result in high errors, whereas choosing a detergent composition in which this T_m value is much higher than 30°C can limit the detection of thermostabilizing mutants. Both of these scenarios will result in more false positives and/or the omission of thermostabilizing mutations when using the two-point assay format for screening.

In theory, the problem of false positives arising from the two-point assay screen can be resolved by determining full seven-point thermostability curves for all mutants. However, seven-point curves use 3.5-fold more reagents and time. The expression, harvesting, and assessment of a library of point mutants such as the one made for this study already takes approximately 6 months for one person to complete. Therefore, while seven-point stability curves are more accurate than the data generated by two-point assays, determining more than 400 thermostability curves is exceedingly onerous for one person. Furthermore, the two-point assay format is generally sufficient for successfully identifying a subset of thermostabilizing point mutations. These mutants must then be assessed using seven-point curves to quantify the extent of thermostabilization, so that they can be ranked from most thermostabilizing to least thermostabilizing.

It has been reported previously that for a typical Ala scan of the transmembrane domain of a GPCR, between 5% and 9% of point mutants are thermostabilizing (Tate, 2012), although it is also known that other amino acid residues can also be thermostabilizing (Serrano-Vega et al., 2008). Therefore, in addition to mutating 104 residues to Ala or Leu, we also mutated them to four other residues, namely, Glu, Arg, Val, and Trp. These residues were chosen due to their different properties, i.e., negatively charged, positively charged, medium hydrophobic, and large hydrophobic, respectively. The thermostabilization methodology worked regardless of the amino acid to which the residues were mutated, with all types of mutations (except the six Leu mutants that were made) yielding some mutants that were more thermostable than the wild-type receptor (Fig. 3). We defined good mutants that are worth assessing with seven-point thermostability curves as those that displayed a relative stability of >60% (where stability for the wild-type covering all assays was ~40% ± 10%) and had at least 40% of the ^{125}I-sauvagine binding prior to heating observed for the wild-type receptor. Based on these constraints, 8.5% of the 416 mutants appeared more stable than the wild-type receptor, which is comparable to an Ala/Leu scan. The Ala mutants appeared to be the most useful since they have the highest number of mutants fulfilling these criteria. However, as Fig. 3 illustrates, half of these mutants are actually in the range of 60%–65% relative stability while other types of mutations indicated much larger improvements in relative stability. The best 36 mutations identified from the two-point thermostability assay comprised 13 Ala mutants, 8 Glu mutants, 6 Arg mutants, 7 Val mutants, and 2 Trp mutants. These were all reassayed using the seven-point thermostability curves and 17 of the mutants were between 1.5 and 5.1°C more thermostable than the wild-type receptor. Within these 17 mutants, there were 5 Glu mutants, 5 Val mutants, 4 Ala mutants, 2 Arg mutants, and 1 Trp mutant. Interestingly, the top three most thermostable mutants were all Glu residues. Therefore, at least in this instance, both Val and Glu outperformed Ala in producing useful thermostabilizing mutations.

Another interesting observation from this work was that there were several positions whose mutation improved thermostability when changed to any one of the four different amino acid residues. Fig. 4 shows two representative examples of improvements in thermostability of the ^{125}I-sauvagine–CRF$_1$R–mini-G$_s$ complex. The two mutations shown were both made at the same position in CRF$_1$R and are both thermostabilizing. This figure also highlights that seven-point curves do not only yield accurate apparent T_m values for the mutants, but also provide information on changes

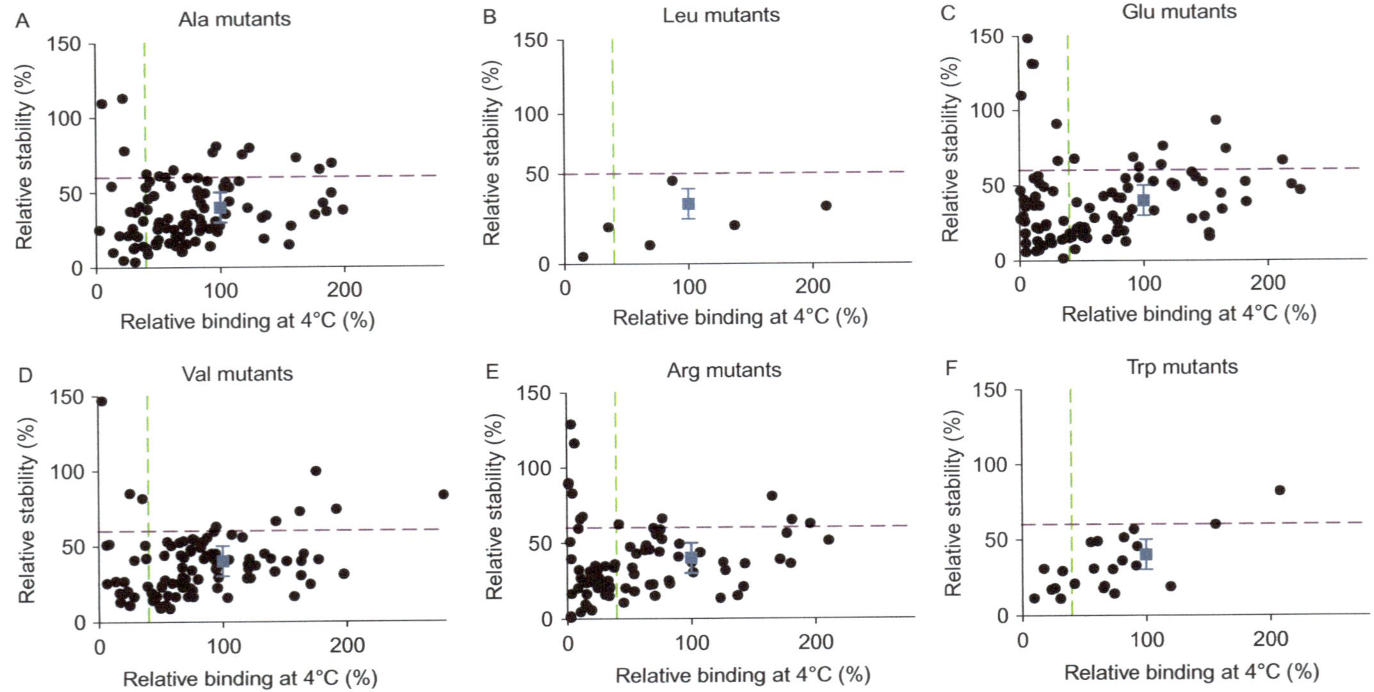

Fig. 3 See legend on opposite page.

in agonist binding under these assay conditions. A mutant that achieves higher agonist binding as the wild-type receptor under the same conditions suggests better coupling between the receptor and the agonist, indicating that the receptor is being stabilized in an active conformation.

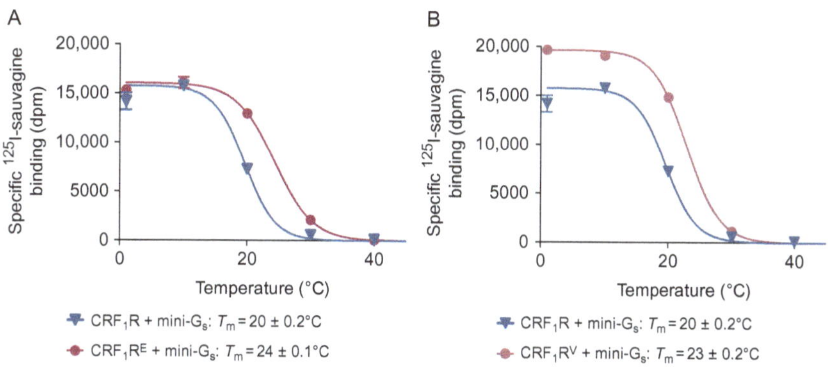

Fig. 4 Representative examples of thermostability shifts obtained as a result of mutating the same position to different amino acids. Both Glu and Val gave rise to relatively large improvements in apparent T_m at this position, and the Val mutant at this position also binds more agonist than the wild-type protein, indicating that it may have a higher B_{max}. This illustrates that amino acids other than Ala can be used for the thermostabilization of GPCR–G protein complexes.

Fig. 3 The stability of 416 point mutants was assessed after expression in HEK293 GnTI$^-$ cells using the two-point thermostability assay. Amino acid residues in CRF$_1$R were mutated to the following: (A) Ala; (B) Leu; (C) Glu; (D) Val; (E) Arg; and (F) Trp. After incubation with the agonist ^{125}I-sauvagine on ice, the cells were solubilized, mini-G$_s$ was bound and the resulting agonist–receptor–mini-G$_s$ complexes were heated at 21°C for 30 min (super-plus mini-G stability assay). Receptor survival was calculated as a percentage from the amount of receptor-bound ^{125}I-sauvagine after heating compared to the unheated control for each mutant. This value is plotted on the *y*-axis of the scatter graphs shown here and indicates a measure of stability relative to the wild-type (*blue square*; 40% ± 10%). For further characterization, we only considered mutants that displayed a significantly higher receptor survival (at least 60%) than the wild type. ^{125}I-sauvagine binding to receptor-G protein complexes incubated on ice relative to the wild-type control measured on the same day is plotted on the *x*-axis. This can be considered a measure of functional expression (provided the affinity for ligand binding is unchanged), since mutants that do not express and/or are not folded in a functional active conformation will display low agonist binding. We excluded mutants that had low specific binding of ^{125}I-sauvagine before heating, since this is an indicator of low functional expression or reduced affinity for the agonist, neither of which are desirable features for overexpression and crystallization of an agonist-bound ternary complex.

In conclusion, we describe the methodology for the super-plus mini-G protein format of the conformational thermostabilization strategy, showing that mini-G proteins are extremely useful tools for the thermostabilization of GPCRs in fully active conformations. Furthermore, our data resulting from the RAVE scan showed that, at least in this instance, Glu and Val scans identified more highly thermostabilizing mutants than the Ala scan. The main advantages of performing a scan with amino acids other than Ala is that different amino acid positions will be identified as thermostabilizing and the mutants may be more stable than Ala mutants. It is also notable that performing an Ala scan and then mutating thermostable mutants to Glu or Val would not have identified the best Glu and Val mutants we found by performing complete scans with these amino acid residues.

ACKNOWLEDGMENTS

This work was supported by core funding from the Medical Research Council (MC_U105197215) and a grant from Heptares Therapeutics Ltd. We thank R. Nehmé for helpful comments on the manuscript.

REFERENCES

Carpenter, B., Nehme, R., Warne, T., Leslie, A. G., & Tate, C. G. (2016). Structure of the adenosine A(2A) receptor bound to an engineered G protein. *Nature*, *536*(7614), 104–107. http://dx.doi.org/10.1038/nature18966.

Carpenter, B., & Tate, C. G. (2016). Engineering a minimal G protein to facilitate crystallisation of G protein-coupled receptors in their active conformation. *Protein Engineering, Design & Selection*, *29*(12), 583–594. http://dx.doi.org/10.1093/protein/gzw049.

Congreve, M., Langmead, C. J., Mason, J. S., & Marshall, F. H. (2011). Progress in structure based drug design for G protein-coupled receptors. *Journal of Medicinal Chemistry*, *54*(13), 4283–4311. http://dx.doi.org/10.1021/jm200371q.

Dore, A. S., Okrasa, K., Patel, J. C., Serrano-Vega, M., Bennett, K., Cooke, R. M., et al. (2014). Structure of class C GPCR metabotropic glutamate receptor 5 transmembrane domain. *Nature*, *511*(7511), 557–562. http://dx.doi.org/10.1038/nature13396.

Dore, A. S., Robertson, N., Errey, J. C., Ng, I., Hollenstein, K., Tehan, B., et al. (2011). Structure of the adenosine A(2A) receptor in complex with ZM241385 and the xanthines XAC and caffeine. *Structure*, *19*(9), 1283–1293. http://dx.doi.org/10.1016/j.str.2011.06.014.

Flock, T., Ravarani, C. N., Sun, D., Venkatakrishnan, A. J., Kayikci, M., Tate, C. G., et al. (2015). Universal allosteric mechanism for Galpha activation by GPCRs. *Nature*, *524*(7564), 173–179. http://dx.doi.org/10.1038/nature14663.

Foord, S. M., Bonner, T. I., Neubig, R. R., Rosser, E. M., Pin, J. P., Davenport, A. P., et al. (2005). International Union of Pharmacology. XLVI. G protein-coupled receptor list. *Pharmacological Reviews*, *57*(2), 279–288.

Fredriksson, R., & Schioth, H. B. (2005). The repertoire of G-protein-coupled receptors in fully sequenced genomes. *Molecular Pharmacology*, *67*(5), 1414–1425.

Hollenstein, K., Kean, J., Bortolato, A., Cheng, R. K., Dore, A. S., Jazayeri, A., et al. (2013). Structure of class B GPCR corticotropin-releasing factor receptor 1. *Nature*, *499*(7459), 438–443. http://dx.doi.org/10.1038/nature12357.

Hopkins, A. L., & Groom, C. R. (2002). The druggable genome. *Nature Reviews. Drug Discovery*, *1*(9), 727–730. http://dx.doi.org/10.1038/nrd892.

Jazayeri, A., Dore, A. S., Lamb, D., Krishnamurthy, H., Southall, S. M., Baig, A. H., et al. (2016). Extra-helical binding site of a glucagon receptor antagonist. *Nature*, *533*(7602), 274–277. http://dx.doi.org/10.1038/nature17414.

Lagerstrom, M. C., & Schioth, H. B. (2008). Structural diversity of G protein-coupled receptors and significance for drug discovery. *Nature Reviews. Drug Discovery*, 7(4), 339–357. http://dx.doi.org/10.1038/nrd2518.

Lebon, G., Bennett, K., Jazayeri, A., & Tate, C. G. (2011). Thermostabilisation of an agonist-bound conformation of the human adenosine A(2A) receptor. *Journal of Molecular Biology*, *409*(3), 298–310. http://dx.doi.org/10.1016/j.jmb.2011.03.075.

Lebon, G., Warne, T., Edwards, P. C., Bennett, K., Langmead, C. J., Leslie, A. G., et al. (2011). Agonist-bound adenosine A2A receptor structures reveal common features of GPCR activation. *Nature*, *474*(7352), 521–525. http://dx.doi.org/10.1038/nature10136.

Lebon, G., Warne, T., & Tate, C. G. (2012). Agonist-bound structures of G protein-coupled receptors. *Current Opinion in Structural Biology*, *22*, 482–490. http://dx.doi.org/10.1016/j.sbi.2012.03.007.

Lee, S., Mao, A., Bhattacharya, S., Robertson, N., Grisshammer, R., Tate, C. G., et al. (2016). How do short chain nonionic detergents destabilize G-protein-coupled receptors? *Journal of the American Chemical Society*, *138*(47), 15425–15433. http://dx.doi.org/10.1021/jacs.6b08742.

Liu, W., Chun, E., Thompson, A. A., Chubukov, P., Xu, F., Katritch, V., et al. (2012). Structural basis for allosteric regulation of GPCRs by sodium ions. *Science*, *337*(6091), 232–236. http://dx.doi.org/10.1126/science.1219218.

Magnani, F., Serrano-Vega, M. J., Shibata, Y., Abdul-Hussein, S., Lebon, G., Miller-Gallacher, J., et al. (2016). A mutagenesis and screening strategy to generate optimally thermostabilized membrane proteins for structural studies. *Nature Protocols*, *11*(8), 1554–1571. http://dx.doi.org/10.1038/nprot.2016.088.

Magnani, F., Shibata, Y., Serrano-Vega, M. J., & Tate, C. G. (2008). Co-evolving stability and conformational homogeneity of the human adenosine A2a receptor. *Proceedings of the National Academy of Sciences of the United States of America*, *105*(31), 10744–10749. http://dx.doi.org/10.1073/pnas.0804396105.

Miller, J. L., & Tate, C. G. (2011). Engineering an ultra-thermostable beta(1)-adrenoceptor. *Journal of Molecular Biology*, *413*(3), 628–638. http://dx.doi.org/10.1016/j.jmb.2011.08.057.

Miller-Gallacher, J. L., Nehme, R., Warne, T., Edwards, P. C., Schertler, G. F., Leslie, A. G., et al. (2014). The 2.1 A resolution structure of cyanopindolol-bound beta1-adrenoceptor identifies an intramembrane Na+ ion that stabilises the ligand-free receptor. *PLoS One*, 9(3), e92727. http://dx.doi.org/10.1371/journal.pone.0092727.

Nehmé, R., Carpenter, B., Singhal, A., Strege, A., Edwards, P. C., White, C. F., et al. (2017). Mini-G proteins: Novel tools for studying GPCRs in their active conformation. *PLoS One*, *12*, e0175642. http://dx.doi.org/10.1371/journal.pone.0175642.

Oswald, C., Rappas, M., Kean, J., Dore, A. S., Errey, J. C., Bennett, K., et al. (2016). Intracellular allosteric antagonism of the CCR9 receptor. *Nature*, *540*(7633), 462–465. http://dx.doi.org/10.1038/nature20606.

Rasmussen, S. G., DeVree, B. T., Zou, Y., Kruse, A. C., Chung, K. Y., Kobilka, T. S., et al. (2011). Crystal structure of the beta2 adrenergic receptor-Gs protein complex. *Nature*, *477*(7366), 549–555. http://dx.doi.org/10.1038/nature10361.

Robertson, N., Jazayeri, A., Errey, J., Baig, A., Hurrell, E., Zhukov, A., et al. (2011). The properties of thermostabilised G protein-coupled receptors (StaRs) and their use in drug discovery. *Neuropharmacology*, *60*(1), 36–44. http://doi:10.1016/j.neuropharm.2010.07.001.

Rosenbaum, D. M., Rasmussen, S. G., & Kobilka, B. K. (2009). The structure and function of G-protein-coupled receptors. *Nature*, *459*(7245), 356–363. http://dx.doi.org/10.1038/nature08144.

Serrano-Vega, M. J., Magnani, F., Shibata, Y., & Tate, C. G. (2008). Conformational thermostabilization of the beta1-adrenergic receptor in a detergent-resistant form. *Proceedings of the National Academy of Sciences of the United States of America*, *105*(3), 877–882. http://dx.doi.org/10.1073/pnas.0711253105.

Shibata, Y., Gvozdenovic-Jeremic, J., Love, J., Kloss, B., White, J. F., Grisshammer, R., et al. (2013). Optimising the combination of thermostabilising mutations in the neurotensin receptor for structure determination. *Biochimica et Biophysica Acta*, *1828*(4), 1293–1301. http://dx.doi.org/10.1016/j.bbamem.2013.01.008.

Shibata, Y., White, J. F., Serrano-Vega, M. J., Magnani, F., Aloia, A. L., Grisshammer, R., et al. (2009). Thermostabilization of the neurotensin receptor NTS1. *Journal of Molecular Biology*, *390*(2), 262–277. http://dx.doi.org/10.1016/j.jmb.2009.04.068.

Tate, C. G. (2010). Practical considerations of membrane protein instability during purification and crystallisation. *Methods in Molecular Biology*, *601*, 187–203. http://dx.doi.org/10.1007/978-1-60761-344-2_12.

Tate, C. G. (2012). A crystal clear solution for determining G-protein-coupled receptor structures. *Trends in Biochemical Sciences*, *37*(9), 343–352. http://dx.doi.org/10.1016/j.tibs.2012.06.003.

Venkatakrishnan, A. J., Deupi, X., Lebon, G., Heydenreich, F. M., Flock, T., Miljus, T., et al. (2016). Diverse activation pathways in class A GPCRs converge near the G-protein-coupling region. *Nature*, *536*(7617), 484–487. http://dx.doi.org/10.1038/nature19107.

Venkatakrishnan, A. J., Deupi, X., Lebon, G., Tate, C. G., Schertler, G. F., & Babu, M. M. (2013). Molecular signatures of G-protein-coupled receptors. *Nature*, *494*(7436), 185–194. http://dx.doi.org/10.1038/nature11896.

Warne, T., Moukhametzianov, R., Baker, J. G., Nehme, R., Edwards, P. C., Leslie, A. G., et al. (2011). The structural basis for agonist and partial agonist action on a beta(1)-adrenergic receptor. *Nature*, *469*(7329), 241–244. http://dx.doi.org/10.1038/nature09746.

Warne, T., Serrano-Vega, M. J., Baker, J. G., Moukhametzianov, R., Edwards, P. C., Henderson, R., et al. (2008). Structure of a beta1-adrenergic G-protein-coupled receptor. *Nature*, *454*(7203), 486–491. http://dx.doi.org/10.1038/nature07101.

White, J. F., Noinaj, N., Shibata, Y., Love, J., Kloss, B., Xu, F., et al. (2012). Structure of the agonist-bound neurotensin receptor. *Nature*, *490*(7421), 508–513. http://dx.doi.org/10.1038/nature11558.

Zhang, C., Srinivasan, Y., Arlow, D. H., Fung, J. J., Palmer, D., Zheng, Y., et al. (2012). High-resolution crystal structure of human protease-activated receptor 1. *Nature*, *492*(7429), 387–392. http://dx.doi.org/10.1038/nature11701.

Zhang, X., Stevens, R. C., & Xu, F. (2015). The importance of ligands for G protein-coupled receptor stability. *Trends in Biochemical Sciences*, *40*(2), 79–87. http://dx.org.doi.10.1016/j.tibs.2014.12.005.

AUTHOR INDEX

Note: Page numbers followed by "*f*" indicate figures, and "*t*" indicate tables.

C

D

L

T

SUBJECT INDEX

Note: Page numbers followed by "*f*" indicate figures, and "*t*" indicate tables.

T

FSC
www.fsc.org
MIX
Papier | Fördert gute Waldnutzung
FSC® C083411

Kolibri 360 GmbH
Ferdinand-Jühlke-Straße 7
99095 Erfurt, Deutschland
produktsicherheit@kolibri360.de